T0250363

Insect-Plant Interactions

Volume III

Elizabeth Bernays, Ph.D.
Professor and Head
Department of Entomology
University of Arizona
Tucson, Arizona

CRC Press
Taylor & Francis Group
Boca Raton London New York

CRC Press is an imprint of the
Taylor & Francis Group, an **informa** business

First published 1991 by CRC Press
Taylor & Francis Group
6000 Broken Sound Parkway NW, Suite 300
Boca Raton, FL 33487-2742

Reissued 2018 by CRC Press

© 1991 by Taylor & Francis
CRC Press is an imprint of Taylor & Francis Group, an Informa business

No claim to original U.S. Government works

This book contains information obtained from authentic and highly regarded sources. Reasonable efforts have been made to publish reliable data and information, but the author and publisher cannot assume responsibility for the validity of all materials or the consequences of their use. The authors and publishers have attempted to trace the copyright holders of all material reproduced in this publication and apologize to copyright holders if permission to publish in this form has not been obtained. If any copyright material has not been acknowledged please write and let us know so we may rectify in any future reprint.

Except as permitted under U.S. Copyright Law, no part of this book may be reprinted, reproduced, transmitted, or utilized in any form by any electronic, mechanical, or other means, now known or hereafter invented, including photocopying, microfilming, and recording, or in any information storage or retrieval system, without written permission from the publishers.

For permission to photocopy or use material electronically from this work, please access www. copyright.com (http://www.copyright.com/) or contact the Copyright Clearance Center, Inc. (CCC), 222 Rosewood Drive, Danvers, MA 01923, 978-750-8400. CCC is a not-for-profit organiza-tion that provides licenses and registration for a variety of users. For organizations that have been granted a photocopy license by the CCC, a separate system of payment has been arranged.

Trademark Notice: Product or corporate names may be trademarks or registered trademarks, and are used only for identification and explanation without intent to infringe.

A Library of Congress record exists under LC control number: 88035918

Publisher's Note
The publisher has gone to great lengths to ensure the quality of this reprint but points out that some imperfections in the original copies may be apparent.

Disclaimer
The publisher has made every effort to trace copyright holders and welcomes correspondence from those they have been unable to contact.

ISBN 13: 978-1-138-50598-8 (hbk)
ISBN 13: 978-1-138-56038-3 (pbk)
ISBN 13: 978-0-203-71169-9 (ebk)

Visit the Taylor & Francis Web site at http://www.taylorandfrancis.com and the CRC Press Web site at http://www.crcpress.com

PREFACE

Insect-Plant Interactions is a series devoted to reviews across the breadth of the topic from biochemistry to ecology and evolution. Topics are selected from areas of particular current interest or subjects that would especially benefit from a new review. It is hoped that the interdisciplinary selection in each volume will help readers to enter new fields of insect-plant interactions.

Volume 3 contains six contrasting articles. Lindroth provides an up-to-date review of insect detoxification mechanisms, and demonstrates that this arena is expanding fast. The new information will help us to understand not just the biochemistry of how insects deal with plant secondary compounds, but also the adaptations of insects to their particular diets and diet breadths. The differential lability of different enzyme systems is an important aspect of insect toxicology.

Mitter and Farrell give a very comprehensive account of evolution among plant-feeding insects from a comparative systematics perspective. The study of plant and animal clades can tell us about the likelihood and degree of coevolution and the relative importance of phylogenetic constraints in determining host affiliations of herbivores.

Van Loon brings a physiologist's approach to nutritional indices. He demonstrates the spectrum of problems associated with the commonly used utilization parameters and points out the need to take some new approaches. This is an important article for ecologists who may not appreciate the physiological details that impact their need to measure food value to insect herbivores.

Aphid feeding behavior in relation to probing is reviewed by Montllor, who provides an account of how plant chemicals influence the details of probing behavior. The need for further work and some new approaches to measurement are emphasized.

Using a modeling approach in combination with field-collected observations, Morris and Kareiva demonstrate very convincingly that host-plant finding can be effective with a mixture of random movement and orientation mechanisms, and that the random elements are most probably much more important than usually thought.

Finally, Sheehan discusses insect-plant relationships in connection with how parasitoids attack and in what prey species they are found. He uses very large data sets on lepidopteran parasitoids to examine such prey features as diet breadth, host plant type, and host plant family.

CONTRIBUTORS

Dr. S. Courtney
Department of Biology
University of Oregon
Eugene, Oregon

Mr. Brian D. Farrell
Deparment of Entomology
University of Maryland
College Park, Maryland

Prof. Peter M. Kareiva
Department of Zoology
University of Washington
Seattle, Washington

Dr. Richard Lindroth
Department of Entomology
University of Wisconsin
Madison, Wisconsin

Dr. Charles Mitter
Department of Entomology
University of Maryland
College Park, Maryland

Dr. Clytia Montllor
Department of Entomology
University of California
Berkeley, California

Dr. William F. Morris
Department of Zoology
University of Washington
Seattle, Washington

Dr. Peter W. Price
Department of Biological Sciences
Northern Arizona University
Flagstaff, Arizona

Dr. Martine Rowell-Rahier
Department of Zoology
University of Basel
Basel, Switzerland

Dr. William Sheehan
Insect Biology and Population Management Research Laboratory
USDA/ARS
Tifton, Georgia

Dr. Joop J. A. van Loon
Department of Entomology
Wageningen Agricultural University
Wageningen, The Netherlands

ADVISORY BOARD

Dr. M. Berenbaum
Department of Entomology
University of Illinois
Urbana, Illinois

Dr. R. F. Chapman
Department of Entomology
University of California
Berkeley, California

Dr. Fred Gould
Department of Entomology
North Carolina State University
Raleigh, North Carolina

Dr. Michael M. Martin
Department of Biology
University of Michigan

Dr. L. M. Schoonhoven
Department of Entomology
Agricultural University
Wageningen, The Netherlands

Dr. Árpád Szentesi
Department of Zoology
Plant Protection Institute
Hungarian Academy of Sciences
Budapest, Hungary

THE EDITOR

Elizabeth Bernays, Ph.D., is Professor and Head of the Department of Entomology, and Professor of Ecology and Evolutionary Biology at the University of Arizona, Tucson. After receiving a B.S. with honors in 1962 at the University of Queensland, Brisbane, Australia, she traveled in Europe and taught biology before studying for the M.Sc. and Ph.D. at the University of London, England. Her Ph.D. was a study of the hatching mechanism in desert locust involving morphology, physiology, and behavior.

From 1970 to 1983, Dr. Bernays was scientist, senior scientist, and then principal scientist in the British Government Research Institute (now the Natural Resources Institute), involved with research on insect pests in developing countries. During this period, she worked in the laboratory on the physiological regulation of feeding behavior in grasshoppers and locusts, and on the effects of various plant compounds on behavior and physiology of plant-feeding insects. In the field, she worked on cassava pest biology in Nigeria, and cereal resistance to insects in India. In 1978, with Dr. R. F. Chapman, she organized the Fourth International Symposium on Plant-Insect Interactions and edited the resultant book.

From 1983 to 1989 Dr. Bernays was a Professor at the University of California Berkeley, where she developed new evolutionary approaches in her study of plant-insect interactions. She continues research on the causes and functions of different diet breadths in insects.

TABLE OF CONTENTS

1

Differential Toxicity of Plant Allelochemicals to Insects: Roles of Enzymatic Detoxication Systems

Richard L. Lindroth
Department of Entomology and
Environmental Toxicology Center
University of Wisconsin
Madison, Wisconsin

TABLE OF CONTENTS

I. INTRODUCTION

Since 1970, research on the mechanisms underlying interactions of plants and insects has focused on plant chemistry. That plant secondary compounds in particular have played dominant roles in the evolution of such interactions is now widely accepted. Many allelochemicals have been shown to be differentially toxic to insects; results have been interpreted to explain ecological phenomena as diverse as dietary specialization, population dynamics, and community organization.

That insects have not been regarded as innocent bystanders in the evolution of chemically mediated interactions is evidenced by the militaristic jargon (e.g., "arms race") used to describe such associations. Indeed, insect adaptations to evade or overcome plant toxins were accorded pivotal roles in early and influential theoretical discussions of plant-insect evolution.[48,49,95] In contrast, however, to research on plant chemistry, empirical studies of insect adaptations have been few in number.

Insect adaptations to potentially toxic chemicals are categorized into three classes.[14,30] *Behavioral resistance mechanisms* reduce exposure to toxins, typically through a variety of changes in feeding behavior. *Physiological resistance mechanisms* reduce uptake and transport of toxins to active sites and include, for example, sequestration and enhanced excretion rates. *Biochemical resistance mechanisms* are further subdivided into target site insensitivity and enzymatic detoxication. The latter is considered the most generally important set of adaptations to plant allelochemicals,[30] and is the subject of this chapter. By focusing on enzymatic detoxication mechanisms, however, the author's intent is not to diminish the importance of other adaptations, which for any particular plant-insect association may play critical roles.

The classic paper of Krieger et al.[61] strongly shaped scientific opinion of the importance of enzymatic detoxication systems in plant-insect interactions. These researchers measured polysubstrate monooxygenase (PSMO) activities in numerous species of Lepidoptera and found that as predicted, activity levels were correlated with the degree of polyphagy. Further support was provided by the early work of Brattsten,[26,34] who showed that PSMO activities were elevated in southern armyworms (*Spodoptera eridania*) following consumption of a variety of plant secondary compounds, and that larvae were subsequently more resistant to the alkaloid nicotine. By the early 1980s, perspectives on the importance of insect detoxication systems vis à vis plant allelochemicals had, on the basis of relatively little empirical evidence, attained the status of ecological dogma. Uncritical enthusiasm was tempered, however, by several researchers, most notably Gould,[53] who argued that available evidence poorly supported some of the widely held notions. Gould's criticism was written to "foster rigorous research", and by all appearances did just that. By the late 1980s, a number of biochemists and chemical ecologists had focused their research efforts on the ecological relevance of insect detoxication systems. Studies appeared that clearly implicated biochemical detoxication as a central feature in many plant-insect interactions.

The time is now appropriate for a review of the role played by enzymatic detoxication systems in the comparative toxicity of secondary plant compounds to insects. This chapter will address the interface of plant chemistry and insect biochemistry as a means of understanding the comparative toxicity of allelochemicals, and implications for the ecology of plant-insect interactions. It will not extensively review the biochemistry of enzymatic detoxication systems; simply catalog known detoxication pathways for various plant secondary compounds. Other recent and informative articles on these topics can be found in References 5, 28, 47, 57, and 79.

II. ENZYMATIC DETOXICATION SYSTEMS

The primary role of the numerous enzyme systems in insects is the conversion of lipophilic foreign compounds (xenobiotics) into hydrophilic products, thereby enhancing rates of solubilization and excretion. This transformation from lipid- to water-soluble product reduces the tendency to penetrate cell membranes and accumulate in lipid-rich tissues, thereby decreasing the biological activity of compounds.[5] Therefore, these reactions are typically regarded as "detoxication" reactions, although in rare cases they may actually increase the biological activity of natural compounds.

A. GENERAL CHARACTERISTICS

Insect detoxication enzymes are differentially distributed in various organs and tissues. The site of highest activity is usually the midgut, a location that accords well with the function of these enzymes as defenses against plant secondary compounds. The fat body and Malpighian tubules are also often sites of high activity. Other tissues may exhibit high activity in individual cases.

Several properties of these enzyme systems are of considerable importance with respect to their ability to metabolize allelochemicals. First, the enzymes exhibit broad substrate specificity — they catalyze the metabolism of a wide variety of compounds. This nonspecificity may be due in part to the fact that

Table 1
MAJOR CLASSES OF INSECT ENZYMES INVOLVED IN METABOLISM OF ALLELOCHEMICALS[5]

Class	Enzymes	Alternative names
Oxidases	PSMOs	Mixed function oxidases (MFOs); cytochrome P-450-dependent monooxygenases
Reductases	Carbonyl reductases	Aldehyde-ketone (AK) reductases
Hydrolases	Glycosidases	
	Esterases	
	Epoxide hydrolases	Epoxide hydratases
Transferases	Glutathione transferases	
	UDP-glucosyl transferases	Phenol β-glucosyl transferases

most, if not all, of these enzymes occur in multiple isoenzymic forms, each of which may accept as substrates somewhat different sets of chemicals. A second feature is that some of these enzyme systems, most notably the PSMOs, catalyze a remarkable variety of biochemical reactions. This, too, may be a consequence of the enzymes occurring in multiple forms. Third, these enzymes can be rapidly and dramatically induced. At least for PSMO enzymes, induction appears to involve synthesis of new enzymes rather than activation of pre-existing enzymes.[1] Induction of activities up to eightfold higher than control levels is not unusual,[41,124] although induction capacity appears to vary among the major enzyme systems. PSMOs exhibit a striking capacity for induction; esterases, in contrast, exhibit nominal capacity.[124]

B. MAJOR CLASSES OF DETOXICATION ENZYMES

A host of enzyme systems are known to contribute to the metabolism of xenobiotics in insects. Most of these can be conveniently classified into one of four categories, determined by the type of reaction catalyzed: oxidation, reduction, hydrolysis, or group transfer (conjugation) (Table 1). Moreover, toxicologists recognize two (not always distinct) phases of metabolism. In phase I, enzymes attack lipophilic compounds directly, often introducing hydroxyl or carboxyl groups into the molecules. PSMOs, reductases, esterases, and at times, glutathione transferases, catalyze phase I reactions. Products of phase I metabolism typically exhibit decreased biological activity. In some cases, however, usually resulting from oxidations, the products are more toxic than their precursors. In phase II, group transferases conjugate chemicals with highly water-soluble, endogenous compounds such as sugars, amino acids, and peptides. Substrates for these enzymes are often products of phase I reactions, but may also be previously unaltered xenobiotics. Epoxide hydrolases are important in the conversion of epoxides (usually the products of PSMO metabolism) to compounds that in turn can be conjugated, so these enzymes are also considered part of the phase II complex.[5] Because phase II reactions increase the hydrophilic nature of chemicals, and do so through conjugation with endogenous compounds, they almost invariably reduce the biological activity of substrates.

Enzymatic systems that play significant roles in the metabolism of allelochemicals are reviewed below. The discussion is admittedly brief; Ahmad et al.[5] provide an excellent general review of these and additional enzyme systems.

1. Polysubstrate Monooxygenases

Polysubstrate monooxygenases are considered the most generally important of all enzymes involved in the metabolism of xenobiotics by insects.[27] The monooxygenases are associated with the endoplasmic reticulum which, upon homogenization, forms small vesicles known as microsomes. Hence, these enzymes are sometimes referred to as microsomal oxidases.

PSMO-catalyzed reactions require NADPH and molecular oxygen as in the following general scheme:

$$RH + NADPH + O_2 + H^+ \rightarrow ROH + NADP^+ + H_2O$$

The terms "monooxygenation" and "mixed-function oxidation" refer to the fact that one atom of oxygen is inserted into the substrate, whereas the second is reduced to form water. Reducing equivalents are provided by NADPH, which transfers electrons to the terminal cytochrome P-450 via another enzyme, NADPH cytochrome P-450 reductase (also known as cytochrome *c* reductase). Cytochrome P-450 is located in the terminal catalytic site of the monooxygenase enzyme.

The polysubstrate monooxygenases catalyze a tremendous variety of reactions. These include, for example, a variety of hydroxylations, *N*-, *O*-, and *S*-dealkylations, epoxidations, and *N*- and *S*-oxidations. Most of these reactions are known in insects only for insecticides; the roles of relatively few are known with respect to plant allelochemicals. Still, because of their well-known involvement in the metabolism of many xenobiotics, the roles of PSMO enzymes in the detoxication of allelochemicals have received more attention than those of any other enzyme system.

2. Carbonyl Reductases

Metabolic reduction of xenobiotics is much less common than oxidation, although in particular instances it may be of considerable importance. Reduction reactions are more frequently employed by microorganisms than by higher organisms; indeed, gut microorganisms may oftentimes be responsible for the appearance of reduced metabolites in animals.[101,102]

Reductase enzymes occur both freely (cytosolic) and in membrane-bound form, as both soluble and microsomal fractions of insect enzyme preparations exhibit activity.[65,68,125] Of the various types of reductases present in animals, the carbonyl reductases are the most important with respect to the metabolism of allelochemicals. Many plant secondary compounds occur as lipophilic aldehydes or ketones. Reductases convert such compounds to alcohols, which then may be conjugated or excreted directly.

The general reaction is

$$R_1R_2CO + NADPH + H^+ \rightarrow R_1R_2CHOH + NADP^+$$

3. Glycosidases

The hydrolases are a large class of enzymes that catalyze the splitting of covalent bonds with the aid of a water molecule.[5] A subgroup, the glycosidases, cleave glycosidic bonds to *O*-, *N*-, or *S*- functional groups, releasing sugar (most commonly glucose) and an aglycone. Aglycones are less water soluble and thus usually more biologically active than the corresponding glycoside. For this reason, glycosidases are not usually considered detoxication enzymes; in fact, they probably accomplish the opposite. Because of their potential importance in differential metabolism of secondary compounds, however, they are considered here. Plants produce a variety of allelochemical glycosides (e.g., glucosinolates, glycoalkaloids, cyanogenic, phenolic and iridoid glycosides) against which the activity of glycosidases is largely unknown. Because most of these are β-linked *O*-glycosyl compounds, however, it is likely that β-glucosidases play important roles in their metabolic transformation.

β-Glucosidases occur primarily in the cytosol, and are concentrated in the midgut.[127] Like other enzymes involved in the metabolism of allelochemicals, these are inducible.[64] β-Glucosidases hydrolyze compounds without the aid of high-energy cofactors such as NADPH:

$$\text{R-O}'\text{-Glucose} + H_2O \rightarrow \text{R-OH} + \text{HO}'\text{-Glucose}$$

4. Esterases

Esterases, which catalyze hydrolysis of compounds containing ester linkages, comprise one of the largest sets of hydrolytic enzymes. Esterases occur in multiple isoenzymic forms, with broad substrate specificities. Two major subcategories, aryl esterases and carboxylesterases, are generally recognized; these catalyze the hydrolysis of aromatic and aliphatic esters, respectively. Considerable overlap exists, however, in the substrate preferences of the two groups.[86] Esterases occur in both microsomal and cytosolic forms, but the latter accounts for the largest proportion of total esterase activity.[65-67]

Many plant natural products contain esters and lactones, and esterases are known to hydrolyze a variety of aromatic, aliphatic, hetero-, and polycyclic esters and lactones.[3] Relatively little is known about the natural roles of esterases against either endogenous compounds or allelochemicals. Given their unusually high activity, their occurrence in multiple forms, and the prevalence of potential substrates, their function is likely to be very important.

The general reaction results in formation of an acid and alcohol:

$$RCOO'R' + H_2O \rightarrow RCOOH + HO'R'$$

5. Epoxide Hydrolases

Epoxide hydrolases hydrolyze cyclic ether bonds by the stereospecific addition of water.[5] Like esterases, they are present in both microsomal and cytosolic fractions, and occur in multiple isoenzymic forms. Some individual forms exhibit specificity for cis or trans configurations. Epoxide hydrolase activities are inducible by standard PSMO inducers.[80]

Epoxides are strong electrophiles, and react readily with macromolecules such as proteins and nucleic acids to produce a variety of toxic effects. These compounds may be generated from the PSMO-mediated metabolism. Moreover, hundreds of epoxides have been identified in higher plants,[80] and these allelochemicals may also be detoxified by epoxide hydrolases.

The hydrolysis of epoxides results in the formation of diols:

$$\text{R-}\overset{\overset{\text{O}}{\wedge}}{\text{C-C}}\text{-R}' + H_2O \rightarrow \text{R-}\overset{\text{HO}}{\overset{|}{\text{C}}}\text{-}\overset{\text{OH}}{\overset{|}{\text{C}}}\text{-R}'$$

6. Glutathione Transferases

Glutathione transferases are probably the most important of the various conjugating enzymes in insects. These enzymes catalyze the conjugation of reduced glutathione (a tripeptide) to electrophilic portions (e.g., α,β-unsaturated carbonyls) of lipophilic compounds. Although the reaction does not require high-energy intermediates, energy is needed to produce the tripeptide and to maintain it in the reduced state. Thus, Ahmad et al.[5] consider glutathione not only as the conjugating agent but as an atypical high-energy intermediate. Glutathione transferases occur primarily as cytosolic but also as membrane-bound enzymes. They are inducible and occur in multiple isoenzymic forms.[3,124,128] In mammals, glutathione-conjugated compounds are further metabolized and excreted as mercapturic acid derivatives. The presence of a similar pathway in insects is not yet known, but the required enzymes do occur in some insects.[5]

Glutathione transferases are important in phase II metabolism of some reactive products formed by PSMO-catalyzed reactions. Their involvement in the metabolism of allelochemicals is still largely unknown, but they do appear critical to the detoxication of thiocyanates[120] and allelochemicals containing α,β-unsaturated carbonyls.[65,119] Given the frequency with which such compounds occur in plants, future research on this enzyme system will most likely only further establish its importance.

7. UDP-Glucosyltransferases

Animals conjugate xenobiotics with a variety of hexoses; of these, glucose is the most common sugar used by insects. In reactions approximately the reverse of those catalyzed by β-glucosidases, glucosyl transferases link glucose to xenobiotics, rendering them much more water soluble and readily excreted. Although glucose transferase activity has been measured in a variety of insects, few details are known about the enzyme system. The enzymes have been associated with soluble and microsomal cell fractions, and appear to exist in isoenzymic forms.[5]

Little information is available on the metabolism of allelochemicals by glucosyl transferases. These enzymes do catalyze the conjugation of glucose to a variety of phenols, aromatic carboxylic acids, alcohols, amines, and sulfhydryls. Ahmad et al.[5] thus argue that they probably play important roles in the elimination of allelochemicals or deactivation of such prior to sequestration.

In insects, glucose conjugation is preceded by the activation of glucose by UTP (catalyzed by UDPG-pyrophosphorylase) to form UDP-α-D-glucose. Conjugation via glucosyl transferase then occurs as:

$$\text{ROH} + \text{UDP-}\alpha\text{-D-Glucose} \rightarrow \text{RO-}\beta\text{-D-Glucose}$$

III. ENZYMATIC DETOXICATION AND INTERSPECIFIC VARIATION IN SUSCEPTIBILITY TO ALLELOCHEMICALS

This section highlights the manner in which enzymatic detoxication systems provide a mechanistic basis for observed differences in the toxicity of plant secondary compounds to insects. For this chapter, the word "toxicity" is used in the typical broad sense, i.e., capacity of a compound to cause deleterious effects in organisms.[39] For insects, deleterious effects include any that negatively influence the performance parameters generally associated with fitness, e.g., growth, development, survival and reproduction.

Roles of particular enzyme systems in the detoxication of specific allelochemicals can be inferred from several lines of evidence. One approach is to assess the toxicity of a secondary compound when administered in the presence and absence of chemical inhibitors of specific enzyme systems. This procedure is widely used to investigate pathways of metabolic degradation of insecticides.[92] For example, piperonyl butoxide and DEF (*S,S,S*-tributylphosphorotrithioate) selectively inhibit PSMO and esterase enzymes, respectively. Augmentation of biological activity in the presence of such specific inhibitors provides a clue to the route of detoxication of a compound. A second line of evidence involves isolation and identification of the metabolic products of an allelochemical. Roles of particular enzyme systems can be inferred from knowledge of the biochemical reactions required to transform the compound into its metabolic product(s). For example, isolation of glucosyl or glutathione conjugates would indicate metabolism via glucosyl or glutathione transferases. A third line of evidence includes a variety of enzymatic techniques. The preferred approach is to monitor enzymatic turnover of the allelochemical of interest in a semipurified enzyme preparation (usually from the midgut).

Addition or elimination of inhibitors (e.g., piperonyl butoxide) or cofactors (e.g., NADPH) to the reaction mixture helps to identify specific enzyme systems involved in the metabolism of an allelochemical. A host of technical problems (e.g., lability of the allelochemical in aqueous solution, absence of a suitable chromophore for spectrophotometric assay), however, usually precludes use of the allelochemical of interest as a substrate because enzymatic assays must be fast, convenient, and highly sensitive for routine laboratory practice. The latter characteristics are afforded by the use of model substrates; however, results from such studies must be interpreted with caution, because high levels of activity against a model compound do not necessarily correlate with activity against a substrate of interest. Enzyme-binding studies may also be employed to identify involvement of a particular enzyme system with the metabolism of an allelochemical. For example, spectral perturbation of cytochrome P-450 is often used as a means of confirming affinity of the enzyme for a compound. Binding, however, does not indicate rates of catalytic activity. Piperonyl butoxide, a PSMO inhibitor, binds readily to cytochrome P-450 but is metabolized very slowly.[5] Clearly, employing several lines of evidence is best in assessing detoxication of an allelochemical by a particular enzyme system. Determinations ideally should be made on the basis of both *in vivo* and *in vitro* studies.

A. POLYSUBSTRATE MONOOXYGENASES

The most thoroughly studied role of PSMO enzymes in conferring differential tolerance to plant secondary compounds involves the detoxication of furanocoumarins by various Lepidoptera. Furanocoumarins are tricyclic compounds derived from the shikimic acid pathway, and are characteristic of the plant families Umbelliferae (Apiaceae) and Rutaceae. The allelochemicals are also photoactive. Excited state compounds may interact directly with macromolecules such as nucleic acids, proteins, or lipids, or may react with oxygen to produce oxygen radicals, which in turn may react with macromolecules.[16]

The furanocoumarin xanthotoxin has been shown to be toxic to a variety of polyphagous Lepidoptera, dramatically increasing mortality rates of *Spodoptera eridania, S. frugiperda,* and *Heliothis virescens* at levels as low as 0.1 to 1.5% diet weight.[12,13,19] In contrast, performance of the black swallowtail (*Papilio polyxenes*) is actually improved at these levels.[13] The black swallowtail feeds exclusively on members of the Umbelliferae and Rutaceae, and thus regularly encounters furanocoumarins in its diet.[16]

A highly active PSMO enzyme system is responsible for detoxication of furanocoumarins by black swallowtail larvae.[38,41,59] Metabolism occurs in the midgut, thereby reducing uptake and transport of the toxins throughout body tissues. Ivie et al.[59] showed that after administration of a standard dose, the levels of xanthotoxin in the body tissues of *S. frugiperda* were nearly 60-fold higher than in *P. polyxenes.* PSMO activities, determined as *O*-demethylation of the model substrate *p*-nitroanisole, are 9 to 250 times higher in *P. polyxenes* than in xanthotoxin-susceptible Lepidoptera (Table 2).

A similar pattern holds for differential PSMO activity and susceptibility to xanthotoxin even within the Papilionidae. *Battus philenor* and *Papilio glaucus*

Table 2
PSMO ACTIVITY (*O*-DEMETHYLATION OF *p*-NITROANISOLE; $\overline{X} \pm 1$. S.E.) IN *PAPILIO POLYXENES* AND THREE POLYPHAGOUS LEPIDOPTERAN SPECIES

Species	Diet	Enzyme activity ($pmol \cdot min^{-1} \cdot mg\ prot^{-1}$)	Ref.
Papilio polyxenes	Wild parsnip	4266 ± 1651	81
Spodoptera frugiperda	Artificial diet	56 ± 9	123
	Soybean	45 ± 3	123
	Corn	257 ± 13	123
Spodoptera eridania	Artificial diet	17 ± 1—483 ± 43[a]	32
Heliothis zea	Artificial diet	<30—80 ± 18	82
	Artificial diet	479 ± 71	126

[a] Temperature-dependent variation.

Table 3
PSMO ACTIVITY (O-DEMETHYLATION OF *p*-NITROANISOLE; $\overline{X} \pm 1$ S.E.) AND EFFECTS OF XANTHOTOXIN (0.1%) ON LARVAL SURVIVAL OF PAPILIONIDS[16,81]

Species	Enzyme activity ($pmol \cdot min^{-1} \cdot mg\ prot^{-1}$)	Survival (%)	
		Control diet	Xanthotoxin diet
Papilio polyxenes	4266 ± 1651	85	95
Papilio cresphontes	1174 ± 1174	—	—
Papilio glaucus	307 ± 114	94	22
Battus philenor	88 ± 20	100	0

Note: *Papilio polyxenes* and *P. cresphontes* larvae used for enzyme assays were reared on plants containing furanocoumarins.

do not normally feed on plants containing furanocoumarins. These species perform poorly on diets containing 0.1% xanthotoxin, and exhibit low PSMO activities (Table 3). In contrast, *P. polyxenes* and *P. cresphontes* (a Rutaceae specialist) normally consume plants with furanocoumarins, and have high to extremely high PSMO activity. Some of the documented difference in enzyme activities, however, may be attributed to induction; both *P. polyxenes* and *P. cresphontes* were reared on plants containing furanocoumarins.

Another species that exhibits PSMO-based tolerance to furanocoumarins is the parsnip webworm (*Depressaria pastinacella*). This insect feeds exclusively on wild parsnip (*Pastinaca sativa*) and a few closely related umbellifer species.[20] Its performance is unaffected by the furanocoumarin, xanthotoxin, which is rapidly metabolized by midgut PSMOs.[87] This enzyme-based resistance is not as effective, however, against bergapten, another furanocoumarin. Berenbaum et al.[20] found that bergapten reduced growth and digestion efficiency of *D. pastinacella.* A seemingly minor difference in chemical structure may inhibit detoxication of bergapten by PSMOs.

One of the most interesting cases of PSMO-mediated detoxication of allelochemicals is that of monoterpene metabolism by bark beetles (Scolytidae). Successful colonization of their coniferous hosts requires mass attack by these insects, which itself is facilitated by the production of aggregation pheromones.[91] The pheromones, produced from host monoterpenes, are typically monoterpene alcohols. These oxygenated products are likely to come from PSMO-catalyzed reactions, and that such is the case for *Dendroctonus ponderosae* was confirmed by Hunt and Smirle.[58] Bark beetles are not, however, uniformly tolerant of conifer terpenes. The fir engraver (*Scolytus ventralis*) is repelled by host (grand fir) monoterpenes, whereas *D. ponderosae* is not strongly repelled by any host (lodgepole pine) compounds. Acute toxicity tests showed that *S. ventralis* is also much more susceptible to monoterpenes than is *D. ponderosae.*[91] Raffa and Berryman[91] argue that high constitutive concentrations of monoterpenes in pines (as opposed to levels in fir) have led to the evolution of mechanisms affording tolerance in *D. ponderosae.* That differences in susceptibility of *S. ventralis* and *D. ponderosae* are due to differences in their PSMO enzyme systems is likely, but currently unconfirmed.

Because metabolism by PSMOs can occasionally increase the biological activity of compounds, susceptibility of insects to some allelochemicals may at times be inversely correlated with PSMO activity. Gunderson et al.[54] propose that such a phenomenon may explain differential toxicity of the mint monoterpene, pulegone, to *Spodoptera eridania* and *S. frugiperda.* Pulegone is metabolized by PSMOs in both species of armyworms; one of the hydroxylation products may rearrange spontaneously to form menthofuran. Both pulegone and menthofuran are differentially toxic to *S. eridania* and *S. frugiperda.* In acute toxicity trials pulegone and menthofuran were 4.2- and 3.5-fold more toxic to *S. eridania* than *S. frugiperda*, respectively. Moreover, *S. eridania* has significantly higher cytochrome P-450 content and PSMO activities than does *S. frugiperda.* Gunderson et al.[54] suggest that the product(s) of PSMO metabolism of pulegone and menthofuran may be more reactive than the parent compounds, which may explain why *S. frugiperda* is more tolerant of the monoterpenes than is *S. eridania.* The reaction products were not identified, however, so this explanation is tentative.

B. CARBONYL REDUCTASES

Reduction reactions appear to play relatively minor roles in the detoxication of xenobiotics by animals, and the enzymes catalyzing such reactions have accordingly been the subject of few empirical studies. They have been implicated as important factors, however, in several plant-insect associations.

A phytochemical trait characteristic of the milkweeds (*Asclepias* sp.) is the production of complex mixtures of cardenolides, C_{23} steroid glycosides. Monarch (*Danaus plexippus*) caterpillars are adapted to feed on milkweeds; some of the cardenolides are sequestered while others are metabolized and excreted. The compounds are toxic and deterrent to some vertebrate[35,36] and insect[18] predators. Marty and Krieger[73] investigated metabolism of the cardenolide, uscharidin, by *D. plexippus.* They found that uscharidin was metabolized by reduction of a keto group on the glycosidic portion of the molecule to form the enantiomers, calactin and calotropin. Carbonyl reductases are probably involved in the transformation, but activity of this enzyme system and the extent to which it confers tolerance to cardenolides in *D. plexippus* are unknown.

The roles played by carbonyl reductases in the comparative toxicity of allelochemicals are best (although still poorly) known with respect to quinone metabolism. Juglone, a 1,4-naphthoquinone, commonly occurs in members of the Juglandaceae and is deterrent and toxic to a variety of insects at concentrations as low as 0.01% diet weight.[52,68,125] Juglone is metabolized by carbonyl reductases in several species of Lepidoptera,[65,68,125] and variation in the susceptibility of these insects to juglone appears to be linked to reductase activity (Table 4). In contrast to Lepidoptera unadapted to juglone-containing plants, *Actias luna* larvae exhibit exceptionally high levels of reductase activity. This high activity may at least partially explain why luna larvae perform well on a variety of members of the Juglandaceae.[65]

C. GLYCOSIDASES

β-Glucosidases catalyze the hydrolysis of both nutrients (e.g., oligosaccharides) and allelochemical glycosides in insects. The latter function often gives rise to products more biologically active than their precursors, and detrimental to insects. Yu[127] showed that a variety of allelochemicals were more toxic than their corresponding glycosides to *Spodoptera frugiperda* larvae. Thus, the variation in β-glucosidase-mediated metabolism of plant glycosides may provide insight into the differential toxicity of these compounds to insects.

The work of Reilly et al.[93] provides an interesting example. Growth of the peachtree borer (*Synanthedon exitiosa*) is improved by amygdalin, a cyanogenic glycoside. In contrast, growth of the lesser peachtree borer (*S. pictipes*) tends to decline in the presence of amygdalin. β-Glucosidase activities of *S. exitiosa* collected from infested trees were ninefold greater than those of *S. pictipes.* Moreover, *S. exitiosa* exhibited measurable β-cyanoalanine synthase (enzyme responsible for the detoxication of cyanide released upon hydrolysis of cyanogenic glycosides) activity, whereas *S. pictipes* did not. Thus, the combination of high glucosidase and synthase activities in *S. exitiosa* ostensibly enables this species to utilize the products of cyanogenic glycoside metabolism for improved growth. Reilly et al. suggest differential host and tissue selection behaviors of *S. exitiosa* and *S. pictipes* may be influenced by interspecific variation in the metabolism of cyanogenic glycosides.

Differences in β-glucosidase activity between subspecies of the eastern tiger swallowtail (*Papilio glaucus*) may partially explain differential susceptibility of the two subspecies to phenolic glycosides. Several phenolic glycosides from aspen are toxic at naturally occurring concentrations to *P. g. glaucus,* but

Table 4
CARBONYL REDUCTASE ACTIVITY,[a] GLUTATHIONE TRANSFERASE ACTIVITY,[b] AND SURVIVAL ON JUGLONE-CONTAINING DIETS FOR VARIOUS SPECIES OF LEPIDOPTERA[65,68,125,126]

Species	Reductase activity		Transferase activity	Larval survival (%)
	Soluble	Microsomal		
Spodoptera frugiperda	26.9 ± 1.5	53.2 ± 2.9	692 ± 30	0, 50[c]
Heliothis zea	—	34.6 ± 0.8	971 ± 60	—
H. virescens	—	46.4 ± 1.7	2147 ± 143	—
Anticarsia gemmatalis	—	51.1 ± 2.3	61 ± 5	—
Lymantria dispar	40.5 ± 4.4—68.7 ± 5.6	62.2 ± 14.4—99.4 ± 9.8	475 ± 51—715 ± 58	<10[d]
Actias luna	47.0 ± 2.0—61.3 ± 2.7	173.9 ± 9.6—288.3 ± 17.3	1016 ± 86—2867 ± 378	92—98[e]

Note: Survival rates given for larvae fed foliage containing juglone or artificial diets containing natural levels (0.1—0.2% fresh weight, Hedin et al.[55]) of juglone. Enzyme activities given as $\overline{X} \pm 1$ S.E.

[a] nmd . $min^{-1} \cdot mg \cdot prot^{-1}$, measured as NADPH oxidation.
[b] nmd . $min^{-1} \cdot mg \cdot prot^{-1}$, measured as conjugation of 1-chloro-2,4-dinitrobenzene.
[c] Artificial diet containing 0.1 or 0.2% juglone, respectively; survival from egg hatch to adult emergence.
[d] Artificial diet containing 0.1 or 0.2% juglone; survival from egg hatch through third larval stadium.
[e] Survival of first instars on black walnut, butternut, and shagbark hickory foliage.

not to the aspen-adapted *P. g. canadensis.*[71] These compounds are hydrolyzed by β-glucosidases in each subspecies, but rates of hydrolysis are significantly lower in *P. g. canadensis* than in *P. g. glaucus.*[64] Lower activity reduces the rate of production of the putative toxic aglycone, and thus may represent one form of adaptation of *P. g. canadensis* to phenolic glycosides in its host plants.

D. ESTERASES

Roles of esterases in conferring differential tolerance to ecologically relevant allelochemicals have, until recently, been virtually unknown. Several studies have now indicated that they play pivotal roles in interactions between tiger swallowtails and their host plants. The species *Papilio glaucus* is comprised of three subspecies in North America. *P. g. canadensis* occurs throughout most of Canada and the northern tier of states east of the Great Plains in the U.S. This subspecies feeds extensively on members of the Salicaceae, whose most characteristic phytochemical feature is phenolic glycosides. *P. g. glaucus* occurs throughout the eastern U.S., as far south as Florida. *P. g. australis* has a narrowly restricted range in central and southern Florida. The latter two subspecies do not survive to pupation when reared on members of the Salicaceae. Phenolic glycosides, particularly salicortin and tremulacin, are responsible for differential host use abilities; these compounds deter feeding and cause degenerative gut lesions in unadapted insects.[70,71,108]

Particularly high midgut esterase activity appears to provide *P. g. canadensis* with tolerance to phenolic glycosides in its salicaceous hosts. Esterase activity is threefold greater in *P. g. canadensis* than in *P. g. glaucus*, and is (moderately) inducible as a consequence of consumption of phenolic glycosides in the former but not the latter subspecies.[67] Scriber et al.[108] showed with a variety of *P. glaucus* hybrids and backcrosses that the ability to grow on diets containing phenolic glycosides was closely correlated with midgut esterase activity. These results have been corroborated by *in vivo* toxicity trials using detoxication enzyme inhibitors. When *P. g. canadensis* larvae are fed diets containing phenolic glycosides and DEF, an esterase inhibitor, survival rates plummet.[67] Development of an efficacious esterase detoxication system has apparently played a central role in the evolution of host range among tiger swallowtails.[16,67,107]

E. EPOXIDE HYDROLASES

As of this writing, no examples exist for a role of epoxide hydrolases in the differential toxicity of allelochemicals to insects. This is due in part to the fact that the enzyme system is not amenable to several standard toxicological techniques. For example, the lack of effective enzyme inhibitors and the poor availability of radiolabeled substrates have limited both *in vivo* and *in vitro* approaches.[80]

Several lines of evidence suggest, however, that epoxide hydrolases may be important in conferring resistance to plant secondary compounds or products of PSMO-catalyzed metabolism of such compounds. Epoxide hydrolase activity tends to be lower in predatory insects than in herbivorous insects[126] and lower in phloem-sucking insects than in leaf-chewing insects,[79] patterns consistent with the relative exposure of these insects to plant allelochemicals. In addition, epoxide hydrolases appear to be involved in differential resistance to synthetic toxicants such as insecticides. For example, pioneer populations of *Diabrotica virgifera virgifera* in central Pennsylvania are much more resistant to aldrin than is an endemic population of *D. barber. D. v. virgifera* exhibit much higher aldrin epoxidase and epoxide hydrolase activities than *D. barberi.*[111] Epoxide hydrolases are likely to confer differential resistance to allelochemicals as well as to insecticides in phytophagous insects.

F. GLUTATHIONE TRANSFERASES

Glutathione transferases were recently implicated as responsible for the differential toxicity of glucosinolates to Lepidoptera. Glucosinolates are common constituents of members of the Cruciferae. When plant tissues are damaged, thioglucosidase enzymes hydrolyze glucosinolates to form organic products, which in turn rearrange to produce isothiocyanates (most commonly), thiocyanates, and nitriles.[28]

Wadleigh and Yu[120] investigated the toxicity and metabolism of allyl isothiocyanate in two generalist (*Spodoptera frugiperda* and *Trichoplusia ni*) and a specialist (*Anticarsia gemmatalis*) Lepidoptera. These species are adapted to cruciferous plants in the order *T. ni* > *S. frugiperda* > *A. gemmatalis*. The authors showed that allyl isothiocynate is twice as toxic to *S. frugiperda* than it is to *T. ni* and *A. gemmatalis*, and that conjugation of allyl isothiocyanate by glutathione transferases is sixfold greater in *T. ni* than *S. frugiperda*, but not detectable in *A. gemmatalis*. Clearly, more is involved in the differential susceptibility of these species to glucosinates than can be explained by transferase-catalyzed metabolism. Wadleigh and Yu suggest, however, that differential toxicity of thiocyanates to *S. frugiperda* and *T. ni* may be explained by differences in the glutathione transferase activity of the two species.

Glutathione transferases are involved in the metabolism of a variety of allelochemicals containing α,β-unsaturated carbonyls.[119] This structural variety is commonly found in alkaloids, coumarins, quinones, terpenoids and simple phenols. For example, glutathione transferases may provide a second line of defense against juglone in *Actias luna*. Transferase activities in this species are exceptionally high (Table 4) and induced 2 to 2.8-fold in larvae feeding on foliage containing juglone.[65]

G. UDP-GLUCOSYL TRANSFERASES

Glycosylation reactions in insects are probably less common than in plants, but the process serves similar purposes in both organisms. Glycosylated products are more water soluble and thus more easily transported; they also exhibit reduced biological activity and thus serve as storage forms.

UDP-glucosyl transferases are likely involved in conferring tolerance to cucurbitacins in some insects. Cucurbitacins are oxygenated tetracyclic triterpenes characteristic of plants in the family Cucurbitaceae. These compounds are highly toxic to mammals[45,78] and a deterrent to many arthropods.[72] Little information exists on direct toxicity of the compounds to insects, although they reduce growth and fecundity of *Epilachna borealis*.[114,115] This species is adapted to feeding on cucurbits, but avoids encountering high (induced) concentrations of mobile cucurbitacins by "trenching" leaves and feeding distally to the trench.

Other phytophagous beetles rely on metabolic detoxication as an adaptation for cucurbit feeding. Cucurbitacins are not only nontoxic to beetles in the genera *Diabrotica* and *Acalymma*, but actually serve as feeding stimulants for these insects.[78] Moreover, Ferguson and Metcalf[50] showed that *Diabrotica* and *Acalymma* that fed on cucurbit fruits rich in cucurbitacins were rejected as food by the praying mantis, *Tenodera aridifolia sinensis*. A significant portion of ingested curcurbitacin is stored as a polar conjugate in hemolymph and tissues of *Diabrotica* and *Acalymma*.[51] Andersen et al.[8] fed cucurbit fruits and purified cucurbitacins to *Diabrotica virgifera virgifera* and *D. undecimpunctata howardi*, and subsequently isolated four metabolic products from beetle tissues. All were curcurbitacin *O*-glucosides. Evidence indicated that the cucurbitacins had undergone hydrogenation, desaturation, and acetylation reactions. Measurements of UDP-glucosyl transferase activity were not made, but these enzymes are most likely responsible for glucosylation of the cucurbitacins.

UDP-glucosyl transferases have also been implicated in conferring cycasin tolerance to caterpillars of the moth, *Seirarctia echo*. These larvae are adapted to feeding on plants in the genus *Cycas*, which produce the glycoside, cycasin. Methylazoxymethanol, the aglycone of cycasin, is a strong alkylating compound, with hepatotoxic and carcinogenic effects in mammals.[102] Teas[116] showed that *S. echo* larvae have highly active gut β-glucosidases, which convert ingested cycasin to methylazoxymethanol. β-Glucosidase activity is low or absent in other body tissues and hemolymph. The question arose as to how larvae avoided toxicity from the aglycone. Teas then fed purified methylazoxymethanol to larvae and recovered cycasin in small amounts from the gut but in high concentrations from hemolymph, fat body, and Malpighian tubules. Thus, *S. echo* larvae avoid methylazoxymethanol toxicity by glucosylating the compound, probably via UDP-glucosyl transferases. Cycasin distributed in body tissues and hemolymph may serve as a chemical defense against predators of the larvae. The question remains as to why larvae exhibit high gut β-glucosidase activity, and the likely answer is that the enzymes are required for hydrolysis of dietary constituents other than cycasin.

H. MISCELLANEOUS ENZYME SYSTEMS

A variety of other enzyme systems, not readily classified according to the categories described previously, also metabolize xenobiotics in insects. In a few instances, these have been shown to play important roles in the detoxicaton of plant secondary compounds.

1. Detoxication of Cyanogenic Glycosides

The production of hydrogen cyanide upon crushing of leaf tissue is widespread in the plant kingdom, occurring in representatives of 100 plant families.[43] The cyanide precursors are typically cyanogenic glycosides, which undergo hydrolysis by plant enzymes upon tissue disruption, or by animal enzymes upon ingestion, to form cyanohydrins. These in turn dissociate (enzymatically or nonenzymatically) to release hydrogen cyanide. Cyanide is considered a potent toxin to most animals; it binds to the terminal cytochrome oxidase in the mitochondrial electron transport chain and thereby inhibits cellular respiration. Few studies have tested the effects of purified cyanogenic glycosides on insects, but cyanogenic foliage is known to deter feeding, for example, of *Locusta migratoria,*[121,122] and to reduce growth and increase mortality of the sawflies *Strongylogaster impressata* and *S. multicincta.*[104]

Insensitivity to cyanide in some insects appears to be due (in addition to potential target site insensitivity[15]) to detoxication by β-cyanoalanine synthase.[5,28] This enzyme catalyzes the first in a series of reactions that channel cyanide into the amino acid pool.[28,46] As mentioned earlier, the cyanogenic glycoside amygdalin promotes growth of *Synanthedon exitiosa*, but has no, or slightly negative, effect on the growth of *S. pictipes.* Reilly et al.[93] attributed this difference partially to β-cyanoalanine synthase activity, which is higher in *S. exitiosa* than in *S. pictipes.* The southern armyworm (*Spodoptera eridania*) provides another example. This polyphagous herbivore appears to prefer cyanogenic plants. Brattsten et al.[33] showed that cyanide stimulated feeding and promoted growth of all instars of *S. eridania.* Brattsten[31] later reported that this adaptation is probably due to high β-cyanoalanine synthase activity in *S. eridania*, although no data were provided.

2. Detoxication of Nonprotein Amino Acids

One of the most thoroughly documented cases of insect biochemical adaptation to plant defenses involves predation by bruchid beetles on canavanine-containing seeds of plants in the family Fabaceae. Canavanine, a nonprotein amino acid, has been detected in hundreds of species of legumes[11] and occurs in concentrations as high as 5 to 10% of seed weight.[24] Canavanine is a structural analog of arginine, and is thus utilized by nonadapted insects in protein synthesis, causing production of malfunctional proteins and in some cases lethal morphological malformations.[97,98,100]

The bruchid beetle *Caryedes brasiliensis* avoids canavanine toxicity by means of target site insensitivity; its modified arginyl-tRNA synthetase does not bind canavanine. Indeed, Rosenthal et al.[99] found that the substitution error frequency (frequency with which canavanine replaces arginine in newly synthesized proteins) was 1 in 360 for *C. brasiliensis* but as high as 1 in 6 for the canavanine-unadapted *Manduca sexta.*

In addition to insensitive target sites, bruchid beetles employ enzymatic mechanisms to catabolize canavanine and recycle the nitrogenous products for use in amino acid biosynthesis. Nitrogen utilization begins with the enzyme arginase, which catalyzes hydrolysis of canavanine to canaline and urea. Canaline itself is potentially neurotoxic; it is converted in *C. brasiliensis* to ammonia and the nontoxic nonprotein amino acid, homoserine, both of which may be recycled into the amino acid pool.[99] Urease metabolizes urea to produce carbon dioxide and ammonia, the latter which may be reused in amino acid synthesis. Bleiler et al.[24] measured arginase (with arginine and canavanine as substrates) and urease activity in three species of canavanine-feeding and nine species of noncanavanine-feeding insects. They found no difference between the two groups in their ability to catabolize canavanine relative to their ability to catabolize arginine. They did find, as predicted, exceptionally high urease activity in canavanine-feeding insects and low to undetectable activity in noncanavanine-feeding insects.

3. Detoxication of Reactive Oxygen Species

Although dioxygen (O_2) is essential for all higher life forms, it also serves as a precursor for the production of a variety of potentially toxic oxygen species. Ground-state molecular oxygen ($^3O_2$) can be the recipient of energy transfer from activated, triplet state molecules to form singlet oxygen ($^1O_2$). Alternatively, $^3O_2$ may undergo one-electron reduction to produce superoxide ($O^{\cdot -}_2$), which in turn may lead to the formation of hydroperoxy and hydroxyl radicals, and hydrogen peroxide.[6,63] Superoxide and hydrogen peroxide are not directly reactive toward most biological molecules,[63] but oxygen radicals generated from these compounds react readily with macromolecules, causing, for example, the deleterious peroxidation of lipids and DNA.

Potentially toxic forms of oxygen may be generated in insects from interactions with endogenous compounds or with prooxidants of plant origin. Examples of the latter include furanocoumarins, which are photochemically activated and produce both superoxide and singlet oxygen, and compounds such as flavonoids and quinones, which are metabolically activated and produce superoxide, hydrogen peroxide, and hydroxyl radicals.[88]

Biochemical adaptation to exposure to allelochemicals that function as prooxidants involves a suite of antioxidant enzymes.[4,7,88,90] Superoxide dismutase catalyzes the dismutation of superoxide to form hydrogen peroxide and ground state oxygen. In turn, catalase promotes the conversion of hydrogen peroxide to oxygen and water. Sequential action of these two enzyme systems appears to be the major antioxidant defense system in insects. Glutathione peroxidases metabolize hydrogen peroxide and other hydroperoxides in vertebrates, but their activity in insects is very low. This low activity may be offset by high peroxidase activity of glutathione transferases in insects, which may provide another form of defense against oxidants. Finally, glutathione reductase may be important for the provision of reduced glutathione for use by glutathione transferases.

Pardini et al.[88] reviewed research on the role of these antioxidant enzymes in differential toxicity of xanthotoxin (a furanocoumarin) and quercetin (a flavonoid) to *Trichoplusia ni, Spodoptera eridania,* and *Papilio polyxenes.* The authors argue that these three species experience low, intermediate and high exposure, respectively, to prooxidants in their host plants. *T. ni* is much more susceptible to both allelochemicals than is *S. eridania,* which in turn is more susceptible than *P. polyxenes* to xanthotoxin. (Toxicity of quercetin to *P. polyxenes* was not reported.) Antioxidant enzyme profiles generally corresponded with susceptibility to the prooxidants. Third, fourth, and fifth instar *P. polyxenes* exhibited higher superoxide dismutase and glutathione reductase activity than did *T. ni,* although catalase activity was higher only in fifth instars. *S. eridania* had intermediate superoxide dismutase, glutathione reductase, and low catalase activities. Moreover, when diethyldithiocarbamate (DETC) was used to inhibit superoxide dismutase activity in *P. polyxenes* and *S. eridania,* toxicity of quercetin increased substantially.[6] Berenbaum[16] reviewed data on antioxidant enzyme activities in *P. polyxenes* and *P. glaucus.* Activities of four enzyme systems were markedly higher in *P. polyxenes* than in *P. glaucus,* results that correspond to the relative exposure of these species to prooxidants. Tolerance to furanocoumarins in *P. polyxenes* thus appears to be due to both extraordinarily high PSMO activity (see Section III.A) and to a highly active suite of antioxidant enzymes.

IV. FACTORS INFLUENCING INTRASPECIFIC VARIATION IN ENZYME ACTIVITY AND SUSCEPTIBILITY TO ALLELOCHEMICALS

A host of factors is known to influence the activity of detoxication enzymes in animals, and hence, susceptibility to xenobiotics. Some of these, such as intraspecific genetic variation, sex, reproductive condition, and disease, have been well documented in vertebrates,[44] but remain virtually unknown with respect to the metabolism of allelochemicals by insects. These factors are clearly deserving of study. Others have received at least a modicum of investigation and are briefly reviewed here.

A. DIET

Nutritional factors have been shown in numerous studies to alter the activity of insect detoxication systems. Both allelochemicals and nutrients have been implicated, although surprisingly few studies have addressed the latter.

1. Effects of Allelochemicals

Consumption of allelochemicals may effect changes in metabolic detoxication such that toxicity of the same or other allelochemicals is markedly altered in subsequent exposures. One of the primary mechanisms responsible is that of enzyme induction, which has been documented to occur in all the major detoxication systems.[117,124] A second mechanism is enzyme inhibition due to exposure to phytosynergists (plant compounds that function as synergists of simultaneously occurring toxins).

Many studies have shown that the activity of detoxication enzymes increases when insects are fed particular allelochemicals or host plants. Indeed, induced activity has often been interpreted as evidence for the involvement of a specific enzyme system in the metabolism of a substrate of interest. Such conclusions, however, are best considered tentative; not all enzyme substrates are inducers, and not all inducers are substrates.[53]

The number of cases in which allelochemicals have been shown to induce enzymes that are responsible for their own metabolism is relatively small, but sufficient to indicate that this biochemical response is important in insect-plant interactions. Cohen et al.[41] showed that PSMO-mediated metabolism of xanthotoxin in *Papilio polyxenes* increased 7.6-fold, relative to control values, in larvae fed diets containing 0.5% xanthotoxin. Induction was less at a lower concentration (0.1%) of dietary xanthotoxin, but no greater at a higher concentration (1%), indicating a limit to which activity could be induced. Similarly, xanthotoxin metabolism was induced by 48% in response to feeding on xanthotoxin-containing diets by *Depressaria pastinacella*.[87] Wadleigh and Yu[120] found that feeding isothiocyanates to *Spodoptera frugiperda* and *Trichoplusia ni* enhanced the capacities of both species to detoxify the compounds via conjugation with glutathione. Glutathione transferase and carbonyl reductase are likely to be responsible for the metabolism of juglone in *Actias luna*, and activities of both were induced when larvae were reared on juglone-containing foliage.[65] Finally, esterases afford protection from phenolic glycosides to *Papilio glaucus canadensis,* and general esterase activity was elevated 22% after consumption of a phenolic glycoside diet.[67] The induction capacity of hydrolytic enzyme systems (e.g., esterases, epoxide hydrolases) is generally marginal in comparison to that of PSMOs and glutathione transferases. Higher maintenance costs associated with the enzyme-coenzyme complexes of the PSMO and glutathione transferase systems may have influenced selection for the capacity to alter catalytic activity on an "as needed" basis.[80]

Elevated enzyme activities presumably provide enhanced protection against the inducing allelochemical. As logical as this conclusion may be, it has not been explicitly demonstrated. Brattsten et al.[33] showed that *S. eridania* larvae preexposed to cyanide are less susceptible to the compound than unexposed larvae. They suggested, but did not document, that induction of β-cyanoalanine synthetase activity may be the determining factor. In another study, Brattsten[26] demonstrated that α-pinene induced PSMO activity in *S. eridania,* and larvae were subsequently more tolerant of an unrelated allelochemical, nicotine. That induced activity does provide general protection is indicated by the fact that allelochemical-mediated induction often reduces the susceptibility of insects to insecticides.[31]

Plant tissues typically contain a variety of secondary compounds, some of which may serve as synergists for others.[14] The mode of action of phytosynergists is probably varied, but many are likely to function by inhibiting specific detoxication enzymes. For example, compounds containing methylenedioxyphenyl (MDP) functional groups may bind to the active site of cytochrome P-450 and thereby inhibit enzyme activity against alternative substrates.[5] Such compounds occur in numerous classes of secondary metabolites.[19,83] Myristicin, safrole, isosafrole, and fagaramide are MDP-containing phenyl propanoids that simultaneously occur, at low concentrations, with furanocoumarins in the Umbelliferae. These compounds exhibit no toxicity against *Heliothis zea* at concentrations of 0.01 to 0.1% diet weight, but potentiate the toxicity of xanthotoxin 2.5- to 6.5-fold at the same concentrations.[19,84] The question then arises as to how *P. polyxenes*, an umbellifer specialist, is able to utilize PSMOs to detoxify furanocoumarins when the compounds co-occur with PSMO inhibitors. Apparently *P. polyxenes* has evolved PSMO enzymes with reduced sensitivity to MDP-containing inhibitors. Neal and Berenbaum[85] showed that *P. polyxenes* is 46- and 10-fold less sensitive to myristicin and safrole, respectively, than is *H. zea*.

2. Effects of Nutrients

Nutrient imbalances, particularly deficiencies, are well known for eliciting significant changes in the susceptibility of vertebrates to xenobiotics.[25,40,44,89,113] Such changes may occur as a consequence of general pathology, or due to alterations in specific detoxication systems. In mammals, inadequate quantity or quality of dietary protein generally reduces cytochrome P-450 concentrations and PSMO activity. Protein-deficient diets also influence conjugation enzymes, but the effect is less consistent; both increases and decreases have been observed. Inadequate dietary carbohydrates and lipids may also lower PSMO activity, although direct and indirect effects are difficult to differentiate. For example, inadequate dietary carbohydrate may lead to the catabolism of protein as an energy source, and this loss of protein may in turn influence detoxication activity.

Surprisingly, no published information exists on the influence of nutrient concentrations on insect detoxication systems. It is reasonable to expect, however, that the effects documented for vertebrates occur in insects as well. The performance of phytophagous insects, especially tree-feeders, is generally believed to be nutrient limited. The most important limiting nutrient is protein, but inadequate carbohydrates and minerals may also reduce performance.[74,77,106] Thus, the capacity of insects to deal with allelochemicals in their food plants is likely to be influenced by nutrient-mediated changes in biochemical detoxication systems.

Several recent experiments in the author's laboratory assessed the effects of dietary nutrients on detoxication enzymes and susceptibility to allelochemicals in the gypsy moth (*Lymantria dispar*). One study investigated the interactive effects of juglone and protein on larval performance and detoxication systems.[68] Low protein resulted in increased esterase and carbonyl reductase (soluble enzymes) activities, but decreased glutathione transferase activities. Reduced transferase activities were not surprising, given that protein deficiencies may have limited the availability of glutathione (a tripeptide) for conjugation *in vivo*. Especially low glutathione reductase activity may have contributed to the very poor performance of larvae fed a low protein, high juglone diet.

A second study addressed the consequences of protein, mineral, and vitamin deficiencies for the activity of four enzyme systems (Table 5). Deficient diets all reduced larval growth rates, compared to those of control animals, but few significant effects and no clear-cut patterns of enzyme response to nutrient limitation were observed. PSMO and glutathione transferase activities were not significantly altered. (The lack of a protein effect on transferase activity contradicts earlier results, and may be a consequence of using a suboptimal, low wheat germ control diet in the second study.[69]) Soluble carbonyl reductase and esterase activities tended to increase in larvae fed a protein-deficient diet, but decreased in larvae fed a vitamin-deficient diet. In short, enzyme activities were surprisingly resistant to diet-induced suppression. Conceivably, insects adapted to feeding on allelochemical-rich but nutrient-poor host plants may have evolved biochemical/physiological mechanisms that help maintain effective enzyme function under conditions of nutrient deficiency. A comparison of

Table 5
EFFECTS OF NUTRIENT DEFICIENCIES ON GROWTH PERFORMANCE (FOURTH INSTARS) AND DETOXICATION ENZYME ACTIVITY (FIFTH INSTARS) IN *LYMANTRIA DISPAR* ($\bar{X} \pm 1$ S.E.[69])

		Enzyme activity[b]					
		PSMO		Carbonyl reductase			
Diet	Relative growth rate[a]	NADPH oxidation	Cytochrome *c* reductase	Soluble	Microsomal	Esterase	Glutathione transferase
Control	0.22 ± 0.01	1.96 ± 0.27	32.1 ± 5.2	40.2 ± 1.0	84.4 ± 7.4	1474 ± 28	372.7 ± 38.9
Low protein	0.15 ± 0.01	1.74 ± 0.30	60.9 ± 14.0	60.4 ± 9.7	87.0 ± 8.7	1906 ± 94	321.3 ± 27.4
Low mineral	0.16 ± 0.02	1.55 ± 0.31	70.7 ± 18.9	51.6 ± 4.8	59.1 ± 13.2	1622 ± 86	293.9 ± 40.0
Low vitamin	0.16 ± 0.01	2.79 ± 0.22	74.3 ± 5.3	25.1 ± 1.9	68.5 ± 2.8	1218 ± 49	349.6 ± 26.3

[a] Units are mg · mg^{-1} · d^{-1}.
[b] Units are nmol · min^{-1} · mg $prot^{-1}$.

the effects of nutrient stress on detoxication activity in insects adapted to relatively nutrient-rich (e.g., forb foliage) vs. nutrient-poor (e.g., tree foliage) foods would be interesting.

B. AGE AND DEVELOPMENT

PSMO activity varies considerably among and within developmental stages, and the pattern is fairly consistent among different insect species.[2,3,27,29,57,60] Typically, activity is low in the egg stage, increases with each larval or nymphal instar, then declines to zero at pupation. In phytophagous adults, activity increases after emergence, then later declines with senescence. Within a stadium, PSMO activity is initially low, rises as insects begin feeding, then declines prior to the next molt. Epoxide hydrolases are the only other insect detoxication system with which age-activity relationships have been documented; within-stadium variation corresponds to that described for PSMOs.[80] In general, enzyme activity corresponds closely with feeding activity, a relationship used to support early arguments about the role of these enzymes in detoxifying allelochemicals.[3,27,53] Exceptions to these general patterns, however, do occur.[53]

Developmental changes in detoxication enzyme activities may parallel concomitant changes in the amounts and diversity of plant allelochemicals to which insects are exposed. For example, relative (weight-adjusted) feeding rates and, consequently, exposure to allelochemicals in *Lymantria dispar* larvae increase with successive larval stadia.[3] Moreover, diet breadth of individual *L. dispar* often increases with age, so larvae consume a greater variety of secondary compounds.[9,62,110] Indeed, reduced growth efficiencies of Lepidoptera reared on multi- rather than single-species diets have been attributed to the metabolic costs of detoxication,[103,105] although no evidence yet exists to support this view.[82] In summary, the potential for ontogenetic changes in detoxication systems to mediate insect-plant interactions is intriguing, but largely unexplored.

C. TEMPERATURE

Biochemical and physiological processes are strongly temperature dependent, so changes in ambient temperatures can markedly influence performance parameters of ectotherms. Within a nonlethal range, temperature affects development, growth, and feeding rates of insects, and effects may vary with different food plants.[109,112] Moreover, insect resistance to insecticides is also temperature dependent, although both positive and negative correlations of toxicity with temperature have been observed.[32,56,118] Thus, temperature-regulated activity of detoxication enzymes may influence insect susceptibility to host allelochemicals.

Few studies have considered effects of temperature on insect detoxication systems. Brattsten et al.[32] found that cytochrome P-450 content and PSMO activity were higher in *Spodoptera eridania* reared at 15°C than in larvae reared at 30°C. In addition, larvae reared under warmer temperatures were 3.5-fold more susceptible to the insecticide carbaryl. No published information exists on temperature modulation of enzyme activities which in turn alter susceptibility to allelochemicals.

Clearly, we are only beginning to understand whether and how many endogenous and exogenous factors influence insect detoxication systems and, consequently, plant-insect interactions. That such effects occur is likely, but their relevance for ecological phenomena such as host selection and insect population dynamics remains to be seen. For example, Mattson and Haack[75,76] suggested that drought conditions improve nutrition and temperature regimes for insects, which in turn enhance the efficacy of detoxication systems and promote population outbreaks. Relevant data are too sketchy at this time, however, to regard this theory as more than speculative.

V. ESTABLISHING A THEORETICAL FRAMEWORK: PATTERNS IN METABOLIC DETOXICATION VIS À VIS PLANT CHEMISTRY

The notion that one of the primary evolutionary responses of plants to herbivory has been the development and diversification of a plethora of secondary compounds is now generally accepted. During the 1970s and 1980s, ecologists witnessed a meteoric rise in interest in the chemical nature of insect-plant interactions. Much of this interest was fueled by compelling, well-articulated theories purporting to explain patterns in the diversity, distribution, and abundance of plant secondary compounds in relation to herbivory. Most notable among these were plant apparency theory,[49,95] optimal defense theory,[94] and more recently, resource availability theory.[10,37,42] Although not completely satisfactory, these theories provided explanatory power in interpreting recognized patterns, and were of considerable heuristic value in providing a framework for additional research.

In contrast, research on insect biochemistry (in relation to host use) has suffered for lack of a guiding theoretical framework from which to derive testable hypotheses. This situation is a consequence of several factors. First is the small number of thoroughly investigated cases from which patterns and principles may be derived;[30] we may only now be approaching the critical mass of observational data required for theory development. A second factor is the problems inherent in comparisons and contrasts of enzyme activities assessed for various insect species in different laboratories. Many different techniques are used, and most rely on indices of activity (i.e., they use model substrates) rather than on measurements of activity against allelochemicals of interest. Given, however, that chemistry is a primary defense of plants against insects, that biochemical detoxication is a primary defense of insects against plant chemistry, and that the evolution of plants and insects has been closely coupled, then broad patterns in host use by phytophagous insects should be at least partially explicable on the basis of enzymatic detoxication metabolism.

If any principle has served as a framework for hypothesis testing with respect to insect detoxication systems and host use, it has been the notion that PSMO activity is inversely correlated with feeding specialization. This view was first promoted by Krieger et al.,[61] who found that PSMO activities in 35 species of Lepidoptera were correlated with the degree of polyphagy. They argued that

PSMOs provide polyphagous insects with an effective defense against the diversity of secondary compounds encountered in their food plants. Subsequent research, however, has failed to support the theory. Rose[96] surveyed PSMO activities in 58 species of Australian Lepidoptera and found no correlation with diet breadth. PSMO activity is not correlated with degree of polyphagy in the Papilionidae; in fact, it is highest in *Papilio polyxenes,* a relative specialist.[16] In short, PSMO activity appears to be associated more with the type of hosts than the number of hosts on which an insect feeds.

Within the last several years, other patterns have begun to emerge, linking enzyme activities to various facets of insect nutritional ecology. Berenbaum[16] has proposed that specialists and generalists differ in the inducibility and range of responsiveness of their enzyme systems. She suggests that as a consequence of diversifying selection to maintain flexibility of response, generalists should exhibit greater induction capacity and less substrate specificity than specialists. Preliminary data on induction of various enzyme systems in seven species of Lepidoptera appear to fit the suggested pattern; range of variation in activity of several enzyme systems tends to increase with insect host range, although the correlation is not tight.[16] Additional research is necessary to determine if this is a general pattern for phytophagous insects.

Another pattern, described by Berenbaum and Isman,[7] associates differences in detoxication enzyme activity with differences in developmental constraints among insect orders. The authors suggest that lower growth rates and efficiencies of Orthoptera in comparison to Lepidoptera may be explained by the substantial loss of cuticular protein and carbohydrate experienced by orthopterans at each molt. Moreover, the peritrophic membrane in orthopterans is unlike that of lepidopterans in that it forms a sheath around fecal pellets and is continuously excreted, thereby exacting additional cuticular loss. Berenbaum and Isman[17] also state that higher gut permeability of lepidopterans, relative to orthopterans, may improve nutritional efficiency, but also increases exposure to allelochemicals. Correspondingly, Lepidoptera have more highly active and inducible PSMO enzymes than do Orthoptera.

Finally, patterns in insect detoxication systems may be associated with particular feeding strategies. For instance, leaf-chewing insects are exposed to higher concentrations of allelochemicals than are phloem-sucking insects, so should have better developed metabolic detoxication systems. Mullin[80] argues that this is indeed the case; aphids have lower PSMO and epoxide hydrolase activities and are more susceptible to toxicants, than are phytophagous Coleoptera and Lepidoptera. Moreover, because leaf-chewers and phloem-imbibers are exposed to different chemical milieu, they may rely on different enzymatic adaptations. For example, because of the relative availability of compounds for use in conjugation reactions, aphids should rely more on glucosyl transferases and less on glutathione transferases than do leaf-chewers.[79] Tree leaves generally have lower protein levels than do forb leaves, so a similar pattern of differential reliance on glutathione transferases may hold for tree and forb feeders.

Differences in insect detoxication systems may also be related to different food processing strategies. Bernays and Janzen[23] showed that sphingid and saturniid caterpillars from Costa Rica process their food in different ways. Sphingids shred their food well, whereas saturniids ingest large, uniformly sized foliage particles. Consequently, both nutrient extraction efficiency and exposure to host allelochemicals may be greater in sphingids than in saturniids. Biochemical detoxication enzymes are likely to exhibit higher activities in sphingids than in saturniids.

VI. CONCLUSIONS

The enzymatic detoxication systems of insects have long been regarded as critical adaptations to host plant chemistry, but until recently little empirical evidence existed to support that view. Research since the mid-1980s, however, has provided a number of examples of detoxication of plant secondary compounds by a variety of enzyme systems. Moreover, differential toxicity of particular allelochemicals to phytophagous insects can now be explained on the basis of differences in the enzyme activity of insects. Differences in enzyme activity may be genetically linked (e.g., interspecific differences), but may also occur due to changes in individual insects as a consequence of a host of intrinsic and extrinsic factors.

We are only now approaching the point where the number of thoroughly investigated cases is sufficient to begin to identify patterns in the relationship of insect biochemistry to plant allelochemistry. Additional research should clarify the patterns and enable the development of a theoretical framework, from which testable hypotheses can be derived.

Many avenues of research are deserving of attention. The role of particular enzyme systems in the detoxication and comparative toxicity of specific allelochemicals needs further study. This is true of all enzyme systems, but especially so of the nonoxidative enzymes. We know almost nothing about the effects of intrinsic factors (e.g., developmental stage, sex, reproductive condition, nutritional status, disease) on the efficacy of detoxication systems. We also know little about how these systems are altered by extrinsic factors such as diet, or stress due to heat or other forms of radiation. Research in these directions can only serve to improve the understanding of mechanisms underlying the evolution and dynamics of insect-plant interactions.

Finally, a note of caution about the perils of reductionism is appropriate. The author has argued that investigation of insect biochemistry will provide substantial insight into the ecology of insect-plant interactions. The dynamics of such interactions are, however, driven by much more than plant chemistry and insect biochemistry. For example, behavioral specialization due to ecological factors such as predator avoidance[21,22] may lead to secondary loss of host tolerance at the biochemical level. We must be careful about impuning cause and effect in studies of biochemical mediation of insect-plant interactions.

REFERENCES

1. **Agosin, M.,** Role of microsomal oxidations in insecticide degradation, in *Comprehensive Insect Physiology, Biochemistry and Pharmacology*, Vol. 12, Kerkut, G. A. and Gilbert, L. I., Eds., Pergamon Press, New York, 1985, 647.
2. **Ahmad, S.,** Mixed-function oxidase activity in a generalist herbivore in relation to its biology, food plants, and feeding history, *Ecology*, 64, 235, 1983.
3. **Ahmad, S.,** Enzymatic adaptations of herbivorous insects and mites to phytochemicals, *J. Chem. Ecol.*, 12, 533, 1986.
4. **Ahmad, S., Beilstein, M. A., and Pardini, R. S.,** Glutathione peroxidase activity in insects: a reassessment, *Arch. Insect Biochem. Physiol.*, 12, 31, 1989.
5. **Ahmad, S., Brattsten, L. B., Mullin, C. A., and Yu, S. J.,** Enzymes involved in the metabolism of plant allelochemicals, in *Molecular Aspects of Insect-Plant Associations*, Brattsten, L. B. and Ahmad, S., Eds., Plenum Press, New York, 1986, 73.
6. **Ahmad, S. and Pardini, R. S.,** Antioxidant defense of the cabbage looper, *Trichoplusia ni*: enzymatic responses to the superoxide-generating flavonoid, quercetin, and photodynamic furanocoumarin, xanthotoxin, *Photochem. Photobiol.*, 51, 305, 1990.
7. **Ahmad, S., Pritsos, C. A., and Pardini, R. S.,** Insect responses to pro-oxidant plant allelochemicals, in *Insects and Plants '89*, Szentesi, A. and Jermy, T., Eds., Akademia Kiado, Budapest, in press.
8. **Andersen, J. F., Plattner, R. D., and Weisleder, D.,** Metabolic transformations of cucurbitacins by *Diabrotica virgifera virgifera* Leconte and *D. undecimpunctata howardi* Barber, *Insect Biochem.*, 18, 71, 1988.
9. **Barbosa, P., Martinat, P. and Waldvogel, M.,** Development, fecundity and survival of the herbivore *Lymantria dispar* and the number of plant species in its diet, *Ecol. Entomol.*, 11, 1, 1986.
10. **Bazzaz, F. A., Chiariello, N. R., Coley, P. D., and Pitelka, L. F.,** Allocating resources to reproduction and defense, *Bioscience*, 37, 58, 1987.
11. **Bell, E. A., Lackey, J. A., and Polhill, R. M.,** Systematic significance of L-canavanine in the Papilionoideae (Faboideae), *Biochem. Syst. Ecol.*, 6, 201, 1978.
12. **Berenbaum, M.,** Toxicity of a furanocoumarin to armyworms: a case of biosynthetic escape from insect herbivores, *Science*, 201, 532, 1978.
13. **Berenbaum, M.,** Effects of linear furanocoumarins on an adapted specialist insect (*Papilio polyxenes*), *Ecol. Entomol.*, 6, 345, 1981.
14. **Berenbaum, M.,** Brementown revisited: interactions among allelochemicals in plants, *Recent Adv. Phytochem.*, 19, 139, 1985.
15. **Berenbaum, M. R.,** Target site insensitivity in insect-plant interactions, in *Molecular Aspects of Insect-Plant Associations*, Brattsten, L. B. and Ahmad, S., Eds., Plenum Press, New York, 1986, 257.
16. **Berenbaum, M. R.,** Comparative allelochemical processing in the Papilionidae (Lepidoptera), *Arch. Insect Biochem. Physiol.*, in press.
17. **Berenbaum, M. R. and Isman, M. B.,** Herbivory in holometabolous and hemimetabolous insects: contrasts between Orthoptera and Lepidoptera, *Experientia*, 45, 229, 1989.
18. **Berenbaum, M. R. and Miliczky,**

E., Mantids and milkweed bugs: efficacy of aposematic coloration against invertebrate predators, *Am. Midl. Nat.*, 111, 64, 1984.

19. **Berenbaum, M. and Neal, J. J.,** Synergism between myristicin and xanthotoxin, a naturally cooccurring plant toxicant, *J. Chem. Ecol.*, 11, 1349, 1985.
20. **Berenbaum, M. R., Zangerl, A. R., and Lee, K.,** Chemical barriers to adaptation by a specialist herbivore, *Oceologia*, 80, 501, 1989.
21. **Bernays, E. A. and Cornelius, M. L.,** Generalist caterpillar prey are more palatable than specialists for the generalist predator *Iridomyrmex humilis*, *Oceologia*, 79, 427, 1989.
22. **Bernays, E. A. and Graham, M.,** On the evolution of host specificity in phytophagous insects, *Ecology*, 69, 886, 1988.
23. **Bernays, E. A. and Janzen, D. H.,** Saturniid and sphingid caterpillars: two ways to eat leaves, *Ecology*, 69, 1153, 1988.
24. **Bleiler, J. A., Rosenthal, G. A., and Janzen, D. H.,** Biochemical ecology of canavanine-eating seed predators, *Ecology*, 69, 427, 1988.
25. **Boyd, J. N. and Campbell, T. C.,** Impact of nutrition on detoxication, in *Biological Basis of Detoxication*, Caldwell, J. and Jakoby, W. B., Eds., Academic Press, New York, 1983, 287.
26. **Brattsten, L. B.,** Ecological significance of mixed-function oxidations, *Drug Metab. Rev.*, 10, 35, 1979.
27. **Brattsten, L. B.,** Biochemical defense mechanisms in herbivores against plant allelochemicals, in *Herbivores. Their Interaction with Secondary Plant Metabolites*, Rosenthal, G. A. and Janzen, D. H., Eds., Academic Press, New York, 1979, 199.
28. **Brattsten, L. B.,** Fate of ingested plant allelochemicals in herbivorous insects, in *Molecular Aspects of Insect-Plant Associations*, Brattsten, L. B. and Ahmad, S., Eds., Plenum Press, New York, 1986, 211.
29. **Brattsten, L. B.,** Metabolic insecticide defenses in the boll weevil compared to those in a resistance-prone species, *Pest. Biochem. Physiol.*, 27, 1, 1987.
30. **Brattsten, L. B.,** Enzymic adaptations in leaf-feeding insects to host-plant allelochemicals, *J. Chem. Ecol.*, 14, 1919, 1988.
31. **Brattsten, L. B.,** Potential role of plant allelochemicals in the development of insecticide resistance, in *Novel Aspects of Insect-Plant Interactions*, Barbosa, P. and Letourneau, D., Eds., John Wiley & Sons, New York, 1988, 313.
32. **Brattsten, L. B., Gunderson, C. A., Fleming, J. T., and Nikbahkt, K.N.,** Temperature and diet modulate cytochrome P-450 activities in southern armyworm, *Spodoptera eridania* (Cramer), caterpillars, *Pest. Biochem. Physiol.*, 25, 346, 1986.
33. **Brattsten, L. B., Samuelian, J. H., Long, K. Y., Kincaid, S.A., and Evans, C.K.,** Cyanide as a feeding stimulant for the southern armyworm, *Spodoptera eridania*, *Ecol. Entomol.*, 8, 125, 1983.
34. **Brattsten, L. B. and Wilkinson, C. F.,** Induction of microsomal enzymes in the southern armyworm (*Prodenia eridania*), *Pest. Biochem. Physiol.*, 3, 393, 1973.
35. **Brower, L. P. and van Zandt Brower, J.,** Birds, butterflies and plant poisons: a study in ecological chemistry, *Zoologica,* 49, 137, 1964.
36. **Brower, L. P., Seiber, J. N., Nelson, C. J., Lynch, S. P., and Tuskes, P. M.,** Plant-determined

variation in the cardenolide content, thin-layer chromatography profiles, and emetic potency of monarch butterflies, *Danaus plexippus* reared on milkweed, *Asclepias eriocarpa* in California, *J. Chem. Ecol.*, 8, 579, 1982.
37. **Bryant, J. P., Chapin, F. S., III, and Klein, D. R.,** Carbon/nutrient balance of boreal plants in relation to herbivory, *Oikos*, 40, 357, 1983.
38. **Bull, D. L., Ivie, G. W., Beier, R. C. et al.,** *In vitro* metabolism of a linear furanocoumarin (8-methoxypsoralen, xanthotoxin) by mixed-function oxidases of larvae of black swallowtail butterfly and fall armyworm, *J. Chem. Ecol.*, 12, 885, 1986.
39. **Casarett, L. J. and Bruce, M. C.,** Origin and scope of toxicology, in *Casarett and Doull's Toxicology. The Basic Science of Poisons*, Doull, J., Klaasen, C. D., and Amdur, M. O., Eds., Macmillan, New York, 1980, 3.
40. **Chhabra, R. S.,** Effect of dietary factors and environmental chemicals on intestinal drug metabolizing enzymes, *Toxicol. Environ. Chem.*, 3, 173, 1981.
41. **Cohen, M. B., Berenbaum, M. R., and Schuler, M. A.,** Induction of cytochrome P450-mediated detoxification of xanthotoxin in the black swallowtail, *J. Chem. Ecol.*, 15, 2347, 1989.
42. **Coley, P. D., Bryant, J. P., and Chapin, F. S., III,** Resource availability and plant antiherbivore defense, *Science*, 230, 895, 1985.
43. **Conn, E. E.,** Cyanide and cyanogenic glycosides, in *Herbivores. Their Interaction with Secondary Plant Metabolites*, Rosenthal, G. A. and Janzen, D. H., Eds., Academic Press, New York, 1979, 387.
44. **Dauterman, W. C.,** Physiological factors affecting metabolism of xenobiotics, in *Introduction to Biochemical Toxicology*, Hodgson, E. and Guthrie, F. E., Eds., Elsevier, New York, 1980, 133.
45. **David, A. and Vallance, D. K.,** Bitter principles of cucurbitaceae, *J. Pharm. Pharmacol.*, 7, 295, 1988.
46. **Davis, R. H. and Nahrstedt, A.,** Cyanogenesis in insects, in *Comprehensive Insect Physiology, Biochemistry and Pharmacology*, Vol. 11, Kerkut, G. A. and Gilbert, L. I., Eds., Pergamon Press, New York, 1985, 635.
47. **Dowd, P. F., Smith, C. M., and Sparks, T. C.,** Detoxification of plant toxins by insects, *Insect Biochem.*, 5, 453, 1983.
48. **Ehrlich, P. R. and Raven, P. H.,** Butterflies and plants: a study in coevolution, *Evolution*, 18, 586, 1964.
49. **Feeny, P. P.,** Plant apparency and chemical defense, *Rec. Adv. Phytochem.*, 10, 1, 1976.
50. **Ferguson, J. E. and Metcalf, R. L.,** Cucurbitacins: plant derived defense compounds for the Diabroticites, *J. Chem. Ecol.*, 11, 311, 1985.
51. **Ferguson, J. E., Metcalf, R. L., and Fischer, D. C.,** Disposition and fate of cucurbitacin B in five species of diabroticites, *J. Chem. Ecol.* 11, 1307, 1985.
52. **Gilbert, B.L., Baker, J.E., and Norris, D.M.,** Juglone (5-hydroxy-1,4-naphthoquinone) from *Carya ovata*, a deterrent to feeding by *Scolytus multistriatus*, *J. Insect Physiol.*, 13, 1453, 1967.
53. **Gould, F.,** Mixed function oxidases and herbivore polyphagy: the devil's advocate position, *Ecol. Entomol.*, 9, 29, 1984.
54. **Gunderson, C. A., Brattsten, L. B., and Fleming, J. T.,** Microsomal oxidase and glutathione transferase as factors influencing the effects of pulegone in southern and fall armyworm larvae, *Pest. Biochem. Physiol.*, 26, 238, 1986.

55. **Hedin, P. A., Langhans, V. E., and Graves, C. H., Jr.,** Identification of juglone in pecan as a possible factor of resistance to *Fusicladium effusum*, *J. Agric. Food Chem.*, 27, 92, 1979.
56. **Hinks, C. F.,** The influence of temperature on the efficacy of three pyrethroid insecticides against the grasshopper, *Melanoplus sanguinipes* (Fab.) (Orthoptera: Acrididae), under laboratory conditions, *Can. Entomol.*, 117, 1007, 1985.
57. **Hodgson, E.,** Microsomal monooxygenases, in *Comprehensive Insect Physiology, Biochemistry and Pharmacology*, Vol. 11, Kerkut, G. A. and Gilbert, L. I., Eds., Pergamon Press, New York, 1985, 225.
58. **Hunt, D. W. A. and Smirle, M. J.,** Partial inhibition of pheromone production in *Dendroctonus ponderosae* (Coleoptera: Scolytidae) by polysubstrate monooxygenase inhibitors, *J. Chem. Ecol.*, 14, 529, 1988.
59. **Ivie, G. W., Bull, D. L., Beier, R. C., Pryor, N. W., and Oertli, E. H.,** Metabolic detoxification: mechanism of insect resistance to plant psoralens, *Science*, 221, 374, 1983.
60. **Krieger, R. I., Wilkinson, C. F., Hicks, L. J., and Taschenberg, E. F.,** Aldrin epoxidation, dihydroisodrin hydroxylation, and *p*-chloro-N-methylaniline demethylation in six species of saturniid larvae. *J. Econ. Entomol.*, 69, 1, 1976.
61. **Krieger, R., Feeny, P. P., and Wilkinson, C.,** Detoxification enzymes in the guts of caterpillars: an evolutionary answer to plant defenses?, *Science*, 172, 579, 1971.
62. **Lance, D. L. and Barbosa, P.,** Host tree influences on the dispersal of late instar gypsy moths, *Lymantria dispar*, *Oikos*, 38, 1, 1982.
63. **Larson, R. A.,** The antioxidants of higher plants, *Phytochemistry*, 27, 969, 1988.
64. **Lindroth, R. L.,** Hydrolysis of phenolic glycosides by midgut β-glucosidases in *Papilio glaucus* subspecies, *Insect Biochem.*, 8, 789, 1988.
65. **Lindroth, R. L.,** Chemical ecology of the luna moth: effects of host plant on detoxification enzyme activity, *J. Chem. Ecol.*, 15, 2019, 1989.
66. **Lindroth, R. L.,** Host plant alteration of detoxication activity in *Papilio glaucus glaucus*, *Entomol. Exp. Appl.*, 50, 29, 1989.
67. **Lindroth, R. L.,** Biochemical detoxication: mechanism of differential tiger swallowtail tolerance to phenolic glycosides, *Oecologia*, 81, 219, 1989.
68. **Lindroth, R. L., Anson, B. D., and Weisbrod, A. V.,** Effects of dietary protein and juglone on the gypsy moth: growth performance and detoxication enzyme activity, *J. Chem. Ecol.*, 16, 2533, 1990.
69. **Lindroth, R. L., Barman, M. A., and Weisbrod, A. V.,** Nutrient deficiencies and the gypsy moth *Lymantria dispar*: effects on larval performance and detoxication enzyme activities, *J. Insect Physiol.*, 37, 45, 1991.
70. **Lindroth, R. L. and Peterson, S. S.,** Effects of plant phenols on performance of southern armyworm larvae, *Oecologia*, 75, 185, 1988.
71. **Lindroth, R. L., Scriber, J. M., and Hsia, M. T. S.,** Chemical ecology of the tiger swallowtail: mediation of host use by phenolic glycosides, *Ecology*, 69, 814, 1988.
72. **Mabry, T. J. and Gill, J. E.,** Sesquiterpene lactones and other terpenoids, in *Herbivores. Their Interaction with Secondary Plant Metabolites*, Rosenthal, G. A. and Janzen, D. H., Eds., Academic

Press, New York, 1979, 501.

73. **Marty, M. A. and Krieger, R. I.,** Metabolism of uscharidin, a milkweed cardenolide, by tissue homogenates of monarch butterfly larvae, *Danaus plexippus* L., *J. Chem. Ecol.*, 10, 945, 1984.
74. **Mattson, W. J., Jr.,** Herbivory in relation to plant nitrogen content, *Annu. Rev. Ecol. Syst.*, 11, 119, 1980.
75. **Mattson, W. J. and Haack, R. A.,** The role of drought in outbreaks of plant-eating insects, *Bioscience,* 37, 110, 1987.
76. **Mattson, W. J. and Haack, R. A.,** The role of drought stress in provoking outbreaks of phytophagous insects, in *Insect Outbreaks*, Barbosa, P. and Schultz, J. C., Eds., Academic Press, New York, 1987, 365.
77. **Mattson, W. J. and Scriber, J. M.,** Nutritional ecology of insect folivores of woody plants: nitrogen, water, fiber, and mineral considerations, in *Nutritional Ecology of Insects, Mites, and Spiders*, Slansky, F., Jr. and Rodriguez, J. G., Eds., John Wiley & Sons, New York, 1987, 105.
78. **Metcalf, R. L., Metcalf, R. A., and Rhodes, A. M.,** Cucurbitacins as kairomones for diabroticite beetles, *Proc. Natl. Acad. Sci. U.S.A.*, 77, 3769, 1980.
79. **Mullin, C. A.,** Adaptive divergence of chewing and sucking arthropods to plant allelochemicals, in *Molecular Aspects of Insect-Plant Associations*, Brattsten, L. B. and Ahmad, S., Eds., Plenum Press, New York, 1986, 175.
80. **Mullin, C. A.,** Adaptive relationships of epoxide hydrolase in herbivorous arthropods, *J. Chem. Ecol.*, 14, 1867, 1988.
81. **Neal, J. J.,** Ecological Aspects of Insect Detoxication Enzymes and Their Interaction with Plant Allelochemicals, Ph.D. thesis, University of Illinois, Urbana, 1987.
82. **Neal, J. J.,** Metabolic costs of mixed-function oxidase induction in *Heliothis zea*, *Entomol. Exp. Appl.*, 43, 175, 1987.
83. **Neal, J. J.,** Methylenedioxyphenyl-containing alkaloids and autosynergism, *Phytochemistry*, 28, 451, 1989.
84. **Neal, J. J.,** Myristicin, safrole, and fagaramide as phytosynergists of xanthotoxin, *J. Chem. Ecol.*, 15, 309, 1989.
85. **Neal, J. J. and Berenbaum, M.,** Decreased sensitivity of mixed-function oxidases from *Papilio polyxenes* to inhibitors in host plants, *J. Chem. Ecol.*, 15, 439, 1989.
86. **Neal, R. A.,** Metabolism of toxic substances, in *Casarett and Doull's Toxicology. The Basic Science of Poisons*, Doull, J., Klaasen, D. D., and Amdur, M. O., Eds., Macmillan, New York, 1980, 56.
87. **Nitao, J. K.,** Enzymatic adaptation in a specialist herbivore for feeding on furanocoumarin-containing plants, *Ecology*, 70, 629, 1989.
88. **Pardini, R. S., Pritsos, C. A., Bowen, S. M., Ahmad, S., and Blomquist, G. J.,** Adaptations to plant pro-oxidants in a phytophagous insect model: enzymatic protection from oxidative stress, in *Oxygen Radicals in Biology and Medicine*, Simic, M. G., Taylor, K. A., Ward, J. F., and von Sonntag, G., Eds., Plenum Press, New York, 1989, 725.
89. **Parke, D. V. and Ioannides, C.,** The role of nutrition in toxicology, *Annu. Rev. Nutr.*, 1, 207, 1981.
90. **Pritsos, C. A., Ahmad, S., Bowen, S. M., Elliot, A. J., Blomquist, G. J., and Pardini, R. S.,** Antioxidant enzymes of the black swallowtail butterfly, *Papilio polyxenes*, and their response to the prooxidant allelochemical, quercetin, *Arch. Insect Biochem.*

Physiol., 8, 101, 1988.

91. **Raffa, K. F. and Berryman, A. A.,** Interacting selective pressures in conifer-bark beetle systems: a basis for reciprocal adaptations?, *Am. Nat.*, 129, 234, 1987.
92. **Raffa, K. F. and Priester, T. M.,** Synergists as research tools and control agents in agriculture, *J. Agric. Entomol.*, 2, 27, 1985.
93. **Reilly, C. C., Gentry, C. R., and McVay, J. R.,** Biochemical evidence for resistance of rootstocks to the peachtree borer and species separation of peachtree borer and lesser peachtree borer (Lepidoptera: Sesiidae) on peach trees, *J. Econ. Entomol.*, 80, 338, 1987.
94. **Rhoades, D. F.,** Evolution of plant chemical defense against herbivores, in *Herbivores. Their Interaction with Secondary Plant Metabolites*, Rosenthal, G. A. and Janzen, D. H., Eds., Academic Press, New York, 1979, 3.
95. **Rhoades, D. F. and Cates, R. G.,** Toward a general theory of plant antiherbivore chemistry, *Recent Adv. Phytochem.*, 10, 168, 1976.
96. **Rose, H. A.,** The relationship between feeding specialization and host plants to aldrin epoxidase activities of midgut homogenates in larval Lepidoptera, *Ecol. Entomol.*, 10, 455, 1985.
97. **Rosenthal, G. A.,** A seed-eating beetle's adaptations to a poisonous seed, *Sci. Am.* 249, 164, 1983.
98. **Rosenthal, G. A.,** Biochemical insight into insecticidal properties of L-canavanine, a higher plant protective allelochemical, *J. Chem. Ecol.*, 12, 1145, 1986.
99. **Rosenthal, G. A., Berge, M. A., Bleiler, J. A., and Rudd, T. P.,** Aberrant, canavanyl protein formation and the ability to tolerate or utilize L-canavanine, *Experientia*, 43, 558, 1987.
100. **Rosenthal, G. A. and Dahlman, D. L.,** L-Canavanine and protein synthesis in the tobacco hornworm *Manduca sexta*, *Proc. Natl. Acad. Sci. U.S.A.*, 83, 14, 1986.
101. **Scheline, R. R.,** Metabolism of foreign compounds by gastrointestinal microorganisms. *Pharm. Rev.*, 25, 451, 1973.
102. **Scheline, R. R.,** *Mammalian Metabolism of Plant Xenobiotics*, Academic Press, New York, 1978.
103. **Schoonhoven, L. M. and Meerman, J.,** Metabolic cost of changes in diet and neutralization of allelochemics, *Entomol. Exp. Appl.*, 24, 689, 1978.
104. **Schreiner, I., Nafus, D., and Pimentel, D.,** Effects of cyanogenesis in bracken fern (*Pteridium aquilinum*) on associated insects, *Ecol. Entomol.*, 9, 69, 1984.
105. **Scriber, J. M.,** Sequential diets, metabolic costs, and growth of *Spodoptera eridania* (Lepidoptera: Noctuidae) feeding upon dill, lima bean, and cabbage, *Oecologia*, 51, 175, 1981.
106. **Scriber, J. M.,** Host-plant suitability, in *Chemical Ecology of Insects*, Bell, W. J. and Carde, R. T., Eds., Chapman & Hall, London, 1984, 159.
107. **Scriber, J. M.,** Tale of the tiger: Beringial biogeography, binomial classification, and breakfast choices in the *Papilio glaucus* complex of butterflies, in *Chemical Mediation of Coevolution*, Spencer, K. C., Ed., Academic Press, New York, 1988, 241.
108. **Scriber, J. M., Lindroth, R. L., and Nitao, J.,** Differential toxicity of a phenolic glycoside from quaking aspen to *Papilio glaucus* butterfly subspecies, hybrids and backcrosses, *Oecologia*, 81, 186, 1989.
109. **Scriber, J. M. and Slansky, F., Jr.,** The nutritional ecology of immature insects, *Annu. Rev. Entomol.*, 26, 183, 1981.
110. **Sheppard, C. A. and Friedman,**

S., Endogenous and induced monooxygenase activity in gypsy moth larvae feeding on natural and artificial diets, *Arch. Insect Biochem. Physiol.*, 10, 47, 1989.

111. **Siegfried, B. D. and Mullin, C. A.,** Influence of alternative host plant feeding on aldrin susceptibility and detoxification enzymes in western and northern corn rootworms, *Pest. Biochem. Physiol.*, 35, 155, 1989.
112. **Slansky, F., Jr. and Scriber, J. M.,** Food consumption and utilization, in *Comprehensive Insect Physiology, Biochemistry, and Pharmacology*, Vol. 4, Kerkut, G. A. and Gilbert, L. I., Eds., Pergamon Press, New York, 1985, 87.
113. **Swick, R. A.,** Hepatic metabolism and bioactivation of mycotoxins and plant toxins, *J. Anim. Sci.*, 58, 1017, 1984.
114. **Tallamy, D. W.,** Behavioral adaptations in insects to plant allelochemicals, in *Molecular Aspects of Insect-Plant Associations*, Brattsten, L. B. and Ahmad, S., Eds., Plenum Press, New York, 1986, 273.
115. **Tallamy, D. W. and Krischik, V. A.,** Variation and function of cucurbitacins in *Cucurbita*: an examination of current hypotheses, *Am. Nat.*, 133, 766, 1989.
116. **Teas, H. J.,** Cycasin synthesis in *Seirarctia echo* (Lepidoptera) larvae fed methylazoxymethanol, *Biochem. Biophys. Res. Commun.*, 26, 686, 1967.
117. **Terriere, L. C.,** Induction of detoxication enzymes in insects, *Annu. Rev. Entomol.*, 29, 71, 1984.
118. **Toth, S. J., Jr. and Sparks, T. C.,** Effect of temperature on toxicity and knockdown activity of *cis*-permethrin, esfenvalerate, and λ-cyhalothrin in the cabbage looper (Lepidoptera: Noctuidae), *J. Econ. Entomol.*, 83, 342, 1990.
119. **Wadleigh, R. W. and Yu, S. J.,** Glutathione transferase activity of fall armyworm larvae toward α,β-unsaturated carbonyl allelochemicals and its induction by allelochemicals, *Insect Biochem.*, 17, 759, 1987.
120. **Wadleigh, R. W. and Yu, S. J.,** Detoxification of isothiocyanate allelochemicals by glutathione transferase in three lepidopterous species, *J. Chem. Ecol.*, 14, 127, 1988.
121. **Woodhead, S. and Bernays, E.,** Changes in release rates of cyanide in relation to palatability of *Sorghum* to insects, *Nature*, 270, 235, 1977.
122. **Woodhead, S. and Bernays, E.,** The chemical basis of resistance of *Sorghum bicolor* to attack by *Locusta migratoria*, *Entomol. Exp. Appl.*, 24, 123, 1978.
123. **Yu, S. J.,** Induction of detoxifying enzymes by allelochemicals and host plants in the fall armyworm, *Pest. Biochem. Physiol.*, 19, 330, 1983.
124. **Yu, S. J.,** Consequences of induction of foreign compound-metabolizing enzymes in insects, in *Molecular Aspects of Insect-Plant Associations*, Brattsten, L. B. and Ahmad, S., Eds., Plenum Press, New York, 1986, 153.
125. **Yu, S. J.,** Quinone reductase of phytophagous insects and its induction by allelochemicals, *Comp. Biochem. Physiol.*, 87B, 621, 1987.
126. **Yu, S. J.,** Biochemical defense capacity in the spined soldier bug (*Podisus maculiventris*) and its lepidopterous prey, *Pest. Biochem. Physiol.*, 28, 216, 1987.
127. **Yu, S. J.,** β-Glucosidase in four phytophagous Lepidoptera, *Insect Biochem.*, 19, 103, 1989.
128. **Yu, S. J.,** Purification and characterization of glutathione transferases from five phytophagous Lepidoptera, *Pest. Biochem. Physiol.*, 35, 97, 1989.

2

Macroevolutionary Aspects of Insect-Plant Relationships

Charles Mitter and Brian Farrell
Department of Entomology
University of Maryland
College Park, Maryland

TABLE OF CONTENTS

I. INTRODUCTION

A fundamental implication of Ehrlich and Raven's[57] essay on coevolution is that the structure of insect-plant communities strongly reflects their evolutionary history. These authors' "escape and radiation" model[192] exemplifies a Simpsonian view of ecological evolution. The hypothesized arms race between insect and plant lineages follows a gradual evolutionary pathway, each step providing the preadaptation for the next, and being sufficiently improbable that new adaptations have time to promote radiation before they are superseded by newer adaptations. The present distribution of insects across plants, and differential diversity and ecological dominance among both insect and plant clades, are taken to reflect that historical sequence.

Until recently, however, such postulates have received little direct scrutiny. Reflecting the climate in evolutionary biology generally,[79] the voluminous work on insect-plant interactions inspired by Ehrlich and Raven's paper has, with notable exceptions (see, for example, References 9, 11, 108, 209), mostly concerned individual- and population-level processes. With the recent resurgence of interest in evolutionary history, stimulated in part by advances in phylogeny reconstruction, the time seems right to reassess the macroevolutionary aspects of Ehrlich and Raven's proposal. Whether or not the escape and radiation model applies in its entirety,[109,190] when broadly viewed, it suggests a variety of plausible evolutionary constraints on insect-plant interactions. This chapter is an attempt to catalog and survey the evidence for the kinds of conservatism and marks of past interactions that contemporary associations might exhibit. First, character conservatism, specifically of insect host plant use, is examined, followed by a treatment of its broader, coevolutionary implications.

To the extent that traits affecting them can evolve freely, the distributions of insects over host plants may be predictably set by natural selection.[153,190] Evolutionary lability of insect host use,[16] for example, is suggested by the common observation of heritable variation in host use traits[43,82,84] and by the rapid accumulation of herbivores on some introduced plant species;[190] however, potential limits on host use evolution are also evident. For example, if the multiple, interdependent traits involved in host use are genetically uncorrelated and subject to strongly divergent selection, the combination of genes required for origin of a new habit may arise only rarely.[81] Host use may change slowly and along restricted evolutionary pathways, as in the escape and radiation model, and rarely attain the highest plausible adaptive peak.

Communities structured by such "constrained" characters might reflect their evolutionary history in several ways.[64a] Limits on the rate or direction of evolution might cause a departure from simple ahistorical models of the distribution of insects over plants.[190] For example, the current size and composition of a plant's insect fauna might depend on its past accessibility to enemies.[183,190,211] Sufficiently conservative interactions could persist through geological time, leaving present-day insect feeding habits to reflect their age of origin, and opening the possibility of long-term coevolution between associated lineages. Finally, as in the escape and radiation model, sufficiently conservative associations may change in prevalence by differential diversification.

Clearly, most interactions have both labile and conservative aspects. Two approaches to distinguishing these might be recognized.[142] One may search first for adaptive patterns of interaction, then investigate departures from these as possible results of phylogenetic constraint. To take a simple example, ecological study shows geographic variation in host use by the butterfly *Euphydryas editha* to be adaptive with respect to plant phenology;[180] however, potential hosts outside the Scrophulariaceae and relatives are never used, though undoubtedly some of these are suitable on phenological grounds. A "constraint" explanation for this restriction is supported by the prevalence of similar predilections among other Euphydryini and related tribes.[1]

The alternative approach, which the authors adopt for the purposes of this text, is to search first for the conservative, hence possibly constrained, aspects of associations in their reconstructed histories, and ultimately in their genetic bases;[80] departures from phylogenetic pattern might then suggest labile adaptation. A limitation exists in that insect-plant interactions leave little fossil record, restricting inference about their histories primarily to comparisons among extant species. We shall digress when necessary to discuss recent advances in comparative methods based on phylogenies.

II. PHYLOGENETIC CONSERVATISM OF INSECT HOST PLANT USE

The Introduction argued that hypotheses ascribing community structure to evolutionary history, such as escape and radiation coevolution, tacitly assume genetic limits on the evolution of interactions. This section attempts to define and review the evidence for some phylogenetic patterns which suggest that the distribution of insects over plants reflects in part genetic limits on the evolution of insect diet. Of course, a firm conclusion about such limits or their absence requires integration of these patterns with study of the genetic and adaptive bases of host use.[43,79,80,133]

The analysis of host use evolution below is based on cladograms (phylogenetic trees), which depict the sequence of divergence of the observed taxa from a common ancestor. Each line segment or branch in a cladogram (see Figure 1) represents the succession of generations within a single lineage, while each branch point or node represents the most recent common ancestor of the lineages it subtends, at the time of speciation. This section assumes the cladogram as established (on grounds apart from host associations), and considers only how it can be used to infer the evolutionary history of host use. Of course, the ability to detect the historical patterns under discussion should be increased by the current rapid development of phylogeny estimation techniques.

In phylogenetics, a "character" is commonly defined as a homologous, heritable descriptor (e.g., number of segmented legs) which can take on

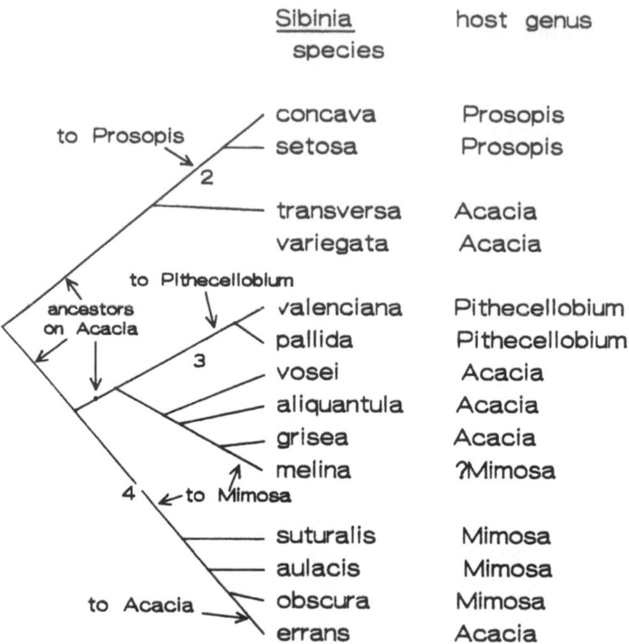

FIGURE 1. Example illustrating inference of host use evolution on a cladogram, extracted from a study of the weevil subgenus *Sibinia Microtychius.*[38] Relationships follow Clark,[38] but only 13 of 133 species are shown and several complexities of diet evolution are omitted. Numbers denote ancestral species. Most parsimonious history of diet is original association with *Acacia,* with subsequent transitions to other host genera as indicated.

different conditions, or states, in different species (e.g., six, the number of legs in insects). Given a cladogram, the evolution of a character is typically inferred by some version of the parsimony criterion; i.e., states are assigned to their most recent common ancestors so as to minimize the number of required character state changes summed over the branches of the tree.[67,73,129,130]

To study the evolution of host use, the parsimony criterion is frequently applied to a "character" whose states are the different host plant taxa or sets thereof used by individual insect species. This heuristic formulation has clear limitations. First, it is strictly appropriate only when new host-taxon associations reflect changes in heritable insect characters; under parallel insect-plant cladogenesis, this need not occur (see below). Second, "host use" is generally treated as a single character, but is actually a compound trait, whose morphological, physiological, and behavioral components may have complex, separate genetic bases.[43,44,82,104] Moreover, "host taxon used" is a quite indirect, possibly artificial, summary of those underlying insect characters; for example, the conclusions can depend on the level at which host taxonomy is considered. Better understanding of the genetic and functional bases of host use will eventually allow its phylogenesis to be analyzed in more precise, mechanistic terms.

The authors' approach is illustrated by the weevil cladogram in Figure 1, which is a deliberate oversimplification of a real example.[38] In Figure 1, the most parsimonious hypothesis of diet evolution is an original (primitive) association with *Acacia*, with a shift to *Prosopis* occurring in ancestral lineage 2, to *Pithecellobium* in ancestor 3, to *Mimosa* both in ancestor 4 and in *Sibinia melina*, and back to *Acacia* in *Sibinia errans*. Reversal from a derived to a more primitive condition (as S. *errans)* is one form of *homoplasy* (independent origin of the same state), the other being parallel origin of the same derived state, as in ancestor 4 and *S. melina*.

Given such reconstructions, one can, in principle, distinguish two categories of potential constraint on host use evolution: limitations on its sequence or direction, and on its rate or frequency.

Genetic constraint on the sequence or direction of evolution is suggested when the estimated sequence or frequency of transition among pairs of states is predictable from the probability of occurrence of the genetic variation necessary for those transitions. The latter probabilities may be measured directly,[43,44,80,135] or inferred indirectly from differences among the hosts. For example, the history of host shifts implied by the weevil cladogram in Figure 1 shows no direct transitions between any of the three host genera apart from *Acacia*. Suppose it were known that the three host genera apart from *Acacia* in Figure 1 were all less similar to each other in secondary chemistry than each was to *Acacia* (in fact, these similarities are not known). One, then, might predict that less genetic change would be required for transfers to and from *Acacia* than between other pairs of hosts. Because the transitions on the cladogram match that expectation, i.e., occur only between *Acacia* and other genera, one might reasonably infer that diet evolution was limited to small genetic steps in weevil traits affected by host secondary chemistry.

The preservation of such genetically predictable orderings or transformation series[129] in extant species would suggest a lack of optimality of some contemporary diets; it is difficult to understand why the optimality of a particular diet should be exactly predictable from the phytophage species' phylogenetic position. Conversely, the lack of any genetically predictable ordering would be consistent with, though hardly establish, diet optimality.

A character's evolution might also be thought "constrained", and the possibility of maladaptation raised, if change occurred rarely or slowly, whatever its sequence or direction. Of course, compelling "constraint" explanations for individual conservative features, such as the highly specialized aphid fundatrix,[141] require detailed study; the authors attempt only to sketch broad patterns of conservatism and their possible implications.

The rate of character evolution is conventionally expressed as amount of change per time per lineage, but the detailed fossil or molecular-clock datings necessary for computing such rates are available for very few insect clades; however, an alternative form of rate comparison is obtainable from cladograms.[167] When evolution in a character is reconstructed by minimizing the number of required changes,[73] the maximum detectable number of changes equals the number of branch points, i.e., speciation events. It therefore seems logical to employ that standard in comparing clades for the frequency of change in host use. For example, in Figure 2 there are 5 apparent changes in host use, as against 13 speciation events — a frequency of 0.38. Patterns in rates so expressed could be obscured by differences in speciation rates unrelated to diet. On the other hand, in addition to being more generally applicable, rates per speciation event might be more appropriate than those based on absolute or generation times, if changes in host use are concentrated at speciation, or if novel feeding habits are evolutionarily evanescent unless protected by reproductive isolation.[78]

The distinction between sequence and rate of diet evolution is in practice often blurred by incomplete life history information and by the indirectness of host classification (taxonomic or otherwise) as a reflection of the insect characters underlying host use. For example, the consistent association of close relatives with the same host genus in Figure 1 might indicate infrequent change in diet; however, if most weevil species proved on close inspection to prefer different host species, host shift might instead be considered frequent, but limited in direction, i.e., to very similar hosts. In the discussion that follows interest is focused on relative conservatism (rate of evolution) with respect to some host property, and there is no attempt to distinguish between these alternatives.

Different aspects of variation among host plants surely affect different insect characters, which in turn vary in their degree and kind of conservatism.[43] For example, some groups "track" the same host taxon across many different habitats, while others are more conservative in habitat than in host taxon choice. Classification and quantification of such variation in dietary conservatism is a necessary step toward full understanding of diet evolution,[109] but has been little attempted.

A. CONSERVATISM WITH RESPECT TO HOST TAXONOMY

Conservatism with respect to plant taxonomy is of particular interest because of its bearing on the question of parallel insect-plant diversification (reviewed below). More generally, because so many plant characteristics are correlated with it, taxonomy offers a convenient measure of the overall lability of host associations; however, it is important to remember that host taxonomy per se gives little clue as to which plant features are constraining evolution in which insect characters. In other words, the apparent lability of the diet as judged from host taxonomy may sometimes mask strong conservatism with respect to some host property poorly correlated with host taxonomy.

Jermy[109] erected four categories of diet-taxonomic patterns and argued that although most phytophage species are oligophagous, groups of related insects most commonly feed on taxonomically disparate sets of hosts. This and many contrary assertions are based on anecdotal evidence, however. A quantitative distribution of diet patterns across the full spectrum of phytophage clades is needed but difficult to assemble, given the dearth of phylogenies and the diverse origins of phytophagy.[136] For many large groups only recently, or never, has there been the prerequisite critical synthesis of feeding habits.[1,4,27,55,57,86,170,202]

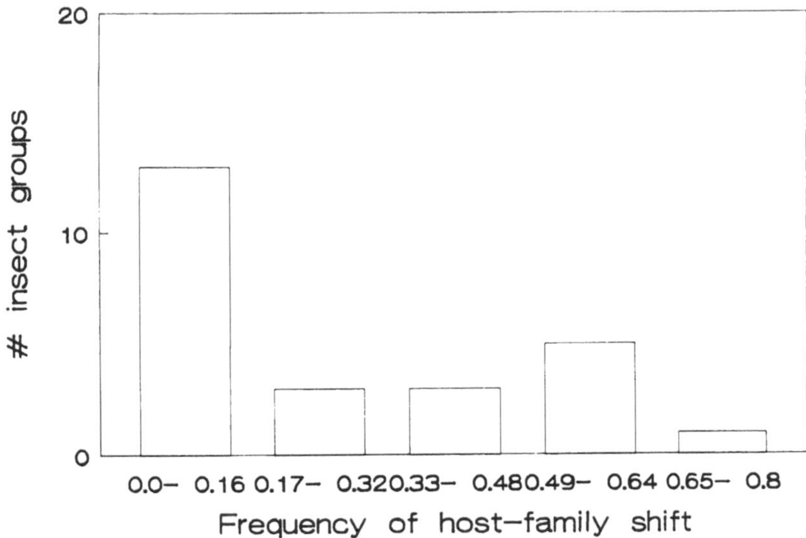

FIGURE 2. Distribution of rates (per speciation event) of shift to new host family, for insect groups in Table 1 plus seven additional groups.[30a,87,91,101,157,158,165]

As a basis for discussion, Table 1 and Figure 2 provide an initial compilation of rates of diet evolution with respect to host taxonomy. The compilation consists of those groups revealed by a search of the literature by the authors, for which both a substantial number of host records and a phylogeny estimate were available. Histories of change in host family and (where possible) host genus association were estimated by the parsimony criterion, as discussed earlier. The rates presented are rough values: even approximate habits are often known for less than half of the species, and the evidential support for the phylogenies is highly heterogeneous. Moreover, it is not clear that the sample is representative for phytophagous insects (or even what that would mean); for example, it includes essentially no polyphagous species. Nonetheless, it provides a first test of Jermy's postulate.

The host family level, frequently the limit of host record precision and a widely used indicator of "major" diet differences,[109] is the focus. Since nearly all species in the sample (except in *Graphium*) use a single family, the values in Table 1 and Figure 2 may be regarded as approximating the fraction of speciation events accompanied by host family shift. (The distribution at the plant genus level has a similar shape.) It appears that this fraction is most typically <20%. The authors interpret this as weak quantitative support for the generalization that related insects most often eat related plants, contrary to Jermy's postulate; however, it is also clear why conflicting generalizations have been made: some groups are remarkably labile, and the rate of host shift varies greatly. For example, if these rates are treated (roughly) as actual proportions of speciation events, they appear to differ significantly even between two weevil genera in the same tribe, *Cleonidius* and *Apleurus* (6/10 vs. 0/8, X_1^2 = 7.2).

Assuming that insect and plant taxonomies are independently derived, the

Table 1
RATES OF HOST-USE EVOLUTION WITH RESPECT TO HOST TAXONOMY, FOR SOME PHYTOPHAGE GROUPS WITH PUBLISHED PHYLOGENY ESTIMATES

	Phytophage group	Fraction of species with known hosts[a]	Shifts per speciation event[b,c] to host plants in new: Genus	Family[d]	Primary hosts	Distribution	Ref.
I.	*Heterobathmia* (Lep.: Heterobathmiidina)	10/10	0.00	0.00	*Nothofagus* Leaf miners	S. America	102
2.	*Lonchophorus* (Col.: Curculionidae)	9/13	—[e]	0.00	Fruits or buds of Bombacaceae	Neotropics	39
3.	Heliconiini (Lep.: Nymphalidae)	60/65	0.05	0.00	Passifloraceae	Neotropics	26
4.	*Apleurus* (Col.: Curculionidae)	8/8	—	0.00	Asteraceae	SW U.S., Mexico	5
5.	*Sibinia* subgen. *Microtychius* (Col.: Curculionidae)	46/126	0.16	0.00	Mimosaceae, buds or seeds	New World	38
6.	*Trioza apicalis* complex (Homop.: Psylloidea)	6/8	—	0.00	Apioid Apiaceae	Palearctic	30

7.	*Dalbulus* (Homop.: Cicadellidae)	11/11	0.20	0.00	*Zea, Tripsacum*	Central America	144
8.	*Phyllobrotica* subgen. *Phyllobrotica* (Col.: Chrysomelidae)	9/17	0.13	0.00	Root worms of Lamiaceae	Holarctic	64
9.	*Ophraella* (Col.: Chrysomelidae)	13/13	0.83	0.00	Asteraceae	N. America	80
10.	*Empoasca* subgen. *Kybos* (Homop.: Cicadellidae	50/63	0.16	0.02	Salicaceae, Betulaceae	Holarctic	166
11.	Danainae Lep.: Nymphalidae)	63/167	—	0.17	Asclepiadaceae, Apocynaceae Moraceae	Pan-tropical	2
12.	*Erythroneura maculata* group (Homop.: Cicadellidae)	125/136	0.19	0.18	Fagales	N. America	166
13.	*Pseudopsallus* (Hemip.: Miridae)	18/20	—	0.35	Herbaceous dicots	Mexico, western N. America	189
14.	*Graphium* subgen. *Graphium* (Lep.: Papilionidae)	12/25	—	0.64	Magnoliidae	Oriental, Indo-Australian, Africa	169

Table 1 (continued)
RATES OF HOST-USE EVOLUTION WITH RESPECT TO HOST TAXONOMY, FOR SOME PHYTOPHAGE GROUPS WITH PUBLISHED PHYLOGENY ESTIMATES

	Phytophage group	Fraction of species with known hosts[a]	Shifts per speciation event[b,c] to host plants in new		Primary hosts	Distribution	Ref.
			Genus	Family[d]			
15.	*Rhagoletis* (Dipt.: Tephritidae)	21/63[f]	0.65	0.50	Fruits of dicots	Cosmopolitan	13
16.	*Cleonidius* (Col.: Curculionidae)	11/19	—	0.60	Dicot herbs and shrubs	North, Central America	5
17.	*Yponomeuta* (Lep.: Yponomeutidae) "Group A"	7/7	0.83	0.50	Woody Rosidae	Western Europe	127
18.	*Squamacoris* + *Ramacoris* (Hemip.: Miridae)	7/9	0.67	0.67	Dicots, herbs, and shrubs	Mexico, western U.S.	188

[a] Records doubted by original authors excluded.
[b] No. speciation events = no. taxa - 1.
[c] Host shifts inferred by "optimizing" host use evolution on cladogram.[73] For species using more than one host taxon of that rank,each host optimized independently; only additions of hosts counted as shifts.
[d] Plant classification follows Cronquist.[45]
[e] Specificity to plant genera not clearly established.
[f] More hosts known, but only 21 species included in analysis.

rate estimates in Table 1 will presumably decrease as those taxonomies are refined. Moreover, focus on the plant genus and family levels leaves unidentified many insect clades showing higher plant-taxonomic associations. For example, numerous clades, e.g., melitaeine nymphalids,[1] sphingine and doliphonotine sphingids,[97] and *Longitarsus* flea beetles,[74] are broadly associated with the subclass Asteridae, despite its tremendous heterogeneity of major secondary compound types. For 11 North American insect clades that include herbivores of Apocynaceae/Asclepiadaceae, about 80% of the other host family associations lie in other asterid orders.[64a,137]

B. OTHER DIMENSIONS OF DIETARY CONSERVATISM

Diet phylogeny syndromes at least partially distinct from host taxonomy are increasingly being described. Groups which "track" convergent plant "defenses" are of especial interest, as their existence is a prediction of the escape and radiation model.[57] Thus, the hosts of *Graphium* swallowtails are taxonomically diverse (Table 1), but are linked by strong secondary-chemical similarities.[132] Berenbaum[11] showed that the diet phylogenies of several herbivore clades using taxonomically disparate hosts may be interpretable as conserved associations with coumarin-bearing plants. It has recently been observed that numerous insect clades, using convergent behavioral counteradaptations,[52,53] feed on taxonomically diverse plants united only by the possession of latex or resin canals.[52,61] For example, related genera and some individual species in the sphingid tribe Dilophonotini feed on Apocynaceae/Asclepiadaceae, Caricaceae, Euphorbiaceae, and Moraceae, distantly related plant families with variable laticifer systems,[91,97,205] while phloem-feeding bark beetles specialize in diverse laticiferous and resinous plants.[6,7,205] Assuming that other "defense"-related diet patterns await description, and that plant-taxonomic conservatism is also chemically based, it seems likely that constraint by plant chemistry has been underestimated.

In contrast, some groups seem to primarily track other host variables, perhaps reflecting conservatism in different steps of the host selection sequence.[43,44] For example, the hosts of the stilt bugs (Berytidae) appear to include taxonomically disparate plants characterized by exceptional hairiness.[201] Some groups seem most conservative with respect to (i.e., have hosts united primarily by occurrence in) certain habitats. For example, a number of groups, such as donaciine leaf beetles,[111] cylindrorhinine weevils, and many nymphuline pyralid moths,[148] are oligophages attacking diverse aquatic plants. Less extreme examples include the many groups (e.g., various weevils[212]) that have shifted among plants characteristic of xeric habitats, such as Chenopodiaceae, Cruciferae, and Asteraceae, and the many groups using diverse plants of wet boreal habitats (e.g., group 10 in Table 1). Quantification of evolutionary rates with respect to host habitat as compared to host taxonomy may be difficult, but should be possible in restricted comparisons, e.g., across a fixed set of habitats. There has been no attempt to do this.

C. VARIATION IN CONSERVATISM

Many hypotheses have been advanced to explain the variation in diet-taxonomic conservatism exemplified in Table 1. Some invoke insect life history. Taxonomically radical host shifts might be especially probable in groups whose adults use the host only as the rendezvous for mating.[47,83] On the other hand, host shift might be discouraged if the adults feed as well as mate on the host before oviposition, as for example in leaf beetles and most homopteran groups, because oviposition on the new host might require colonizing adults to possess appropriate variants of both feeding and host-finding traits. Internal feeders may experience more host-imposed constraints than external feeders, and exhibit greater specificity and conservatism.[190] Another possible constraint is evolved dependency on the host for herbivore defense, invoked to explain the ancient constancy of troidine swallowtail butterflies to Aristolochiaceae,[72] and suggested by apparent atavism in several groups whose adults seek out defensive compounds that may have been acquired from the larval host prior to host shift.[56,128,172]

Other hypotheses invoke differences in the resource used. Groups that specialize in flowers and fruits may experience less selection differential between host taxa because these structures frequently lack the plants' characteristic secondary chemistry.[13] Thus, groups such as heliothine noctuid moths[90,126] and many tephritid fly groups[13] (Table 1), while typically oligophagous, show diet phylogenies of considerable host-taxonomic lability. Similarly, plants in some taxonomic and/or habitat assemblages may show relative chemical homogeneity, presenting fewer barriers to host transfer;[70,75] examples include the characteristically tanniferous hamamelid and rosid angiosperms, dominants of temperate forests, and many monocot groups. Physically rigorous habitats, which may contain few but taxonomically disparate plant species, may also foster host-taxonomic lability by enforcing habitat specificity in insects.

While anecdotal evidence exists for (and against) all of these hypotheses, rigorous tests are largely lacking. One line of evidence is the confirmation of the hypothesized mechanism. For example, an experimentally testable expectation under the "defense-dependence" hypothesis is that use of hosts from which less chemical defense can be acquired should reduce fitness through increased predation;[47] however, any single defense-dependent lineage (e.g. troidine swallowtails) will have its own history and a variety of distinguishing features,[85] in addition to its host relations, undermining the attribution of its dietary conservatism to just its defense syndrome. One control for such confounded influences on the evolution of a single group is to seek replicate rate contrasts between groups which do and do not possess the trait (e.g., use of plant-derived defenses) under test. For true replication, each contrast must represent an independent origin (or loss) of the trait.[31,136] Whether there is enough evidence at present to apply such a test to any of the hypotheses reviewed above is unknown.

D. PHYLOGENETIC ORDERING OF HOST USE

Earlier, it was argued that genetic limitation on diet evolution would be suggested if the frequencies of different types of transitions between hosts, reflected in the orderings of different habits on phytophage cladograms, were predictable from the likely genetic barriers to those transitions. Several kinds of such orderings have been postulated.

Stepwise diet evolution with respect to host taxonomy or chemistry, as has been seen, is a fundamental assumption of the "escape and radiation" coevolution model. The compilation of rates of diet evolution with respect to host taxonomy (Table 1, Figure 2), assuming that relatives feeding on the same host genus or family often differ in specific preference, suggests that most diet transitions occur among related hosts; however, it provides little evidence on the scale or precision of diet evolution constraint by host similarity. At what level, if any, does host similarity become unimportant? Within insect clades specialized to one plant family, for example, might the plant genera be largely of equal evolutionary accessibility from one another? Do evolutionary sequences of habits show only rough correspondence to host similarity, or do they show precise gradation? For most of the groups in the authors' survey, neither host use nor host relatedness is known in sufficient detail to ask such questions. In a few well-studied oligophagous groups (e.g., various butterflies[37,50,181]), close relatives using the same host family show broadly overlapping and locally variable choice of plant genera and species. In one of these groups, the heliconiine butterflies, species groups and genera do differ on average in their use of host genera or subgenera,[9,26] but there is little apparent phylogenetic ordering among those associations.[135] Thus, in many groups, host similarity may place only coarse limits on diet evolution. There are, however, two possible instances of host-taxonomically "stepwise" evolution on a fine scale (Table 1): the sequential occupation of related host genera and species by *Phyllobrotica* (which may actually represent parallel cladogenesis; see below) and a possible trend toward successively more chemically distinctive Asteraceae in *Ophraella*.[80] More such detailed studies are needed.

Another long-postulated form of diet-phylogenetic ordering,[166] consistent with some recent models of host shift,[43,44] is that the genesis of one specialized habit from another occurs most readily through an intermediate stage in which both hosts are used. The array of habits among close relatives frequently suggests such a history, but phylogenetic corroboration of the proposed transformation sequence is lacking. Thus, most species of the leafhopper subgenus *Empoasca Kybos* feed on either *Salix* or *Populus*.[166] The few species that eat both were postulated to represent evolutionary intermediates, but in Ross'[166] phylogeny (doubted by Dworakowska[53a]), they occupy terminal positions rather than lying between the two kinds of specialists, as would be expected under Ross' postulate. The coccinellid *Epilachna pustulosa* uses the hosts of both its more specialized close relatives, but its phylogenetic intermediacy, while suggested by geographic distribution, has not been established.[112]

These examples illustrate a limitation of phylogenies for testing hypotheses of evolutionary mechanisms. The nonintermediate phylogenetic position of the oligophagous leafhoppers has at least three interpretations. New specializations might not actually evolve through an oligophagous intermediate stage. Alternatively, the oligophagous intermediate hypothesis may be correct, but evolution is rapid with respect to speciation, making the hypothesized pathway undetectable by phylogenies; i.e., phylogenetic sequences provide evidence on evolutionary pathways only if the intermediate stages are preserved as interspecific variation. (Conversely, the occurrence of predictable phylogenetic ordering would imply that evolution in that character is at least as infrequent or as slow as speciation.) Finally, a phylogenetic pattern reflecting the oligophagous-intermediate pathway may have been partly obscured by extinction.

This last hypothesis, in contrast to the others, implies that the expected sequence may still be detectable statistically, if multiple host shifts have occurred. A statistical approach seems generally desirable, and statistical tests for transformation sequences are being developed.[49,122,139,140] A limitation to this approach is the difficulty of obtaining a large enough sample of phylogenetic sequences. It is not clear whether pathways of diet transformation can be profitably studied by statistical-phylogenetic comparisons, as opposed to experimentally controlled analysis of single events.

The "no oligophagous intermediate" and "rapid evolution" hypotheses may be distinguishable in comparisons of very recently diverged species or populations, as in the coccinellid example above. From an ecological perspective, however, these hypotheses similarly imply that the present distribution of insects across host plants little reflects the evolutionary constraint embodied in the oligophagous-intermediate hypothesis.

Convergent evolution of analogous diets, if it has strong ecological correlates, is potentially conclusive evidence for local adaptation; an example is the geographic shift to more abundant hosts seen in several unrelated thistle insects;[213] however, some forms of parallel evolution may also reflect genetic constraints.[179] For example, if genes permitting use of a novel host arise so rarely as to limit diet evolution, but are often retained when a lineage subsequently occupies other hosts, reversions to formerly used hosts may occur more readily than adoption of entirely new hosts; this is a version of Dollo's law.[68] Retention of ancestral feeding abilities is known to occur, although no general review exists. For example, the one alder-feeding *Chrysomela* species feeds and develops normally when confined to its clearly ancestral willow hosts, while willow-feeding *Chrysomela* refuse to accept alders.[28] Similarly, when reared on diets incorporating extracts from 48 different plant families, the strictly solanaceous feeding *Leptinotarsa decemlineata* developed normally only on diets with extracts from composites and milkweeds, hosts to its congeners, and to other doryphorine genera.[99,100] Within the ancestrally cucurbit-feeding chrysomelid genus *Diabrotica*, adults of the derived, grass-feeding *virgifera* group will compulsively feed on toxic cucurbitacins, the host-derived defense chemicals utilized by their cucurbit-feeding relatives.[116,128]

A preferential shift to ancestral hosts, which would seem to have no purely adaptive explanation, may prove hard to demonstrate conclusively, but is suggested by some phylogenetic patterns. Thus, the noctuid moth *Catocala* (240+ species) feeds mostly on the dominant north temperate woody angiosperms, among which legumes cannot be counted. Yet, legume feeding, widespread and probably ancestral in the largely tropical Catocalinae, may have arisen more times in *Catocala* than any other habitat.[8,139] Patterns of homoplasy in host use have yet to be rigorously examined.

E. PHYLOGENESIS OF HOST RANGE

Change in the number of host taxa used is of special interest because of the central role of niche width in evolutionary theory.[81] Small expansions and contractions of diet breadth may be incidental to shifts of host preference. In some groups, however, the variation in the number of host taxa used is so great as to suggest that host range per se has been selected on. The high proportion of host specialists among phytophagous species may reflect a prevailing selective advantage,[16] but may also be subject to nonadaptive, phylogenetic influences.

The sensory, physiological, and genetic bases of host range differences, and their adaptive significance, are little understood; however, host range seems likely to involve a complex of adaptations. Host use in polyphagous species may involve qualitatively different characters from that in monophagous species, and may reflect larger life history syndromes.[14-16,81,145,173,199]

If host range involves a complex of adaptations, one may expect it to be evolutionarily conservative. Some very broad phylogenetic patterns are evident, such as the generally greater host ranges in the orthopteroid than hemipteroid or holometabolous orders,[29,35] and in scale insects as contrasted to other Sternorrhyncha,[54] but there is little detailed evidence. In a few cases, host range has been shown to vary among more restricted taxa, in correlation with other traits. Among the bombycoid moths of a Costa Rican forest, for example,[17,106] the Saturniidae are characterized by slow-growing, digestively efficient, well-defended larvae and short-lived, weak-flying adults. They feed on nutrient-poor, mature leaves of tanniferous tree taxa, and tend to have broad host ranges. In contrast, the Sphingidae have fast-growing, cryptic larvae and long-lived, highly vagile adults. They select fast-growing tissues of plant taxa characterized by "toxic" defenses, and are typically oligophagous. Replicates of this single phylogenetic contrast, and studies of the rate of change in diet breadth within such groups, are needed.

If host range is constrained by correlation with other life-history traits, one may further expect its present-day variation to represent stepwise phylogenetic sequences. No such sequences are known, but plausible cases exist. For example, heliothine Noctuidae,[89,90] at one extreme exemplified by the 100+ species of *Schinia*, attack one or a few closely related host species, lay few, large eggs, and are modified specifically for crypsis on the host in both larval and adult stages. The other extreme, exemplified by the corn earworm and relatives (*Helicoverpa*; 17 species[89]), is near-omnivory, accompanied by high fecundity (>1000 eggs in *Helicoverpa zea*) and highly vagile, even migratory adults with unspecialized coloration. Hardwick[89,90] proposed a gradual transformation between these extremes, regarding the paraphyletic genus *Heliothis*, which shows a range of intermediate states of these features, to be transitional between *Helicoverpa* and *Schinia* (see Matthews[126]); however, host range in heliothines appears sometimes to contract rapidly, possibly without change in associated characters. Thus, several derived *Helicoverpa* species are apparently host specific,[89] and polyphagy seems to have given rise to oligophagy a number of times within *Heliothis*, for example, within the morphologically homogeneous *H. virescens* group.[159] Close relatives with greatly differing host ranges are also known in many other groups (see Reference 78), particularly Lepidoptera (see Reference 93). Thus, while detailed studies are needed, restriction of diet breadth, at least, may not be closely constrained by its broad correlation with other characters.

Constraint on the direction of host range evolution could also affect the proportion of host specialists.[72,78,81,110] Specialization to part of an ancestrally broad host range may evolve more readily than the reverse,[167] as the heliothine pattern may suggest, since it does not require adaptation to any novel hosts. Conversely, the occasional polyphages in characteristically oligophagous groups may express multiple retained ancestral feeding abilities, and arise most readily in lineages undergoing rapid diet evolution. Thus, the sphingid moth *Manduca rustica*, which feeds on a variety of families of the subclass Asteridae, appears to occupy a derived position in a clade composed mostly of specialists on those families.[97] Evidence on directionality in host range evolution is lacking.

Finally, the unequal numbers of specialist vs. generalist feeders may reflect differential rates of diversification.[81] An inverse correlation of speciation rates with niche width has been predicted (see References 58 and 160), but there is as yet no evidence that this postulate applies to insect host range.[81] Cladograms offer straightforward potential tests of hypotheses about diversification rates.[136] Sister groups, i.e., those originating from the same branch point in the phylogeny, are equal in age by definition. Therefore, differences in species diversity between sister groups reflect differences in the rate of diversification (speciation minus extinction). Thus, if host specificity promotes diversification, multiple independent contrasts of sister groups which differ on average in host range should show consistently higher diversity of the more host-specific sister group.

III. RELICTUAL AND CODIVERSIFIED ASSOCIATIONS

Sufficiently conservative insect-plant interactions, once established, may persist through subsequent diversification of one or both lineages. Such parallel diversification has two implications. First, it is a likely correlate of long-term coevolution (including escape and radiation) between particular pairs of lineages, although it can occur in the absence of adaptive reciprocity. More generally, it implies a historical cause for current patterns of insect-plant associations: they may reflect in part the sequence of origin of the respective plant and insect groups.

A spectrum of mechanisms can be envisioned for correspondence of insect and plant ages. In extremely specific interactions, new associations may arise only through speciation in those occurring previously, leading to a detailed match of insect and plant phylogenies. A coarser match could result if insect diversification lagged somewhat behind that of plants, but the newly evolved plants were colonized most often by herbivores of their near relatives.[212] Thus, the escape and radiation model does not predict exact parallel cladogenesis, but correlation between the evolutionary sequence of plant defense types and the ages of their respective herbivores.[135] Even if newly evolved hosts were colonized at random, the ages of associates may be correlated,[209] if change in host use were sufficiently infrequent and punctuational that ancestral associations frequently persisted.

Most phytophage species belong to clades that have been phytophagous since the Cretaceous Period, in many cases much longer.[200] Thus, phytophagous insect diversification has been broadly contemporaneous with that of higher plants, especially of angiosperms. It would be surprising if present-day associations showed no traces of this shared history. This section asks how common these relictual or codiversified elements are, on what evolutionary scale or breadth of comparison they are most evident, and how their presence can be rigorously demonstrated, given that many associations, in contrast, undoubtedly reflect recent host shifts. Subsequently, the possible coevolutionary consequences of such enduring interactions are considered.

The most obvious expectation, if particular insect and plant lineages have diversified to some extent in concert, is concordance between their phylogenies; however, recognition and statistical corroboration of such concordance raises many difficulties, given the complexity of most taxonomic distributions of host use. Moreover, concordance itself admits several explanations. Therefore, several lines of evidence, reviewed in turn, must be synthesized in assessing parallel diversification.

A. CONCORDANCE OF INSECT AND PLANT CLADOGRAMS

This subsection attempts to review and quantify the as-yet limited evidence on parallel insect-plant diversification provided specifically by phylogeny estimates.

The problem of quantifying the cladogram concordance attributable to parallel phylogenesis closely resembles that of vicariance biogeography. In the latter, the issue is whether disjunct distributions between near relatives arose simultaneously with the physical barriers separating them, through fragmentation of widespread ancestral species, as opposed to more recent dispersal over barriers that were already there. Under vicariance, the order of separation among geographic areas should correspond to the phylogenetic relationships among their endemic species, like those of host plants and their specialized phytophages under parallel diversification. While there is general agreement that the fit between cladograms under vicariance-like processes should be measured by some analogy to phylogenetic parsimony, conflict over precise methods has arisen,[20,21,115,130,135,147,154,156] resolution of which will require rigorous demonstration of the connection between measures and the basic properties of those processes. Moreover, sampling distributions for vicariance measures are largely unavailable (see, however, References 154 and 178); still, one can take advantage of the better development of methods for a problem similar to but simpler than vicariance, namely the comparison of independently derived trees for the same set of taxa. Thus, the implicit question in this heuristic review is how similar is the actual host plant phylogeny to the host phylogeny implied by the relationships among the herbivores of those hosts?

Table 2 is an attempt to summarize the evidence on concordance for the few associations in which at least provisional phylogenies are available for both insects and host plants. Host phylogeny was examined at the taxonomic level on which feeding habits were most often specialized. Herbivores using multiple host taxa were omitted, as their interpretation is much debated[115,154] (see below), and treatment of convergent host use followed Kluge.[115] The groupings shared between insect- and host plant-derived cladograms, following the usual practice, were summarized in a consensus tree (see Figure 3) and quantified by a consensus index. Of the two widely used consensus tree methods, the authors prefer that of Adams[3] (Figure 4), a method the authors have argued elsewhere[64] as being more realistic for parallel phylogenesis because it counts partially (as well as fully) concordant groupings, which result when parallel diversification is partially obscured by subsequent host shifts. Heuristic significance tests for concordance (against a null model of independent cladogenesis) were based on two slightly different randomization distributions[176,177] for similarity between randomly constructed trees. (For a detailed discussion of these methods, see Farrell and Mitter.[64])

Although the evidence in Table 2 is very uneven, it seems clear that unambigous cladogram concordance, suggesting closely parallel phylogenesis, is rare. For some groups such as *Yponomeuta*, there is strong evidence against concordance, while for others (e.g., Heliconiini) there is limited concordance at best. A few groups show complete agreement but include few taxa, and for only three, *Phyllobrotica*, *Empoasca Kybos*, and carsidarid psyllids, does concordance even approach significance under any of the authors' approximate statistical tests. Examination of the *Phyllobrotica* case will illustrate the complexity of demonstrating parallel phylogenesis.

Phyllobrotica, a galerucine chrysomelid, contains two subgenera. The subgenus *Phyllobrotica* includes 13 North American and 4 Eurasian species. Eight of the nine species with known food plants are strictly monophagous on various species of *Scutellaria* (Lamiaceae), mostly perennial herbs; the other attacks the mint *Physostegia virginiana*. The less studied western North American subgenus *Stachysivora* is associated with the mint genus *Stachys*.[60] The closest relatives of *Phyllobrotica* appear to be the Asian *Hoplasoma* and *Hoplasomoides*, which feed on the verbenaceous genus *Clerodendrum*. *Phyllobrotica* larvae, like all Luperini, are root feeders. The adults, which emerge as the host plants flower, eat both leaves and reproductive parts of the larval host, often nearly destroying the seed crop and constituting the only obvious herbivory.[60]

An estimate of the relationships among the *Phyllobrotica* species of known feeding habits, extracted from a study of adult morphology,[60, 64] is shown in Figure 3, together with an estimate of relationships among their host plants, which was synthesized subsequently from the literature. All the beetle host plants belong to the Lamiales sensu Cantino.[33] The beetles are absent from two large, advanced lamialean clades, the lamiaceous Nepetoideae and the verbe-

Table 2
SYNOPSIS OF EVIDENCE ON PARALLEL DIVERSIFICATION, ON SOME ASSOCIATIONS FOR WHICH BOTH INSECT AND HOST PLANT PHYLOGENY ESTIMATES ARE AVAILABLE

Phytophage group	Plant phylogeny level	Groups resolved in consensus tree	P[b]	Equivalent ages plausible?	Parallel diversification plausible?	Plant phylogeny ref.
1. *Heterobathmia*	Species	0/1	ns/ns	?No: geogr. restr.[c]	No	102
3. Heliconiini	Genus/ subgenus	~3/6	ns/ns	?Yes: geogr.[d]	?Partial	113
8. *Phyllobrotica*	Species	6/8	~0.05[e]/*	?Yes: geogr.	Yes	10, 33, 34, 59; see 64
9. *Ophraella*	Genus/ tribe	1/2 or 0/2[f]	ns/ns	No: allozyme clock[e] geogr. restr.	No	19 vs. 155; See 80
10. *Empoasca*	Genus/ order	2/2	~0.07/*	No: geogr. restr. unoccup. hosts[g]	No	46
13. *Pseudopsallus*	Family	1/3	ns/ns	No geogr. restr. unoccup. hosts	No	46
15. *Rhagoletis*	Family	~3/7	ns/ns	No: allozyme clock unoccup. hosts	No	46
16. *Cleonidius*	Family	0.5/1	ns/ns	No: geogr. restr. unoccup. hosts	No	46

Table 2 (continued)
SYNOPSIS OF EVIDENCE ON PARALLEL DIVERSIFICATION, ON SOME ASSOCIATIONS FOR WHICH BOTH INSECT AND HOST PLANT PHYLOGENY ESTIMATES ARE AVAILABLE

Phytophage group	Plant phylogeny level	Groups resolved in consensus tree	P[b]	Equivalent ages plausible?	Parallel diversification plausible?	Plant phylogeny ref.
17. *Yponomeuta*	Family	0/3	ns/ns	No: geogr. restr. unoccup. hosts	No	46
18. *Squamacoris*	Family	1/3	ns/ns	No: geogr. restr. unoccup. hosts	No	46
Papilionidae (Lep.) tribes[132]	Family	1/2	ns/ns	No: unoccup. hosts	No	46
Graphiini genera (Lep.: Papil.)[132]	Family	1/2	ns/ns	No: unoccup. hosts	No	46
Larinus (Col.: Curcul.) Subgenera[212]	Tribe	1/1	ns/ns	Yes: fossil; allozyme clock geogr.	Yes	See 203
Genera of Carsidaridae (Homop.: Psylloidea)[87]	Family	~1.5/2	ns/[h]	?Yes: geogr.	?	46

Note: Numbered taxa correspond to those in Table 1.*$p < 0.05$.

[a] Number of groups resolved/maximum possible (no. host taxa -2), in Adams consensus tree between actual plant cladogram and plant cladogram(s) best fitting insect cladogram (or average thereof, for multiple solutions) following Kluge.[115] Herbivores not host-specific at level of plant phylogeny excluded.

[b] Randomization distributions of Shao and Rohlf[176]/Shao and Sokal[177] based on "consensus fork" index of consensus tree resolution.[41]
[c] Insect distribution much more restricted than that of plant clade descended from the common ancestor of all hosts used.
[d] Distribution of insects suggests age equivalent to that of host clade.
[e] Test value equals critical value.
[f] Two different host phylogenies current.
[g] Many families, orders, etc., of host clade not used by insect clade.
[h] Multiple solutions; test values for some exceed critical value.

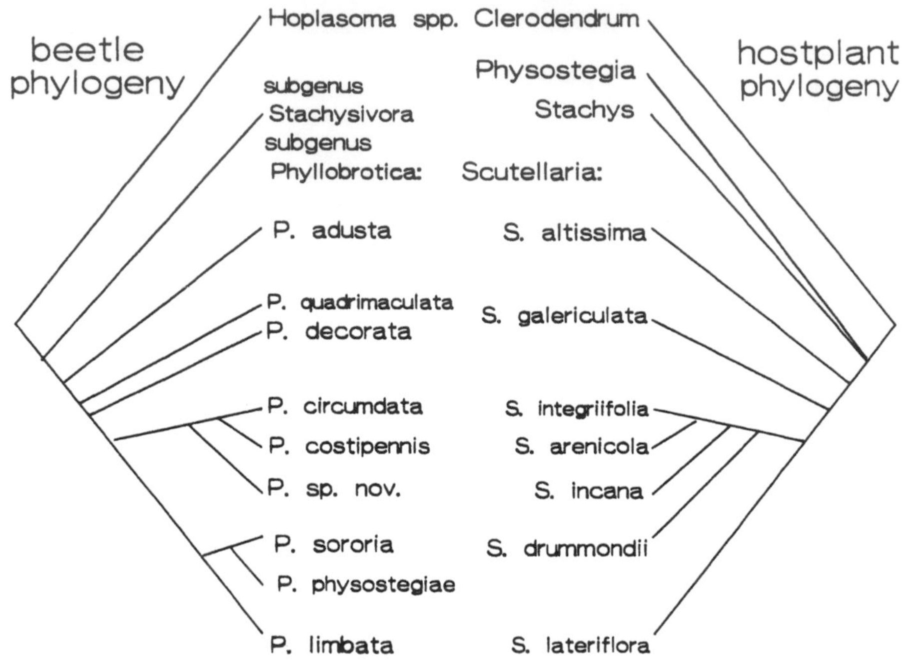

FIGURE 3. Comparison of phylogeny estimates for *Phyllobrotica* leaf beetles (species with known hosts only) and their host plant clade (excluding nonhost taxa; see text), modified from Farrell and Mitter[64] (refer to for plant phylogeny sources). Host names are opposite herbivore names, except that *P. physotegiae* feeds on *Physostegia*.

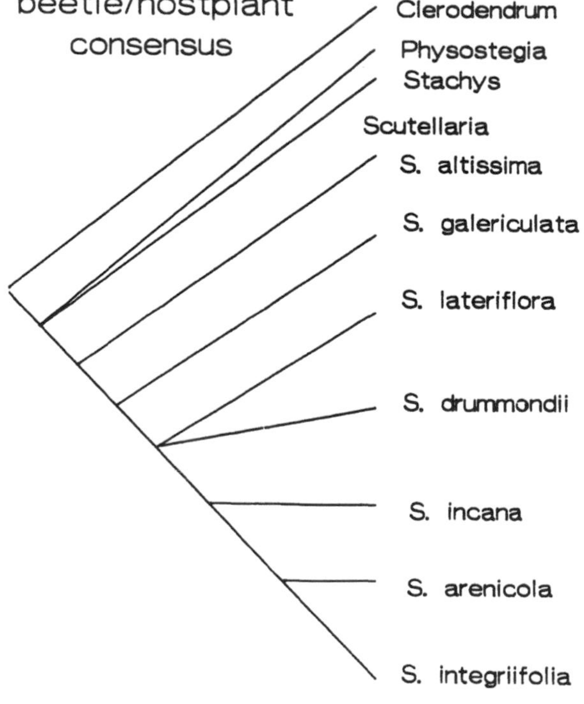

FIGURE 4. Adams' consensus tree[3] representing degree of agreement between phylogeny of host taxa extracted from botanical literature, and host phylogeny implied by cladogram of the *Phyllobrotica* beetles of Figure 3. The consensus tree is resolved for 6 of 8 possible groups (thus, the consensus fork index of Colless[41] = 0.75), indicating that the concordance between the two cladograms is significant or nearly so under the randomization distributions of Shao and Rohlf[176] and Shao and Sokal.[177]

naceous Verbenoideae, and are restricted to the subfamilies Viticoideae (Verbenaceae) and Lamioideae (Lamiaceae), basal (probably paraphyletic) elements of their respective families.[34]

The overall match of cladograms (Figure 3) is significant or nearly so (Figure 4) despite one major disagreement — the placement of the mint *Physostegia*; the adjacent *Scutellaria* sections Lateriflorae and Annulatae are also transposed between the cladograms. While this pattern strongly suggests parallel phylogenesis overlaid with occasional host transfers, other lines of evidence must be considered before dismissing the alternative explanation of colonization that is entirely subsequent to host diversification by herbivores tracking some feature correlated with host phylogeny.[64]

B. EVOLUTIONARY MECHANISMS

Many assemblages may reflect complex mixtures of parallel phylogenesis and host transfer. Delineating these would be easier if one could also independently specify the pattern expected under host transfer, if only in hindsight. In the *Phyllobrotica* example, the unusual ecology of the "anomalous" association of *P. physostegiae* with *Physostegia* strongly suggests host transfer: this beetle and its sister species are unique in the genus in inhabiting dry uplands instead of wet bottomlands; prairie *Scutellaria*, unlike riparian species, are often annual and unpredictably available, which could strongly favor insect host shift to a related, chemically similar perennial host such as *Physostegia*.[37]

Similarly, the interpretation of partial concordances as parallel diversification would be stronger if a mechanism could be identified. One possibility is joint geographic isolation, as suggested by the distributions of some *Phyllobrotica*/host pairs. Thus, *P. costipennis* and its host *Scutellaria arenicola* are both endemic to the biotically distinctive Lake Wales Ridge region of central Florida. They were probably isolated together by Pleistocene sea level fluctuations[198] from their respective sister species to the northwest.[64] Joint vicariance for other insect-plant assemblages has been suggested,[36,98,114,171] but is little investigated like vicariance hypotheses generally (see, for instance, References 117 and 151). In contrast, Zwölfer[211,212] suggests that partial, higher-level congruence between *Larinus* weevils and tribes of thistles (see below) may result simply from gradual restriction of inter- vs. intratribal host shifts as the tribes diverge. This mechanism may be most likely in a species-rich host clade.

More generally, parallel phylogenesis would more credible if variation in its apparent prevalence had a biological rationale. Unlike the animal parasites in which concordant phylogenesis is most evident,[88] transmission of phytophagous insects between host individuals involves a free-living stage, greatly increasing the chance of host shift.[165] It is not surprising that the most detailed match of insect and plant cladograms known involves a group with many features thought to promote specificity and conservatism of feeding habits, including larval endophagy, adult feeding on the larval host, and a "toxic" host chemistry (including most notably iridoid glycosides[18]) on which the apparently aposematic adults may depend for defense. The patchy host distribution and sedentary adult behavior that in other herbivores have been linked to genetic population subdivision[123] may further increase the chance for joint geographic speciation. Closely parallel phylogenesis of insects and host plants, if ever common, is most likely to be found in other assemblages with similar natural history. These may be common in other leaf beetles and their relatives.[36]

Finally, parallel phylogenesis would be more credible if it could be explained why the herbivores are generally much less diverse than, and use a small fraction of, the host clade.[211] Thus, the subgenus *Phyllobrotica* attacks only about $^{1}/_{10}$ of the species of *Scutellaria*, and the *Phyllobrotica* clade as a whole attacks <1% of the Lamiales. Unoccupied hosts present the greatest difficulties when their cladistic position requires postulation of many host "escapes", or of extinct or unsampled associations, particularly if these must have distributions uncorroborated by the cladograms.[22,64,135] Thus, although the host relationships best fitting the cladogram for the Holarctic leaf hopper subgenus *Kybos* (Table 2) agree with plant phylogeny, the two pairs of host genera (*Salix* and *Populus*, Salicaceae, and *Betula* and *Alnus*, Betulaceae) are in different, cosmopolitan subclasses that diverged in the lower Cretaceous (see below). Rather than parallel diversification, it seems much more parsimonious to suppose that the three closely related species on Betulaceae represent a host shift within a common habitat from a large clade ancestrally associated with Salicaceae.[166] More generally, parallel phylogenesis can probably be safely ruled out for many geographically restricted herbivore species and clades, the divergence of whose taxonomically disparate hosts occurred on a much broader scale.

Unoccupied hosts are more readily accomodated if their "escape" can be independently explained. Thus, the absence of *Phyllobrotica* and relatives from the very diverse Nepetoideae and Verbenoideae requires only escapes by the single ancestors of these groups, which for Nepetoideae could been facilitated by the substitution of aromatic oils for iridoid glycosides. Within *Scutellaria*, it is primarily the most widespread species that are used by the beetles, which are often less widespread than their hosts. Thus, the absence of beetles from narrowly endemic skullcap species and from the distinctive South American clade[59] can be plausibly ascribed to host geographic "escape".[64]

C. EQUIVALENCE OF INSECT AND HOST PLANT AGES

A critical expectation, if current associations commonly reflect a joint evolutionary history, is that associated insect and plant lineages will commonly be of equivalent ages. If such equivalence could be disproved, cladogram concordance would need some explanation other than parallel cladogenesis.[88] Conversely, if equivalence of ages were established, it would be difficult to maintain any other explanation for even partial concordance. Moreover, some relictual feeding patterns reflecting herbivore stasis but not concerted cladogenesis may only be detectable by comparison of insect and host ages. At present, evidence on the age of associations comes mostly from fossils, limited by the generally poor record for insects, and from biogeography; in the future, approximate molecular clocks should provide critical additional information.[88]

Fossil or biogeographic ages for particular associations of modern species or genera show that these may be surprisingly old.[65,96,119,121,142,152] Modern genera of insect herbivores are often of mid-Tertiary age. Such observations raise the plausibility of, but do not directly bear on, synchronous diversification.

More direct evidence comes from the few datings available for cladistically concordant associations. While there are no fossils of the *Phyllobrotica/Hoplasoma* clade, the "Arcto-Tertiary" distribution[194] of the beetles would date them to the mid-Tertiary, in agreement with fossil pollen dating for the origin of the Lamiales.[143] Biogeography and allozyme divergence levels in the weevil genus *Larinus* and its sister group[212] are consistent with a late Oligocene origin, similar to that indicated by fossils for their host clade, thistles. Thus, although the partial, higher-level phylogeny concordance exhibited by these associates cannot be statistically corroborated by itself, it seems likely to represent parallel cladogenesis. A similar argument may hold for heliconiine and related butterflies and carsidarid psyllids, which show striking geographic disjunctions.[28,98] Conversely, limited allozyme divergence in *Ophraella* renders parallel phylogenesis very unlikely, despite partial agreement with one view of the phylogeny of its much older asteraceous hosts[80](Table 2). Datings for other examples of partial concordance are needed.

A final, potentially powerful test for synchronous diversification comes from the broad, correlative study insect and host plant ages. Although statistical reanalysis is needed, Zwölfer[209] presents evidence that major phytophage clades originating in successively younger geological eras are associated on average with successively younger plant taxa. This finding parallels the observation that the basal divergences in a number of large phytophage clades, such as aphids and some beetle groups, are roughly concordant with host relationships. Similar patterns may exist at lower levels, but have been little sought. A suggestive observation is that the host taxa used by extant chrysomeloid and curculionoid beetle genera which are known also in Eocene fossils are significantly older than those used by extant genera known only from Oligocene fossils.[64a,65] The interpretation is somewhat complicated by the restricted geographic source of the earlier fossils; at the least, however, many of these associations are extraordinarily conservative.

IV. ESCALATION, COEVOLUTION, AND DIVERSIFICATION

Highly specific, long-term interactions have been regarded as the best candidates for pairwise, reciprocal adaptation (see Reference 84 for a contrasting view), and indeed some of the clearest evidence for reciprocity is seen in the long-standing association of heliconiine butterflies with Passifloraceae; however, similar adaptations are not apparent in any other assemblage the authors have surveyed, including those most likely to have diversified in parallel. Reciprocal adaptation between *Phyllobrotica* and *Scutellaria* has yet to be intensively sought, but is not obvious, and there appears to be no specific reciprocity between thistles and weevils or any other of their insect herbivores.[202] Close, pairwise insect-plant coevolution sensu Janzen[105] clearly seems rare.[76,190,191]

Nonetheless, phytophagous insects and higher plants have diversified in at least broad synchrony, and many associations are old. Insects have certainly formed part of the community of enemies with which plant evolution has had to contend, and should participate in such diffuse coevolution as these two trophic levels have undergone, whether or not they are a dominant selective force on plants.

While coevolution in the sense of reciprocal individual adaptation has received most of the attention, Ehrlich and Raven's hypothesis was macroevolutionary in scope, assigning at least an equal role to differential diversification of lineages resulting from those adaptations.[186] Ecological theories of macroevolution, exemplified by Ehrlich and Raven's essay, were a major theme of the New Synthesis[179] but have been in eclipse.[103] Sceptics doubt that interactions, including the effects of insects on plants,[190] are intense or consistent enough to be major evolutionary forces. Advocates of a "hierarchical" theory of evolution have argued that evolutionary trends may be governed by differential speciation rates unrelated to increased individual adaptation,[185] or that "trends" may even be random.[162]

Reassertions of the ecological view come from marine paleontology.[120,196] Vermeij[197] argues that the Phanerozoic fossil record shows pervasive "escalation", that is, improved individual adaptation for efficiency of resource capture or resistance to enemies and competitors, with attendant increase in diversification rate. Reciprocal escalation, as in Vermeij's example of the elaboration of armor in gastropods, correlated with the rise of shell-breaking predators such as crabs, is equivalent to escape and radiation coevolution.

Vermeij[197] prescribes a rigorous test of the escalation hypothesis for cases in which the interactions themselves are fossilized (e.g., bivalve shells drilled by predators): one directly measures past selection due to interaction, then looks for adaptive change and subsequent diversification in the affected lineage.[197] Difficulties arise because plant-insect interactions rarely fossilize, but comparative approaches based on contemporary species, not yet widely applied, may nonetheless provide strong inferences about their macroevolutionary consequences. The following provides a summary of the scant evidence of "escalation" and radiation, separately for plants and insects, and outlines some directions for further work.

A. ESCALATION OF PLANT DEFENSE

Plant traits affecting insects and other enemies are frequently conservative, characterizing higher taxa. This observation suggested to Ehrlich and Raven[57] and others a major role for defense escalation in higher plant evolution. Thus, Cronquist[45,46] postulates a succession of chemical defenses underlying the radiation of the angiosperm subclasses. For example, the most primitive subclass, Magnoliidae, is typified by isoquinoline alkaloids; the Hamamelidae and Rosidae are typically tanniferous; and the most recent and diverse subclass, Asteridae, displays a profuse variety of acutely toxic or repellent substances. Analogous variation can be seen at lower levels.

A first step in testing any specific escalation hypothesis is documentation of the supposed escalation events. Without fossils or other absolute datings, temporal sequences of adaptation can be inferred only from phylogenetic orderings of extant species; however, if the earlier stages in such a sequence now have low fitness, they may be extinct and unavailable for study. Indeed, their very survival would call for an explanation under the escalation hypothesis; Vermeij[197] argues that they will only persist in "refuge" habitats exerting reduced biotic pressures. Moreover, the progression in "diffuse" adaptation of an entire assemblage might fail to be mirrored in the phylogenies of individual lineages, even though all the intermediate stages were preserved, if the traits involved were sufficiently labile.

Nevertheless, a comparison of present-day plant taxa reveals a number of potential defense escalation sequences; three examples follow. Perhaps the most rigorously stated is Berenbaum's[11] hypothesized sequence of coumarin evolution, from *p*-coumaric acid to hydroxycoumarins, to linear furanocoumarins, to possession of both linear and angular furanocoumarins. Each condition occurs independently in several to many plant families,[11] and is postulated to have reduced loss from enemies adapted to the antecedent defense. The increasing taxonomic restriction of the successive stages and their biosynthetic relationships are consistent with the postulated sequence, although explicit phylogenies are lacking.[132] Within the Capparales, the apparently ancestral glucosinolates are supplemented in several independent lineages by various atypical compounds, such as cardenolides, cucurbitacins, and alkaloids, similarly suggested to represent escalations against glucosinolate-adapted enemies.[71] Within the North American subgenus *Asclepias Asclepias* (Asclepiadaceae), cardenolide composition and tissue distribution, derived from extensive surveys of monarch butterfly host plants,[23-25,146,164] accord well with the morphology-based phylogenetic ordering of milkweed species groups (series) by Woodson.[207,208] The relatively advanced milkweeds (series 5 to 8) are unique in producing cardenolides of the structurally complex labriniformin family. Cardenolides in the more primitive series (1 to 4) are of the simpler calotropogenin type, widespread and presumably primitive in the Asclepiadaceae. Within the advanced group of milkweeds, the levels, within-plant locations, and identities of complex cardenolides can be further arranged into an apparent evolutionary sequence. In series 5, complex cardenolides are present in only trace amounts, increasing to a maximum in series 8 (Roseae). In Roseae, cardenolides are confined principally to the latex, where their deterrent effects on herbivores should be maximal.[146]

The defense escalation suggested by these examples is very hard to prove, however.[76] There are strong alternative hypotheses for the origin of new secondary compounds,[191,200] and evidence for an original defensive function is difficult to gather. In the absence of fossil demographies,[197] past selection due to enemies would be most directly inferred from demonstration in the field that plants bearing the advanced "defenses" (still) suffer consistently less fitness loss due to enemies than relatives bearing the antecedent defense (see below). Unfortunately, such comparisons, even when practical, may not reflect the original selection differential: the biotic environment is likely to have changed,[76,107] especially if the clade bearing the trait is now widespread.

If the plant fitness differential itself has been lost, indirect evidence for an initial advantage would be greater reduction of enemy fitness by the advanced than by the primitive defense, in a test battery of relevant enemies such as those adapted to the primitive defense. In other words, if the enemies attacking the advanced defenses, however diverse and successful they may now be, were derived from a relatively small subset of clades attacking the antecedent defenses, one could infer that plants bearing the new defense had initially shed many of their enemies.[161] Thus, some coumarins and atypical crucifer compounds have been shown to present barriers to some insect herbivores,[11,12,37] and chemically atypical crucifers seem to be consistently avoided by crucifer specialists in the field,[37] although the chemical mechanism for this is not clear.[195]

A final complication is that under some plant defense hypotheses, novel secondary compounds may function not to reduce attack rates overall, but rather to provide the same level of protection at less cost or in new environments.[40,200] Such indirect "escalations" against enemies could nevertheless permit radiation through competitive superiority or invasion of new habitats.

Given the difficulty of resolving the adaptive significance of novel "defenses", it may seem premature to ask whether they promote plant diversification. The authors argue, however, that the test of this second prediction of Ehrlich and Raven's model should proceed simultaneously with the first prediction. It could be independently falsified. Conversely, if "defense" innovations are associated with increased diversification, they can hardly be selectively neutral, and the search for their adaptive meaning becomes all the more compelling.

Berenbaum[11] showed that among the Apiaceae, genera with both angular and linear furanocoumarins have markedly higher average numbers of species than genera bearing linear furanocoumarins only, and that the latter in turn are more diverse than genera that lack coumarins. A difficulty with this analysis, however, is that genera bearing similar coumarins are probably not phylogenetically independent. Angular furanocoumarins in Apiaceae, for example, are restricted to two adjacent tribes[149] and may characterize a single lineage, raising the issue discussed earlier of confounded influences on evolutionary rate or "success".

One solution to this problem, as noted earlier, is statistical control.[136] Cladograms can be used to identify lineages bearing independent origins of the same defense and their sister groups. Since sister groups are by definition of equal age, differences in their diversities must reflect different rates of diversification (origination minus extinction).[185] Under the escalation/adaptive radiation hypothesis, lineages bearing the new defense should be consistently more diverse than their sister groups, when the latter retain the antecedent defense. This approach can be applied in the absence of fossils, although a comprehensive record would provide more accurate diversity comparisons and permit other types of rate comparisons, such as of the same lineage before and after a defense innovation.[61]

Multiple comparisons in some form seem essential for demonstrating that escalation promotes diversification, but raise a dilemma of sample size: it is difficult to statistically establish an effect on diversification of adaptations which, like angular furanocoumarins, have arisen very few times. One solution is to first combine such singular events into broader categories. Thus, although the evidence has not been collated, Berenbaum's[11] compilations suggest that numerous comparisons are available to test the hypothesis that the origin of new coumarin classes in general promotes diversification. Corroboration of the broad hypothesis would then strengthen causal interpretation of individual cases.

Analysis of another broad class of putative defenses, secretory canals bearing latex or resin, provides replicated evidence that escalation of defenses can promote plant radiation.[53,61] Such canals have arisen at least 40 times, ranging across all the angiosperm subclasses in addition to ferns and gymnosperms, and show great variation in the details of their anatomy and of the chemistry of their contents; however, they have such strong functional similarities that they can be considered a distinct syndrome.

Secretory canals are typically tubes, consisting either of living cells (laticifers) or intercellular spaces (resin canals), that ramify extensively through the plant. They are filled with viscous fluids that ooze or spurt out when the canals are severed. Many nondefensive functions have been suggested for secretory canals, but evidence for these is lacking.[53] In contrast, there is much support for a defensive function.[53] Plants in which the canal systems are poorly developed or have been artificially disrupted suffer greater herbivore attack than those with well-developed, intact canal systems.[53] Insects attempting to attack plants with intact canals can be immobilized or engulfed by sticky latex or resin. Moreover, other toxic or repellent substances are often taxonomically coincident with, and physically concentrated in, the secretory canals. The adhesive and toxic/repellent aspects of canal defenses are undoubtedly synergistic: the canal system enormously increases the quantities of latex or resin and its noxious constituents which can be delivered rapidly to a wound site, and may also help to isolate toxic substances from sensitive plant tissues.

Secretory canals in different plant groups, despite their anatomical and chemical diversity, evoke characteristic, convergent counteradaptations from insect herbivores.[52] Thus, folivores from at least 11 families in 3 orders, feeding collectively on many unrelated canal-bearing plant groups, disrupt latex and resin canals by transecting veins or cutting trenches before feeding beyond the cuts;[53] as noted earlier, many of the herbivores attack taxonomically diverse canal-bearing plants.

Sister group comparisons are available for only 16 of the 40+ origins of canal systems,[61] but the canal-bearing lineage is more diverse than its presumed sister group in 14 of these. This statistically significant trend strongly suggests that secretory canals have promoted radiation in the plant groups that have evolved them. Moreover, the diversities of canal-bearers relative to their sister groups were found to be independent of geological age, suggesting that the advantages of canals may still be evident.

Secretory canals are most plausibly interpreted as defenses against herbivores and pathogens, but direct plant fitness comparisons are lacking. Fitness comparisons on single taxa suffer from the same drawbacks as single comparisons of diversity, and ideally should also be replicated across independent origins. There have been few attempts to do this. In an initial survey, canal-bearing taxa within a Peruvian rainforest were found to have consistently higher population densities per species than their near relatives.[66] These data may also bear on the unsolved problem of how increased individual adaptatedness (i.e., escalation) is translated into increased diversification.[103,196] Elevated population densities have also been associated with another convergent "escalation", more efficient growth forms in marine encrusting bryozoans.[120]

Comparative studies of other convergent defenses may establish defense escalation as a pervasive influence on plant diversification, contrasting with the lack of support[95] for at least one alternative hypothesis, the development of biotic dispersal syndromes.[193] Attention must then focus on the mechanisms of such radiation, including the relative importance of insects. These may be best studied in relatively recent, geographically restricted examples whose original circumstances are still evident.

B. INSECTS: NOVEL HOSTS AND ADAPTIVE RADIATION

As escape from herbivory promotes plant diversification under Ehrlich and Raven's model, so radiation onto host plant resources is accorded a major role in insect diversification.[57,160,209] This hypothesis has been elaborated in broadest form by Southwood,[184,190] who points out that while feeding on higher plants is restricted to a minority of the 30 insect orders, most of which are predominantly saprophagous or predaceous, phytophages constitute a majority of insect species. He argues that there are strong barriers to plant feeding, such as increased exposure to predators and to desiccation, low nutrient levels, and active defense by plants, but that the relatively few insect lineages that have overcome these barriers have been able to radiate extensively in an unsaturated adaptive zone. Southwood's postulate is supported statistically by replicated sister group comparisons.[135]

This finding is consistent with the hypothesis that the abundance or diversity of plant resources promotes insect diversification, but an alternative (though complementary) explanation is species selection sensu stricto;[182] i.e., "parasites" in the broad sense (including phytophages[160]) may speciate more rapidly than predators or saprophages due to fragmented population structure and disruptive selection imposed by the discrete, patchy distribution of host species and the differences among them.[32,160] This hypothesis predicts that host shifts may directly foster speciation. Evidence for this is limited. It is not clear what fraction of phytophage speciation events are accompanied by discrete change in hosts (the host information in our compilation of cladograms being rarely detailed enough to tell). In at least some groups, close relatives typically overlap in habits, suggesting that host switch has not been a major factor in speciation.[37,168] A second prediction, however, of the species-selection hypothesis, that host differences can serve as primary isolating mechanisms, is supported by some recent studies.[68,112,124,206]

A third prediction of the species-selection hypothesis is that other types of parasitism sensu lato should also accelerate diversification. This prediction has been tested for insects parasitizing animals.[203] Parasitism of animals has arisen from saprophagy or predation at least several dozen times among insects. (To keep the parasite and phytophagy analyses independent, the few instances of animal parasitism arising from phytophagy or vice versa were excluded; these would not have affected the conclusion). Sister group comparisons are available for 16 lineages of parasites sensu stricto. No trend in relative diversities was apparent, either in the entire sample of comparisons, nor when these were variously subdivided as to the mode of parasitism or the host group attacked. This finding indicates that the parasitic lifestyle and its genetic consequences alone do not always result in enhanced diversification. Elevated phytophage diversity, and the lack thereof for animal parasites, are probably related instead of or in addition to the differing resource bases available to different trophic groups.[160]

If resource diversity underlies phytophagous insect diversity, shifts within phytophagous lineages to new resource "adaptive zones", such as new plant tissues, or new groups of plants defined by taxonomy or secondary chemistry should also enhance diversification.[57] This proposition seems almost self-evident, yet it may be false: the diversity of feeding habits could be incidental to, not a cause of, the diversity of phytophage species. Tests of the hypothesis are essentially nonexistent, and face many problems. As argued earlier, it is difficult to rigorously demonstrate the diversifying effect of any single adaptive shift. To obtain the necessary replication, "adaptive zones" must be defined broadly enough that they have been multiply invaded, yet narrowly enough that they remain meaningful. Given the enormous diversity of phytophage feeding habits, this is not easy. If "adaptive zones" exist and have figured in radiation, they should characterize groups of related species; i.e., they should correspond to the conservative dimensions of resource use, as reviewed earlier, but these clearly vary among phytophage groups. Meaningful tests of the hypothesis may thus be possible only within restricted groups; yet great care must be taken to avoid the circularity of excluding groups occupying the hypothesized adaptive zone simply because they are not diverse. Moreover, resource diversity could underlie phytophage diversity, yet the "adaptive zone" concept lacks utility, if the resources permitting the existence of individual species were unique, forming no obvious categories shared among relatives. Finally, even if phytophage diversification proves consistently associated with resource shifts, many questions of mechanism remain. For example, the release from competition or predation implicit in "adaptive zone" hypotheses seems questionable for phytophagous insects.[190]

Despite these difficulties, several kinds of resource groupings constitute plausible though untested "adaptive zones", at least within restricted phytophage groups. The escape and radiation hypothesis predicts sequential insect radiations following colonization of plant groups bearing successively more advanced defenses. The strongest argument for such a sequence is Berenbaum's[11] analysis of the fauna of coumarin plants. The taxonomic placements of several phytophage groups feeding on apparently advanced coumarin plants suggest that these were ancestrally associated with older coumarin types (although no explicit phylogenies exist[132]). Moreover, within two lepidopteran clades, subgroups whose hosts include coumarin plants are more species-rich on average than subgroups that never use coumarin plants;[11] however, these subgroups are probably not phylogenetically independent, and the hypothesis awaits phylogenetic re-examination.

Use of latex- or resin-canal-bearing plants may similarly constitute a repeatedly invaded adaptive zone: it often characterizes lineages of insects, and is typically associated with behaviors for disabling the canals. Specializations to taxonomically isolated and/or biochemically distinctive plant taxa may similarly constitute adaptive zones.[57] For example, the milkweeds and relatives (Apocynaceae + Asclepiadaceae) are host to a specialized herbivore fauna that includes many large clades.

New modes of plant attack may also represent new adaptive zones.[209] For example, the induction of galls, which may enhance both herbivore nutrition and protection from enemies,[160] seems to be a relatively recent development, appearing first in the late Cretaceous Period, but now characterizes many species-rich lineages.[118,209] Derived specializations to plant tissues may also constitute adaptive zones. Thus, specialization to the relatively rich resources of flowers and fruits has arisen multiple times from groups feeding on vegetative structures and characterizes some large clades, such as heliothine and related noctuids[90] and euchloine pierid butterflies. Other examples are given by Zwölfer.[209]

Finally, adaptive radiation may frequently follow colonization of a new geographically or ecologically defined flora.[64a,197] Thus, the northern temperate zone has been invaded many times by tropical phytophage lineages.[63,64a] Some have remained associated with the few temperate representatives of their tropical host groups, but others have switched to the temperate flora dominants, and may thereby have been able to radiate. A possible example, discussed earlier, is the noctuid *Catocala*, by far the most diverse north temperate catocaline genus, which apparently shifted from an ancestral legume association in the lower latitudes to an array of "Arcto-Tertiary"[194] woody angiosperms. The genus contains many well-marked groups, each restricted to a single plant family,[8,134,138] suggesting a series of radiations onto novel host taxa. Similar patterns occur in many other phytophages.[38,64a,97]

V. SUMMARY AND CONCLUSIONS

1. To the degree that evolution in the traits governing them is constrained, the current structure of insect-plant associations may reflect long-term historical processes, rather than local optimization of host use or host defense. This broad postulate is implicit in Ehrlich and Raven's model of coevolution, but has undergone little direct testing. Drawing on recent advances in phylogenetic methods, the authors surveyed some possible forms of such constraints and their macroevolutionary consequences, and reviewed the evidence that systematics and paleontology provide of their prevalence.
2. Genetic limitation on the evolution and current pattern of insect host use is suggested if host use changes infrequently, or if it shows phylogenetic orderings predictable from the likely genetic barriers to transition among different hosts. Phytophage phylogenies are few, and evolutionary conservation of host use has been rarely quantified. An initial compilation, based on cladograms for 25 insect groups, suggests that while the distribution of rates is broad, the largest category is groups in which <20% of speciation events are accompanied by shift in host family association. Continuing discovery of diet-phylogeny patterns attributable to conver-

gent host "defenses" reinforces Ehrlich and Raven's view of host use as conservative and of its evolution as guided by preadaptation to host chemistry.

3. There are many hypotheses for variation among groups in the kind and degree of conservatism, but there have been no rigorous tests of these, based on phylogenetically independent contrasts among groups in which conservatism has been measured.
4. Orderly ("stepwise") phylogenetic sequences of diet, reflecting, for example, gradients of host similarity, are implicit in the escape and radiation model and would suggest strong constraint on diet evolution. These have been sought rarely, but two apparent instances are known.
5. Although there is little phylogenetic evidence concerning its evolution, diet breadth per se seems sometimes conservative and correlated with other traits; however, changes of host range appear to sometimes evolve rapidly, and may be constrained in a minor way by those correlations.
6. Sufficiently conservative associations may diversify together. Evidence from insect-plant cladogram concordance is scanty and difficult to interpret, given the imprecision of food plant records and the apparent frequent overlap of hosts among related insects; however, a review of 12 cases indicates that strong congruence, suggesting close parallel phylogenesis, is rare. Molecular and biogeographic evidence against equivalent ages of some insect and host plant clades, such as the frequent use of taxonomically disparate hosts by geographically restricted herbivores, speaks against parallel cladogenesis.
7. Statistically defensible cladogram correspondence plausibly attributable to parallel phylogenesis is known for one monophagous leaf beetle genus showing unusually intricate dependence on the host, and should be sought in other assemblages with similar biologies. Partial concordance in several other groups may also represent contemporaneous diversification.
8. Paleontological evidence increasingly suggests that at least on a broad scale, many host-conservative insect groups and their characteristic host taxa are of comparable age. Thus, despite the undoubted prevalence of host transfer, the distribution of insect over plant groups reflects in part their ages of origin.
9. Long-term, pairwise insect-plant coevolution is rare and may not even characterize codiversified associates, although it deserves further scrutiny; however, because phytophagous insects and higher plants as a whole have diversified roughly contemporaneously, radiations of both should reflect such diffuse coevolution as may occur between producer and primary consumer trophic levels.
10. The taxonomic distribution of some hypothesized plant defenses suggests stepwise escalation against (and diversification permitted by reduced loss to) enemies, as hypothesized by Ehrlich and Raven, although no tests have been conducted. Sister group comparisons show that plant clades bearing latex- or resin-canal systems, which have arisen many times and for which experimental evidence strongly supports a defensive

function, have significantly enhanced diversification rates. Additional studies, both of other convergent defenses and of localized, recent defense-associated radiations, are needed.

11. Replicate sister group comparisons strongly suggest that colonization of higher plants has enhanced insect diversification, as argued by Ehrlich and Raven and others, whereas adoption of other "parasitic" habits has not. The causes for the spectacular diversity of phytophagous insects have received only minor clarification. Many subcategories of plant feeding, including the attack of plants with similar defense innovations (e.g., latex canals), may constitute adaptive zones responsible for the radiation of particular groups, but there have been no compelling tests of this postulate.

ACKNOWLEDGMENTS

We thank E. Bernays and S. Courtney for highly useful criticism of the manuscript. Our work has been supported by grants from the USDA-CRGO, the Smithsonian Scholarly Studies program, the Maryland Agricultural Experiment Station-CGP, and the American Museum of Natural History. This is contribution no. 8327, scientific article no. A6159 from the Maryland Agricultural Experiment Station.

REFERENCES

1. **Ackery, P. R.,** Hostplants and classification: a review of nymphalid butterflies, *Biol. J. Linn. Soc.,* 33, 95, 1988.
2. **Ackery, P. R. and Vane-Wright, R. I.,** *Milkweed Butterflies: Their Cladistics and Biology,* Cornell University Press, Ithaca, NY, 1984.
3. **Adams, E. N.,** Consensus techniques and the comparison of taxonomic trees, *Syst. Zool.,* 21, 390, 1972.
4. **Ahmad, L. and Schaefer, C.,** Food plants and feeding biology of the Pyrrhocoroidea (Hemiptera), *Phytophaga,* 1, 75, 1987.
5. **Anderson, R. S.,** Systematics, phylogeny and biogeography of New World weevils traditionally of the tribe Cleonini (Coleoptera: Curculionidae; Cleoninae), *Quest. Entomol.,* 23, 431, 1987.
6. **Atkinson, T. H. and Equihua, A.,** Biology of the Scolytidae and Platypodidae (Coleoptera) in a tropical deciduous forest at Chamela, Jalisco, Mexico, *Fla. Entomol.,* 69, 303, 1986.
7. **Atkinson, T. H. and Equihua-Martinez, A.,** Biology of bark and ambrosia beetles (Coleoptera: Scolytidae and Platypodidae) of a tropical rain forest in southeastern Mexico with an annotated checklist of species, *Ann. Entomol. Soc. Am.,* 79, 414, 1986.
8. **Barnes, W. and McDunnough, J.,** Illustrations of the North American species of the genus *Catocala, Mem. Am. Mus. Nat. Hist.,* 3, 1918.

9. **Benson, W. W., Brown, K. S., and Gilbert, L. E.,** Coevolution of plants and herbivores: Passion flower butterflies, *Evolution,* 29, 659, 1975.
10. **Bentham, G.,** *Labiatarum Genera et Species,* Ridgeway & Sons, London, 1832-1836.
11. **Berenbaum, M.,** Coumarins and caterpillars: a case for coevolution, *Evolution,* 37, 163, 1983.
12. **Berenbaum, M. and Feeny, P.,** Toxicity of angular furanocoumarins to swallowtail butterflies: escalation in the coevolutionary arms race?, *Science,* 212, 927, 1981.
13. **Berlocher, S. H. and Bush, G. L.,** An electrophoretic analysis of *Rhagoletis* (Diptera: Tephritidae) phylogeny, *Syst. Zool.,* 31, 136, 1982.
14. **Bernays, E. A.,** Host specificity in phytophagous insects: selection pressure from generalist predators, *Entomol. Exp. Appl.,* 49, 131, 1988.
15. **Bernays, E. A. and Chapman, R.,** Evolution of plant deterrence to insects, in *Perspectives In Chemoreception and Behavior,* Chapman, R. F., Bernays, E. A., and Stoffolano, J. G., Eds., Springer-Verlag, New York, 1987, 159.
16. **Bernays, E. A. and Graham, M.,** On the evolution of host specificity in phytophagous arthropods, *Ecology,* 69, 886, 1988.
17. **Bernays, E. A. and Janzen, D. H.,** Saturniid and sphingid caterpillars: two ways to eat leaves, *Ecology,* 69, 1153, 1988.
18. **Bowers, M. D.,** Chemistry and coevolution: Iridoid glycosides, plants and herbivorous insects, in *Chemical Mediation of Coevolution,* Spencer, K. C., Ed., Academic Press, San Diego, CA, 1988, 133.
19. **Bremer, K.,** Tribal interrelationships of the Asteraceae, *Cladistics,* 3, 210, 1987.
20. **Brooks, D. R.,** Hennig's parasitological method: a proposed solution, *Syst. Zool.,* 30, 229, 1981.
21. **Brooks, D. R.,** Macroevolutionary comparisons of host and parasite phylogenies, *Annu. Rev. Ecol. Syst.,* 19, 235, 1988.
22. **Brooks, D. R. and Bandoni, S. M.,** Coevolution and relicts, *Syst. Zool.,* 37, 19, 1988.
23. **Brower, L.P., Seiber, J. N., Nelson, C. J., Lynch, S. P., and Tuskes, P. M.,** Plant-determined variation in the cardenolide content, thin-layer chromatography profiles, and emetic potency of monarch butterflies, *Danaus plexippus,* reared on the milkweed *Asclepias eriocarpa* in California, *J. Chem. Ecol.,* 8, 579, 1982.
24. **Brower, L.P., Seiber, J. N., Nelson, C. J., Lynch, S. P., Hoggard, M. P., and Cohen, J. A.,** Plant-determined variation in cardenolide content and thin layer chromatography profiles of monarch butterflies, *Danaus plexippus,* reared on milkweed plants in California. III. *Asclepias californica, J. Chem. Ecol.,* 10, 1823, 1984.
25. **Brower, L.P., Seiber, J. N., Nelson, C. J., Lynch, S. P., and Holland, M. M.,** Plant-determined variation in the cardenolide content, thin-layer chromatography profiles, and emetic potency of monarch butterflies, *Danaus plexippus* L. reared on milkweed plants in California. II. *Asclepias speciosa, J. Chem. Ecol.,* 10, 601, 1984.
26. **Brown, K.S.,** The biology of *Heliconius* and related genera, *Annu. Rev. Entomol.,* 26, 427, 1981.
27. **Brown, R. G. and Hodkinson, I. D.,** *Taxonomy and Ecology of Jumping Plant Lice of Panama,* Entomonographs No. 9, Scandinavian Science Press, Klampenborg, Denmark, 1988.
28. **Brown, W. J.,** The New World species of *Chrysomela* L., *Can.*

Entomol., 88(Suppl.), 1, 1956.

29. **Brues, C. T.,** The specificity of food-plants in the evolution of phytophagous insects, *Am. Nat.*, 54, 313, 1924.
30. **Burckhardt, D.,** Taxonomy and host plant relationships of the *Trioza apicalis* Förster complex (Hemiptera, Homoptera: Triozidae), *Entomol. Scand.*, 16, 415, 1986.

30a. **Burckhardt, D., and Lauterer, P.,** Systematics and biology of the Rhinocolinae (Homoptera: Psylloidea), *J. Nat. Hist.*, 23, 643, 1989.

31. **Burt, A.,** Comparative methods using phylogenetically independent contrasts., *Oxf. Surv. Evol. Biol.*, in press.
32. **Bush, G. L.,** Modes of animal speciation, *Annu. Rev. Ecol. Syst.*, 6, 339, 1975.
33. **Cantino, P. D.,** Affinities of the Lamiales: a cladistic analysis, *Syst. Bot.*, 7, 237, 1982.
34. **Cantino, P.D. and Sanders, R. W.,** Subfamilial classification of Labiatae, *Syst. Bot.*, 11, 163, 1986.
35. **Chapman, R. F.,** Chemoreception: the significance of receptor numbers, *Adv. Insect Physiol.*, 16, 247, 1982.
36. **Chemsak, J. A.,** Taxonomy and Bionomics of the Genus *Tetraopes* (Cerambycidae: Coleoptera), University of California Publ. Entomol. No. 30, University of California Press, Berkeley, 1963.
37. **Chew, F. S.,** Searching for defensive chemistry in the Cruciferae, or, do glucosinolates always control interactions of Cruciferae with their potential herbivores and symbionts? No!, in *Chemical Mediation of Coevolution*, Spencer, K. C., Ed., Academic Press, San Diego, CA, 1988, 81.
38. **Clark, W. E.,** The weevil genus *Sibinia* Germar: natural history, taxonomy, phylogeny, and zoogeography, with revision of the New World species (Coleoptera: Curculionidae), *Quest. Entomol.*, 14, 91, 1978.
39. **Clark, W.E.,** Revision of the weevil genus *Lonchophorus* Chevrolat (Coleoptera: Curculionidae, Anthonominae), *Quest. Entomol.*, 24, 465, 1988.
40. **Coley, P.D., Bryant, J. P., and Chapin, F.S., III,** Resource availability and plant anti-herbivore defense, *Science*, 230, 895, 1985.
41. **Colless, D.H.,** Congruence between morphometric and allozyme data for *Menidia* species: a reappraisal, *Syst. Zool.*, 29, 288, 1980.
42. **Collins, J. L.,** A Revision of the Annulate *Scutellaria* (Labiatae), Ph.D. dissertation, Vanderbilt University, Nashville, 1976.
43. **Courtney, S. P. and Kibota, T. T.,** Mother doesn't know best: selection of hosts by ovipositing insects, in *Insect-Plant Interactions*, Vol. 2, Bernays, E. A., Ed.,. CRC Press, Boca Raton, 1990, 161.
44. **Courtney, S. P., Chen, G. K., and Gardner, A.,** A general model for individual host selection, *Oikos*, 55, 55, 1989.
45. **Cronquist, A.,** On the taxonomic significance of secondary metabolites in angiosperms, *Plant Syst. Evol. Suppl.*, 1, 179, 1977.
46. **Cronquist, A.,** *An Integrated System of Classification of Flowering Plants*, Columbia University Press, New York, 1981.
47. **Denno, R.F., Larsson, S., and Olmstead, K. L.,** Role of enemy-free space and plant quality in host-plant selection by willow beetles, *Ecology*, 71, 124, 1990.
48. **Diehl, S. R. and Bush, G. L.,** The role of habitat preference in adaptation and speciation, in *Speciation and Its Consequences*, Otte, D. and Endler, J., Eds., Sinauer Associates, Sunderland, MA, 1989, 345.
49. **Donoghue, M. J.,** Phylogenies and the analysis of evolutionary

sequences, with examples from seed plants, *Evolution,* 43, 1137, 1989.

50. **Drummond, B. A.,** Herbivory and defense, in *Solanaceae: Biology and Systematics*, D'Arcy, W. G., Ed., Columbia University Press, New York, 1986, 303.
51. **Dussourd, D. E., Ubik, K., Harvis, C., Resch, J., Meinwald, J., and Eisner, T.,** Biparental defensive endownment of eggs with acquired plant alkaloid in the moth *Utetheisa ornatrix*, *Proc. Natl. Acad. Sci. U.S.A.,* 85, 5992, 1988.
52. **Dussourd, D. E., and Denno, R. F.,** Deactivation of plant defense: correspondence between insect behavior and secretory canal architecture, *Ecology*, 72, in press, 1991.
53. **Dussourd, D. E. and Eisner, T.,** Vein-cutting behavior: insect counterploy to the latex defense of plants, *Science,* 237, 898, 1987.

53a. **Dworakowska, I.,** *Kybos* Fieb, subgenus of *Empoasca* Walsh (Auchenorrhyncha, Cicadellidae, Typhlocybinae) in Palaearctic, *Acta Zool. Cracoviensia,* 21, 387, 1976.

54. **Eastop, V. F.,** Diversity of the Sternorrhyncha within major climatic zones, *Symp. R. Entomol. Soc. London,* 9, 71, 1978.
55. **Eastop, V. F.,** Aphid-plant associations, in *Coevolution and Systematics*, Stone, A. R., and Hawksworth, D. L., Eds., Clarendon Press, Oxford, 1986, 35.
56. **Edgar, J.A., Culvenor, C. C., and Pliske, T. E.,** Coevolution of danaid butterflies with their host plants, *Nature,* 250, 646, 1974.
57. **Ehrlich, P.R. and Raven. P. H.,** Butterflies and plants: a study in coevolution, *Evolution,* 18, 586, 1964.
58. **Eldredge, N.,** Differential evolutionary rates, *Paleobiology,* 2, 174, 1976.
59. **Epling, C.,** The American Species of *Scutellaria*, Publ. Botany No. 20, University of California, Berkeley, 1942, 1.
60. **Farrell, B.,** A Biosystematic and Evolutionary Study of *Phyllobrotica* (Coleoptera: Chrysomelidae), Master's thesis, University of Maryland, College Park, 1985
61. **Farrell, B., Dussourd, D., and Mitter, C.,** Escalation of plant defense: do latex/resin canals spur plant diversification?, *Am. Nat.,* 137, in press, 1991.
62. **Farrell, B. D. and Erwin, T. L.,** Leaf-beetle community structure in an Amazonian rainforest canopy, in *Biology of Chrysomelidae*, Jolivet, P., Petitpierre, E., and Hsiao, T. J., Eds., Kluwer Academic, Dordrecht, 1988, 73.
63. **Farrell, B. D., Mitter, C., and Erwin, T. L.,** Origins of latitudinal diversity gradients, or why are there so many species of leaf-beetles?, submitted.
64. **Farrell, B. and Mitter, C.,** Phylogeny of host-affiliation: have *Phyllobrotica* (Coleoptera: Chrysomelidae) and the Lamiales diversified in parallel?, *Evolution,* 44,1389, 1990.

64a. **Farrell, B. D. and Mitter, C.,** Phylogenetic determinants of insect/plant community diversity, in Community Diversity: Historical and Geographical Determinants, Ricklefs, R. E. and Schluter, D. Eds., University of Chicago Press, in press, 1991.

65. **Farrell, B. and Mitter, C.,** Paleontology of insect/plant interactions: new evidence from amber, submitted.
66. **Farrell, B. and Mitter, C.,** Demographic correlates of adaptive radiation: why are latex plants so abundant in the tropics?, submitted.
67. **Farris, J. S.,** Methods for computing Wagner trees, *Syst. Zool.,* 19, 83, 1970.
68. **Farris. J. S.,** Phylogenetic analy-

sis under Dollo's law, *Syst. Zool.*, 26, 77, 1976.

69. **Feder, J. L., Chilcote, C. A., and Bush, G. L.,** Genetic differentiation between sympatric host races of *Rhagoletis pomonella*, *Nature*, 336, 61, 1988.
70. **Feeny, P.,** Biochemical coevolution between plants and their insect herbivores, *in Coevolution of Insects and Plants*, Gilbert, L. E. and Raven, P. H., Eds., University of Texas Press, Austin, 1975, 3.
71. **Feeny, P.,** Defensive ecology of the Cruciferae, *Ann. Mo. Bot. Gard.*, 64, 221, 1977.
72. **Feeny, P. P.,** The roles of plant chemistry in associations between swallowtail butterflies and their host plants, in *Proc. 6th Int. Symp. Insect-Plant Relationships*, W. Junk Publishers, Dordrecht, 1987.
73. **Fitch, W. M.,** Toward defining the course of evolution: minimum change for a specific tree topology, *Syst. Zool.*, 20, 406,1971.
74. **Furth, D. G.,** Wing polymorphism, host plant ecology, and biogeography of *Longitarsus* in Israel (Coleoptera: Chrysomelidae), *Isr. J. Entomol.*, 13, 125, 1979.
75. **Futuyma, D. J.,** Food plant specialization and environmental predictability in Lepidoptera, *Am. Nat.*, 110, 285, 1976.
76. **Futuyma, D. J.,** Evolutionary interactions among herbivorous insects and plants, in *Coevolution*, Futuyma, D. J. and Slatkin, M., Eds., Sinauer Associates, Sunderland, MA, 1983, 207.
77. **Futuyma, D. J.,** Evolution and coevolution in communities, in *Patterns and Processes in the Evolution of Life*, Raup, D. and Jablonski, D., Eds., Springer-Verlag, New York, 1986.
78. **Futuyma, D. J.,** Macroevolutionary consequences of speciation: inferences from phytophagous insects, in *Speciation and Its Consequences*, Otte, D. and Endler, J. A., Eds., Sinauer Associates, Sunderland, MA, 1989, 557.
79. **Futuyma, D. J.,** History and evolutionary processes, in *History and Evolution*, Nitecki, M., Ed., University of Chicago Press, Chicago, in press.
80. **Futuyma, D. J. and McCafferty, S. J.,** Phylogeny and the evolution of host associations in the leaf beetle genus *Ophraella* (Coleoptera, Chrysomelidae), *Evolution*, 44, 1885, 1990.
81. **Futuyma, D. J. and Moreno, G.,** The evolution of ecological specialization, *Annu. Rev. Ecol. Syst.*, 19, 207, 1988.
82. **Futuyma, D. J., and Peterson, S. C.,** Genetic variation in the use of resources by insects, *Annu. Rev. Entomol.*, 30, 217, 1985.
83. **Gilbert, L.,** Development of theory in the analysis of insect- plant interactions, in *Analysis of Ecological Systems*, Horn, D., Mitchell, R., and Stairs, G., Eds., Ohio State University Press, Columbus, 1979, 117.
84. **Gould, F.,** Genetics of pairwise and multispecies plant- herbivore coevolution, in *Chemical Mediation of Coevolution*, Spencer, K. C., Ed., Academic Press, San Diego, CA, 1988, 13.
85. **Gould, S. J. and Calloway, C. B.,** Clams and brachiopods — ships that pass in the night, *Paleobiology*, 6, 383, 1980.
86. **Grehan, J. R.,** Larval feeding habits of the Hepialidae, *J. Nat. Hist.*, 23, 803, 1989.
87. **Guttman, S., Wood, T. K., and Karlin, A. A.,** Genetic differentiation along host plant lines in the sympatric *Enchenopa binotata* Say complex (Homoptera: Membracidae), *Evolution*, 35, 205, 1981.
88. **Hafner, M. S. and Nadler, A.,** Phylogenetic trees support the coevolution of parasites and their hosts, *Nature*, 332, 258, 1988.

89. **Hardwick, D. F.,** The corn earworm complex, *Mem. Entomol. Soc. Can.,* 40, 1, 1965.
90. **Hardwick, D. F.,** A generic revision of the North American Heliothidinae (Lepidoptera: Noctuidae), *Mem. Entomol. Soc. Can.,* 73, 1, 1970.
91. **Harris, P.,** Food-plant groups of the Semanophorinae (Lepidoptera: Sphingidae): a possible taxonomic tool, *Can. Entomol.,* 104, 71, 1972.
92. **Heie, O. E.,** Revision of the aphid genus *Nasonovia* Mordvilko, including *Kakimia* Hottes & Frison, with keys and descriptions of the species of the world (Homoptera: Aphididae), *Entomol. Scand. Suppl.,* 9, 1, 1979.
93. **Heitzman, R. L.,** Systematic Study of the Mature Larvae of the Nearctic Ennomini and Related Tribes (Lepidoptera: Geometridae), Ph.D. dissertation, University of Maryland, College Park, 1985.
94. **Hennig, W.,** *Phylogenetic Systematics,* University of Illinois Press, Urbana, 1966.
95. **Herrera, C. M.,** Seed dispersal by animals: a role in angiosperm diversification?, *Am. Nat.,* 133, 309, 1989.
96. **Hickey, L. J. and Hodges, R. W.,** Lepidopteran leaf mine from the early Eocene Wind River formation of northwestern Wyoming, *Science,* 189, 718, 1975.
97. **Hodges, R. W.,** Sphingoidea, in *The Moths of America North of Mexico,* Fasc. 21, E.W. Classey Ltd. and R.B.D. Publ., Inc., 1971.
98. **Hollis, D.,** A review of the Malvales-feeding psyllid family Carsidaridae (Homoptera), *Bull. Br. Mus. Nat. Hist. (Entomol.),* 56, 87, 1987.
99. **Hsiao, T. H. and Fraenkel, G.,** Selection and specificity of the Colorado Potato Beetle for solanaceous and nonsolanaceous hostplants, *Ann. Entomol. Soc. Am.,* 61, 493, 1968.
100. **Hsiao, T. H. and Hsiao, C.,** Chromosomal analysis of *Leptinotarsa* and *Labidomera* species, (Coleoptera: Chrysomelidae), *Genetica,* 60, 139, 1983.
101. **Huemer, P.,** A taxonomic revision of *Caryocolum* (Lepidoptera: Gelechiidae), *Bull. Br. Mus. Nat. Hist., (Entomol.),* 57, 439, 1988.
102. **Humphries, C. J., Cox, J. M., and Nielsen, E. S.,** *Nothofagus* and its parasites: a cladistic approach to coevolution, in *Coevolution and Systematics,* Vol. 32, Stone, A. R. and Hawksworth, D. L., Eds., Clarendon Press, Oxford, 1986, 55.
103. **Jackson, J,** Does ecology matter? (review of *Evolution and Escalation: An Ecological History of Life,* by G. J. Vermeij), *Paleobiology,* 14, 307, 1988.
104. **Jaenike, J.,** Genetic complexity of host-selection behavior in *Drosophila, Proc. Natl. Acad. Sci. U.S.A.,* 83, 2148, 1986.
105. **Janzen, D.H.,** When is it coevolution?, *Evolution,* 34, 611, 1980.
106. **Janzen, D. H.,** Two ways to be a tropical big moth: Santa Rosa saturniids and sphingids, *Oxf. Surv. Evol. Biol.,* 1, 85, 1984.
107. **Janzen, D. H. and Martin, P. S.,** Neotropical anachronisms: the fruits the gomphotheres left behind, *Science,* 215, 19, 1982.
108. **Jermy, T.,** Insect-host-plant relationship — co-evolution or sequential evolution?, *Symp. Biol. Hung.,* 16, 109, 1976.
109. **Jermy, T.,** Evolution of insect/hostplant relationships, *Am. Nat.,* 124, 609, 1984.
110. **Jermy, T.,** Can predation lead to narrow food specialization in phytophagous insects, *Ecology,* 69, 902, 1988.
111. **Jolivet, P.,** Food habits and food selection of Chrysomelidae, bionomic and evolutionary perspectives, in *Biology of Chrysomelidae,*

Jolivet, P., Petitpierre, E., and Hsiao, T. H., Eds., Kluwer Academic, Dordrecht, 1988, 1.

112. **Katakura, H., Shioi, M., and Kira, Y.,** Reproductive isolation by host specificity in a pair of phytophagous ladybird beetles, *Evolution,* 43, 1045, 1989.
113. **Killip , E. P.,** The American species of Passifloraceae, *Publ. Field Mus. Nat. Hist. (Bot.),* 19, 1, 1938.
114. **Kinsey, A. C.,** The gall wasp genus *Cynips*: a study in the origin of species, *Indiana Univ. Stud.,* 16, 1, 1930.
115. **Kluge, A. G.,** Parsimony in vicariance biogeography: a quantitative method and a Greater Antillean example, *Syst. Zool.,* 37, 315, 1988.
116. **Krysan, J. L. and Smith, R. F.,** Systematics of the *virgifera* species group of *Diabrotica* (Coleoptera: Chrysomelidae: Galerucinae), *Entomography,* 5, 375, 1987.
117. **Ladiges, P. Y., Newnham, M. R., and Humphries, C. J.,** Systematics and biogeography of the Australian "Green Ash" eucalypts, *Cladistics,* 5, 345, 1989.
118. **Larew, H. G.,** The fossil gall record: a brief summary, *Proc. Entomol. Soc. Wash.,* 88, 385, 1986.
119. **Larsson, S. G.,** *Baltic Amber — A Palaeobiological Study*, Entomonographs, No. 1, Scandinavian Science Press, Klampenborg, Denmark, 1978.
120. **Lidgard, S. and Jackson, J. B. C.,** Growth in encrusting cheilostome bryozoans. I. Evolutionary trends, *Paleobiology,* 15, 255, 1989.
121. **Linsley, E. G.,** Bering arc relationships in Cerambycidae and their hostplants, in *Pacific Basin Biogeography,* Gressitt, J. L., Ed., Bishop Museum Press, Honolulu, 1963.
122. **Maddison, W. P.,** A method for testing the correlated evolution of two binary characters: are gains or losses concentrated on certain branches of a phylogenetic tree?, *Evolution,* 44, 539, 1990.
123. **McCauley, D. E. and Eanes, W. F.,** Hierarchical population structure analysis of the milkweed beetle, *Tetraopes tetropthalmus* (Forster), *Heredity,* 58, 193, 1987.
124. **McPheron, B. A., Smith, D. C., and Berlocher, S. H.,** Genetic differences between host races of the apple maggot fly, *Nature,* 336, 64, 1988.
125. **Marsh, N. A., Clark, C. A., Rothschild, M., and Kellet, D. N.,** *Hypolimnas bolina* (L.), a mimic of danaid butterflies, and its model *Euploea core* (Cram.) store cardioactive substances, *Nature,* 268, 726, 1977.
126. **Matthews, M.,** The Classification of the Heliothinae, Ph.D. thesis, University of London, London, 1987.
127. **Menken, S. B. J.,** Biochemical genetics and systematics of small ermine moths (Lepidoptera: Yponomeutidae), *Z. Zool. Syst. Evol. Forsch.,* 20, 131, 1982.
128. **Metcalf, R. L.,** Coevolutionary adaptations of rootworm beetles (Coleoptera: Chrysomelidae) to cucurbitacins, *J. Chem. Ecol.,* 12, 1109, 1986.
129. **Mickevich, M. F.,** Transformation series analysis, *Syst. Zool.,* 31, 461, 1982.
130. **Mickevich, M. F.,** Quantitative phylogenetic biogeography, in *Advances in Cladistics*, Proc. 1st Meeting Willi Hennig Society, Funk, V. A. and Brooks, D. R., Eds., The New York Botanical Garden, New York, 1981, 209.
131. **Mickevich, M. F. and Weller, S. J.,** Evolutionary character analysis: tracing character change on a cladogram, *Cladistics,* in press.
132. **Miller, J. S.,** Host-plant relationships in the Papilionidae (Lepidoptera): parallel cladogenesis or colonization?, *Cladistics,* 3, 105,

1987.

133. **Miller, J. S. and Feeny, P. P.,** Interspecific differences among swallowtail larvae (Lepidoptera: Papilionidae) in susceptibility to aristolochic acids and berberine, *Ecol. Entomol.,* 14, 287, 1989.
134. **Mitter, C.,** On the taxonomic utility of some internal reproductive structures in *Catocala* (Schrank) and related genera, *Ann. Entomol. Soc. Am.,* 81, 10, 1988.
135. **Mitter, C. and Brooks, D. R.,** Phylogenetic aspects of coevolution, in *Coevolution,* Futuyma, D. J. and Slatkin, M., Eds., Sinauer Associates, Sunderland, MA, 1983, 65.
136. **Mitter, C., Farrell, B., and Wiegmann, B.,** The phylogenetic study of adaptive radiation: has phytophagy promoted insect diversification?, *Am. Nat.,* 132, 107, 1988.
137. **Mitter, C. and Farrell, B.,** The asterid affinities of the North American milkweed insect fauna, manuscript in preparation
138. **Mitter, C. and Silverfine, E.,** A first-estimate phylogeny for *Catocala* (Lepidoptera: Noctuidae) with comments on the evolution of diet, manuscript in preparation.
139. **Mitter, C. and Silverfine, E.,** On the systematic position of *Catocala* Schrank, *Syst. Entomol.,* 13, 67, 1988.
140. **Miyazaki, J. and Mickevich, M. F.,** Evolution of *Chesapecten* (Mollusca: Bivalvia, Miocene-Pliocene) and the biogenetic law, *Evol. Biol.,* 15, 369, 1982.
141. **Moran, N. A.,** The evolution of host-plant alternation in aphids: evidence for specialization as a dead end, *Am. Nat.,* 132, 682, 1988.
142. **Moran, N. A.,** A 48-million-year-old aphid-host plant association and complex life cycle: biogeographic evidence, Science, 245, 173, 1989.
143. **Muller, J.,** Significance of fossil pollen for angiosperm history, *Ann. Mo. Bot. Gard.,* 71, 419, 1984.
144. **Nault, L. R.,** Evolutionary relationships between maize leafhoppers and their host plants, in *The Leafhoppers and Planthoppers,* Nault, L. R. and Rodriguez, J. G., Eds., John Wiley & Sons, New York, 1985, 309.
145. **Neimela, P., Hanhimaki, S., and Mannila, R.,** The relationship of adult size in noctuid moths (Lepidoptera, Noctuidae) to breadth of diet and growth form of host plants, *Ann. Entomol. Fenn.,* 47, 17, 1981.
146. **Nelson, C. J., Seiber, J. N., and Brower, L. P.,** Seasonal and intraplant variation of cardenolide content in the California milkweed *Asclepias eriocarpa,* and implications for plant defense, *J. Chem. Ecol.,* 7, 981, 1981.
147. **Nelson, G. and Platnick, N.,** *Systematics and Biogeography: Cladistics and Vicariance,* Columbia University Press, New York, 1981.
148. **Neunzig, H. H.,** Pyralidae (Pyraloidea), in *Immature Insects,* Stehr, F. W., Ed., Kendall/Hunt, Dubuque, IA, 1987, 462.
149. **Nielsen, B. E.,** Coumarin profiles in the Umbelliferae, in *The Biology and Chemistry of the Umbelliferae,* Heywood, V., Ed., Academic Press, London, 1971, 325.
150. **Nishio, S., Blum, M. S., and Takahashi, S.,** Intraplant distribution of cardenolides in *Asclepias humistrata* (Asclepiadaceae), with additional notes on their fates in *Tetraopes melanurus* (Coleoptera: Cerambycidae) and *Rhyssomatus lineaticollis* (Coleoptera: Curculionidae), *Mem. Coll. Agric. Kyoto Univ.,* 122, 43, 1983.
151. **Noonan, G. R.,** Biogeography of North American and Mexican insects, and a critique of vicariance biogeography, *Syst. Zool.,* 37, 366, 1988.
152. **Opler, P. A.,** Fossil lepidopterous

leaf-mines demonstrate the age of some insect-plant relationships, *Science,* 179, 1321, 1973.

153. **Orians, G. H. and Paine, R. T.,** Convergent evolution at the community level, in *Coevolution,* Futuyma, D. J. and Slatkin, M., Eds., Sinauer Associates, Sunderland, MA, 1983, 431.
154. **Page, R. D. M.,** Quantitative cladistic biogeography: constructing and comparing area cladograms, *Syst. Zool.,* 37, 254, 1988.
155. **Palmer, J. D., Jansen, R. K., Michaels, H. J., Chase, M. W., and Manhart, J. R.,** Chloroplast DNA variation and plant phylogeny, *Ann. Mo. Bot. Gard.,* 75, 1180, 1988.
156. **Patterson, C.,** Methods of paleobiogeography, in *Vicariance Biogeography: A Critique,* Nelson G. and Rosen, D. E. Eds.,. Columbia University Press, New York, 1981, 446.
157. **Pitkin, L. M.,** Gelechiid moths of the genus *Mirificarma, Bull. Br. Mus. Nat. Hist. (Entomol.),* 48, 1, 1984.
158. **Pitkin, L. M.,** The Holarctic genus *Telieopsis*: host-plants, biogeography and cladistics (Lepidoptera: Gelechiidae), *Entomol. Scand.,* 19, 143, 1988.
159. **Poole, R., Mitter, C., and Huettel, M.,** A revision and cladistic analysis of the *Heliothis virescens* species group, with a preliminary morphometric analysis of *H. virescens,* submitted.
160. **Price, P. W.,** *Evolutionary Biology of Parasites,* Princeton University Press, Princeton, NJ, 1980.
161. **Price, P. W., and Pschorn-Walcher, H.,** Are galling insects better protected against parasitoids than exposed feeders?: a test using tenthredinid sawflies, *Ecol. Entomol.,* 13, 195, 1988.
162. **Raup, D. M., Gould, S. J., Schopf, T. J. M., and Simberloff, D. S.,** Stochastic models of phylogeny and the evolution of diversity, *J. Geol.,* 81, 525, 1973.
163. **Rensch, B.,** *Evolution Above the Species Level,* Columbia University Press, New York, 1959.
164. **Roeske, C. N., Seiber, J. N., Brower, L. P., and Moffitt, C. M.,** Milkweed cardenolides and their comparative processing by monarch butterflies (*Danaus plexippus*), *Recent Adv. Phytochem.,* 10, 93, 1976.
165. **Roskam, J. C.,** Evolutionary patterns in gall midge-host plant associations (Diptera, Cecidomyiidae), *Tijdskr. Entomol.,* 128, 193, 1985.
166. **Ross, H. H.,** An evolutionary outline of the leafhopper genus *Empoasca* subgenus *Kybos,* with a key to the Nearctic fauna (Hemiptera, Cicadellidae), *Ann. Entomol. Soc. Am.,* 56, 202, 1963.
167. **Ross, H. H.,** An uncertainty principle in ecological evolution, in *A Symposium on Ecosystematics,* Occ. Pap., No. 4, Allen, R. T. and James, F. C. Eds., University of Arkansas, Little Rock, 1972.
168. **Rowell, C. H. F.,** The biogeography of Costa Rican acridid grasshoppers, in relation to their putative phylogenetic origins and ecology, in *Evolutionary Biology of Orthopteroid Insects,* Baccetti, B. M., Ed., Ellis Horwood, Chichester, U.K., 1987, 470.
169. **Saigusa, T., Nakanishi, A., Shima, H., and Yata, O.,** Phylogeny and geographical distribution of the swallow-tail subgenus *Graphium* (Lepidoptera: Papilionidae), *Entomol. Gen.,* 8, 59, 1982.
170. **Schaefer, C. and Ahmad, S.,** The food plants of four pentatomoid families (Acanthosomatidae, Tesoaratomidae, Urostylidae, and Dinidoridae), *Phytophaga,* 1, 21, 1987.
171. **Schlinger, E. I.,** Continental drift, *Nothofagus,* and some ecologi-

cally associated insects, *Annu. Rev. Entomol.*, 323, 1974.

172. **Schneider, D.,** The strange fate of pyrrolizidine alkaloids, in *Perspectives in Chemoreception and Behavior*, Chapman, R. F., Bernays, E. A., and Stoffolano, J. G., Eds., Springer-Verlag, New York, 1987, 159.
173. **Schneider, J. C.,** The role of parthenogenesis and female aptery in microgeographic, ecological adaptation in the fall cankerworm, *Alsophila pometaria* Harris (Lepidoptera: Geometridae), *Ecology,* 61, 1082, 1980.
174. **Scudder, G.G.E. and Duffey, S. S.,** Cardiac glycosides in the Lygaeinae (Hemiptera: Lygaeidae), *Can. J. Zool.,* 50, 35, 1972.
175. **Seeno, T. N. and Wilcox, J. A.,** Leaf beetle genera (Coleoptera: Chrysomelidae), *Entomography,* 1, 1, 1982.
176. **Shao, K. and Rohlf, F. J.,** Sampling distributions of consensus indices when all bifurcating trees are equally likely, in *Numerical Taxonomy,* NATO Advanced Study Institute, Ser. G (Ecological Sciences), No. 1., Felsenstein, J., Ed., Springer-Verlag, Berlin, 1982, 132.
177. **Shao, K. and Sokal, R. R.,** Significance tests of consensus indices, *Syst. Zool.,* 35, 582, 1986.
178. **Simberloff, D.,** Calculating probabilities that cladograms match: a method of biogeographical inference, *Syst. Zool.,* 36, 175, 1987.
179. **Simpson, G. G.,** *The Major Features of Evolution,* Columbia University Press, New York, 1953.
180. **Singer, M. C.,** Evolution of foodplant preference in the butterfly *Euphydryas editha, Evolution,* 25, 383, 1971.
181. **Smiley, J. T.,** Plant chemistry and the evolution of host specificity: new evidence from *Heliconius* and *Passiflora, Science,* 201, 745, 1978.
182. **Sober, E.,** *The Nature of Selection,* MIT Press, Cambridge, MA, 1984.
183. **Southwood, T. R. E.,** The number of species of insect associated with various trees, *J. Anim. Ecol.,* 30, 1, 1961.
184. **Southwood, T. R. E.,** The insect/plant relationship—an evolutionary perspective, in *Insect Plant Relationships,* van Emden, H. F., Ed., Blackwell Scientific, Oxford, 1973, 3.
185. **Stanley, S. M.,** *Macroevolution, Pattern and Process,* W. H. Freeman, San Francisco, 1979.
186. **Stanley, S. M., Van Valkenburgh, B., and Steneck, R.S.,** Coevolution and the fossil record, in *Coevolution,* Futuyma, D.J., and Slatkin, M., Eds., Sinauer Associates, Sunderland, MA, 1983.
187. **Stehr, F. W., Ed.,** *Immature Insects,* Kendall/Hunt, Dubuque, IA, 1987.
188. **Stonedahl, G. M. and Schuh, R. T.,** *Squamocoris* Knight and *Ramentomiris,* New Genus (Heteroptera: Miridae: Orthotylinae). A Cladistic Analysis and Description of Seven New Species from Mexico and the Western United States, No. 2852, American Museum of Novitates, 1986, 1.
189. **Stonedahl, G. M., and Schwartz, M. D.,** Revision of the Plant Bug Genus *Pseudopsallus* Van Duzee (Heteroptera: Miridae), No. 2842, American Museum of Novitates, 1986, 1.
190. **Strong, D.R., Lawton, J. H., and Southwood, T. R. E.,** *Insects on Plants,* Harvard University Press, Cambridge, MA, 1984.
191. **Tallamy, D. W. and Krischik, V. A.,** Variation and function of cucurbitacins in *Cucurbita*: an examination of current hypotheses, *Am. Nat.,* 133, 766, 1989.

192. **Thompson, J. H.,** Concepts of coevolution, *Trends Ecol. Evol.*, 4, 179, 1989.
193. **Tiffney, B. H.,** Seed size, dispersal syndromes, and the rise of the angiosperms: evidence and hypothesis, *Ann. Mo. Bot. Gard.*, 71, 551, 1984.
194. **Tiffney, B. H.,** Perspectives on the origin of the floristic similarity between eastern Asia and eastern North America, *J. Arn. Arb.*, 66, 73, 1985.
195. **Usher, B. F. and Feeny, P.,** Atypical secondary compounds in the family Cruciferae: tests for toxicity to *Pieris rapae*, an adapted crucifer-feeding insect, *Ent. Exp. Appl.*, 34, 257, 1983.
196. **Valentine, J. W., Ed.,** *Phanerozoic Diversity Patterns: Profiles in Macroevolution,* Princeton University Press, Princeton, NJ, 1985.
197. **Vermeij, G. J.,** *Evolution and Escalation: An Ecological History of Life*, Princeton University Press, Princeton, NJ, 1987.
198. **Ward, D. B.,** Plants, in *Rare and Endangered Biota of Florida*, Vol. 5, Pritchard, P. C. H., Ed., University Presses of Florida, Gainesville, 1979.
199. **Wasserman, S. S. and Mitter, C.,** The relationship of body size to breadth of diet in some Lepidoptera, *Ecol. Entomol.*, 3, 155, 1978.
200. **Waterman, P. G. and Mole, S.,** Extrinsic factors influencing production of secondary metabolites in plants, in *Insect-Plant Interactions*, Vol. 1, Bernays, E. A., Ed., CRC Press, Boca Raton, FL, 1989, 107.
201. **Wheeler, A. G. and Schaefer, C.,** Food plants of the Berytidae, *Ann. Entomol. Soc. Am.*, 75, 495, 1982.
202. **White, I. M. and Hodkinson, I. D.,** Nymphal taxonomy and systematics of Psylloidea (Hemiptera), *Bull. Br. Mus. Nat. Hist. (Entomol.)*, 50, 153, 1986.
203. **Wiegmann, B. M., Mitter, C., and Farrell, B.,** Does parasitism sensu stricto promote insect diversification?, manuscript in preparation.
204. **Winder, J. A.,** Ecology and control of *Eryinnis ello* and *E. alope*, important insect pests in the New World, *PANS*, 22, 449, 1976.
205. **Wood, S. L.,** The bark and ambrosia beetles (Coleoptera: Scolytidae) of North and Central America, a taxonomic monograph, *Great Basin Nat. Mem.*, No. 6, 1982.
206. **Wood, T. K. and Guttman, S. I.,** *Enchenopa binotata* complex: sympatric speciation?, *Science*, 220, 310, 1983.
207. **Woodson, R. E.,** The North American Asclepiadaceae. I. Perspective of the genera, *Ann. Mo. Bot. Gard.*, 28, 193, 1941.
208. **Woodson, R. E.,** The North American species of *Asclepias* L., *Ann. Mo. Bot. Gard.*, 41, 1, 1954.
209. **Zwölfer, H.,** Mechanismen und Ergebnisse der Co-evolution von phytophagen und entomophagen Insekten und höheren Pflanzen, *Sonderbd. Naturwiss. Ver. Hamburg*, 2, 7, 1978.
210. **Zwölfer, H.,** Species richness, species packing, and evolution in insect-plant systems, *Ecol. Stud.*, 61, 301, 1987.
211. **Zwölfer, H.,** Evolutionary and ecological relationships of the insect fauna of thistles, *Annu. Rev. Entomol.*, 33, 103, 1988.
212. **Zwölfer, H. and Herbst, J.,** Präadaptation, Wirtskreiserweiterung und Parallel-Cladogenese in der Evolution von phytophagen Insekten, *Z. Zool. Syst. Evol. Forsch.*, 26, 320, 1988.
213. **Zwölfer, H. and Romstöck-Völkl, M.,** Biotypes and the evolution of niches in phytophagous insects on Cardueae hosts, in *Herbivory: Tropical and Temperate Perspectives*, Price, P., Ed., John Wiley & Sons, New York, in press.

3

Measuring Food Utilization in Plant-Feeding Insects — Toward A Metabolic and Dynamic Approach

Joop J. A. van Loon
Department of Entomology
Agricultural University
Wageningen, The Netherlands

TABLE OF CONTENTS

I. INTRODUCTION

A. THE SIGNIFICANCE OF FOOD UTILIZATION

In many studies of the relationships between insects and plants, attempts have been made to quantify the efficiency with which insects exploit their food plants. Together with data on the rate of food ingestion and growth, food utilization efficiency is an important component of herbivore performance.[98] From a nutritional point of view, utilization efficiency reflects the quality of a food. Since it had become clear from classical nutritional studies that the qualitative requirements of insects do not differ fundamentally among species,[26] and since the advances in phytochemistry allowed the conclusion that green plants generally contain all nutritionally essential elements required by insects,[26,32] a quantitative nutritional approach became indispensable to achieve a better understanding of the trophic relationships between herbivores and plants.[87,94] Differences in the efficiencies of digestion and conversion of food may contribute to observed patterns in the specificity of insect-plant interactions. Quantification of postingestive processes renders explanations that are complementary to those deduced from short-term behavioral studies in the preingestive phase. Studies on causal factors responsible for preingestive host plant selection behavior have often been focused on the role of sensory recognition of suitable food plants.[13,27,41,46,77] Indeed, while often difficult, it is important to separate sensory discrimination and optimal food utilization as factors determining insect-plant interactions.[9] Moreover, recent evidence indicates that a functional link between them exists.[23,74,91-93]

B. NUTRITIONAL ECOLOGY

The term "nutritional ecology" has been adopted to emphasize the importance of nutritional factors in understanding the behavioral strategies of phytophagous insects.[87,94] In the past 20 years there has been a rash of ecological studies on insect-plant relationships that have been concerned with quantifying the utilization of plant foods by insect herbivores. Relative suitability of plants as food for phytophagous insects[86] as well as the effects of quality and quantity of specific phytochemicals added to artificial diets[40,71] have been quantified using food utilization efficiency as a criterion. This body of literature is adequately summarized in several recent and extensive reviews.[55,97,98] Important notions from these reviews are the differences between major herbivore guilds, such as tree and herb feeders, in the efficiency with which plant food is utilized and the importance of water and nitrogen as primary determinants of the nutritional quality of plant food.[98]

C. THE MEASUREMENT OF FOOD UTILIZATION: TWO DECADES AFTER WALDBAUER

In the majority of cases, the measurement of food utilization is performed along the directives put forward by Waldbauer in his 1968 review, which set the stage for insect quantitative nutrition. Undoubtedly, subsequent research into insect quantitative nutrition in general, including that of herbivores, has benefited by adopting the standardization of methodology and terminology proposed therein.

Waldbauer[110] defined three parameters of utilization, now commonly termed nutritional indices: approximate digestibility (abbreviated as AD), the efficiency with which digested food is converted to body substance or utilization efficiency (ECD) and the efficiency of conversion of ingested food to body substance (ECI). He analyzed several problems associated with measuring these indices and also discussed different methods that had been applied up to 1968. The gravimetric method has been, and still is, the prevalent method used to quantify food intake and utilization. This also seems justified, as the (albeit few) studies specifically devoted to comparing different methods of measurement advocate that gravimetry is the most reliable of available methods.[48,63,110] At the same time, gravimetric measurements are easy to perform, do not require expensive equipment, and are relatively accurate, which have made them widely used by researchers.

The aims of this chapter are to summarize problems connected with the use of gravimetry to determine utilization efficiencies, to bring together recent criticisms, and to evaluate its methodological validity. In doing so, physiological aspects are stressed. Respirometric measurement of metabolic rate over time is presented as an alternative method that has the potential to alleviate the main problems and limitations inherent to the gravimetric measurement of food utilization. The focus of the chapter is on the utilization of total assimilated dry matter and energy, i.e., their relative allocation to growth as opposed to respiration or heat loss.

II. THE RATIONALE OF THE GRAVIMETRIC METHOD

Prior to a detailed discussion of sources of error, the rationale of measuring nutritional indices is outlined briefly. Central to this rationale is the gravimetric budget, that is, in fact, based on the law of conservation of matter and energy.[17] The budget equation is often given as:[65,79]

$$C = G + R + FU \tag{1}$$

where

- C is the amount of food consumed
- G is insect biomass produced, which includes growth (somatic and reproductive) and several secreted and excreted products that do not actually contribute to somatic growth per se, such as exuviae, silk, and digestive enzymes
- R is respiration (amount of carbon dioxide respired)
- FU is feces, including urinary wastes and other metabolic waste products egested as fecal constituents

Budget items are routinely expressed in dry matter units, because water escapes from food, feces, and the insect body as vapor and the losses via these respective routes are very difficult to quantify. In the course of time and based on a number of considerations, different terms, and concomitant abbreviations, symbols and notations have been employed such as ingestion (I) for consumption, production (P), including both growth and the nongrowth secretions, egesta (E) for feces. For an overview of indices and their notations, including those in use in ecological energetics, the reader is referred to Kogan.[48] To avoid the indiscriminate use of these largely equivalent terms within the scope of this chapter, the author prefers to use the terms and their abbreviations as originally proposed by Waldbauer;[110] the terms "absorption efficiency" as equivalent for AD and "metabolic efficiency" for ECD, as proposed by Woodring et al.,[118] are also utilized. "Metabolic" and "energetic" are considered synonyms within the scope of this text.

Three of the four biologically distinct budget items are measured (C, G, and FU), thus, in theory, the R-term follows by difference:

$$R = C - G - FU \tag{2}$$

The gravimetric measurements of C, G, and FU are performed by determining the dry weights of food, insects, and feces at the start and the end of the experimental period and calculating the differences. C equals:

$$C = F_i - F_e \tag{3}$$

where F_i = the dry weight food offered at t(i), the time point on which the experiment is initiated and F_e = the dry weight of food remains at t(e), the time point on which the experiment ends. The actual calculation is more complicated and introduces a first source of error, as the weight of dry food matter at the start of the experiment cannot be measured directly and must be estimated using an aliquot. This is discussed in more detail further on. Thus,

$$F_i = F_{i,f} \cdot \%dm(F_{i,f}) \tag{4}$$

where $F_{i,f}$ = the fresh weight of food offered at t(i) and % dm($F_{i,f}$) the percentage of dry matter of the fresh food at t(i) as determined from a matched aliquot of food. Equation 3 can then be rewritten as:

$$C = (F_{i,g} \cdot \%dm(F_{i,f})) \cdot F_e \tag{5}$$

Growth is measured as:

$$G = B_e - B_i \tag{6}$$

where B_e = the dry weight of the insect body at t(e) and B_i = the dry weight of the insect body at t(i). The measurement of growth likewise involves an aliquot determination necessary to estimate the dry weight of the insect body at t(i):

$$B_i = B_{i,f} \cdot \%dm(B_{i,f}) \tag{7}$$

where $B_{i,f}$ = the fresh weight of the insect body at t(i) and %dm($B_{i,f}$) = the percentage of dry matter of the insect body of an aliquot group of insects of comparable age and rearing history. Equation 6 can then be rewritten as:

$$G = B_e \cdot (B_{i,f} \cdot \%dm(B_{i,f})) \tag{8}$$

The measurement of the total amount of feces produced during the experimental period is relatively straightforward as it equals the dry weight of feces accumulated at t(e)(FU), assuming that the amount of feces at t(i) is nil and, ideally, that the gut of the experimental insects is completely devoid of food. The latter point relates to the timing of the experimental period. Waldbauer[110] advocated the use of physiological periods, i.e., a period clearly demarcated by physiological events such as moulting or pupation. These normally coincide with a natural emptying of the gut (see below).

Having measured C, G, and FU in the ways described, the nutritional indices can subsequently be calculated as follows:

$$AD = \frac{C - FU}{C} \tag{9}$$

$$ECD = \frac{G}{C - FU} = \frac{G}{G + R} \tag{10}$$

$$ECI = \frac{G}{C} = AD \cdot ECD \tag{11}$$

The author, as mentioned previously, focuses in this chapter on the methodology of measuring utilization, or metabolic, efficiency (ECD). ECD is usually expressed as a ratio of flows of mass (expressed in dry matter units) or energy (expressed in Joules or calories) that reflects the physiological partitioning of assimilated components to either growth or respiration. In theory, this can be done legitimately while disregarding experimental duration and time-related parameters such as rate of food consumption or growth; however, it is noted in advance that differences in utilization efficiency may simply be caused by differences in the experimental periods over which utilization was measured. This indicates that it is not really possible to disregard the time dimension of utilization experiments, and stresses the importance of accurate timing of physiological events such as moulting and pupation and the cessation of feeding normally associated with these.

III. CHECKING THE BUDGET: DISCREPANCIES AND THEIR CAUSES

There is an important theoretical objection against the gravimetric approach for quantifying food utilization. A necessary evaluation of its validity requires the construction of a complete matter balance sheet, i.e., including oxygen uptake and the amounts of carbon dioxide and water produced in oxidative metabolism. The next step is to assess the error that occurs when oxygen uptake and water production are neglected, which is what in fact is done in dry matter balance sheets. The gravimetric approach as it has been commonly applied has skipped the basic evaluation and implicitly assumes that the error possibly involved is unimportant. This is one of the major reasons that the gravimetric method contains a methodological pitfall: the budget is not checked. A check can be achieved by measuring all four items (C, FU, G, and R), preferably with independent methods. The absence of a methodological check and the concomitant risk of unrealistic results is not just an academic problem. The following three lines of evidence are presented to support the reality of the problem.

1. Problems with checking energy budgets have primarily been noted in a number of studies on plant feeding insects performed in the field of ecological energetics.[5,58,114,116] Most of these yielded considerable discrepancies between the R-term resulting from the gravimetric budget (Equation 2) and independent estimates of energy expenditure obtained by respirometric measurement of oxygen consumption. A balancing energy budget requires that the R-term is expressed in energy units. This is generally done by determining the calorific content (in Joules per unit dry weight) of the food, the insect body, and the feces by means of bomb calorimetry,[59,66] multiplying these by C, G, and FU (in dry weight units), and calculating R (in Joules) according to Equation 2. This calculated value of R (R(c)), should then equal, within reasonable limits of inaccuracy, the amount of Joules that corresponds to the respirometrically measured amount of oxygen consumed, R(m). The latter is found by multiplication

with an oxycalorific conversion factor.[47,59] The imbalances reported are considerable, since the ratio R(c)/R(m) ranges from 1.4 to 6.0 for 11 different insect species fed fresh leaves.[58] The main causes for these discrepancies are currently the subject of debate.[1,5,58,109,116] The causes of such incorrect estimations are discussed in more detail below.

2. The studies cited above dealt with fresh plant food, which is inherently variable both between and within plants in composition as well as over time, which may explain the problems encountered; however, a recent sensitivity analysis demonstrated that in artificial diet studies, considerable errors in the measurement of nutritional indices may also occur.[75] The two main sources of error that were identified will be dealt with in more detail below.
3. A related, though indirect, indication of the involvement of inaccuracy in measurements comes from the appreciation of variability of nutritional indices. Mattson[55] found that ECD was considerably more variable than AD by analyzing results from four detailed studies of leaf feeders. Also, several compilations of AD and ECD values from plant-feeding insects[95,97,98,110] likewise demonstrate that ECD for a particular species is much more variable than AD, regardless of the experimental variables that may have contributed to this variation, such as plant species or variety offered as food, developmental stages, or even phase within the instar. It is also noted that due to the default way in which it is calculated, the ECD is not particularly sensitive to absolute differences in R, the item reflecting metabolic investment in dry matter units (Figure 1). Imagine a herbivorous insect that attains a growth of 100 mg of dry matter during its final instar for which R is calculated to be 50 mg. This then results in an ECD of 67%. If R increased by 50%, to 75 mg, and growth remained unchanged (making the assimilated fraction C-FU also 25 mg higher), the ECD would become 57%, only 10 percentage points lower. A second situation that seems more probable would be when R varies while assimilation (A) is kept constant and thus G would covary inversely with R. The latter situation leads to a higher degree of variation in ECD (Figure 1). As stated above, one factor that may cause differences in ECD is a considerable difference in the duration of the utilization experiment (see below).

Relating to Point 1, variability in metabolic efficiency expressed as a ratio of dry matter units is predicted to generate an even higher variability when it is expressed as a ratio of energetic equivalents, the physiologically relevant measure of metabolic intensity. This is due to the differences in energetic content of the different biochemical substrates for oxidative energy metabolism, ranging from approximately 16 J/mg for mono- and disaccharides to approximately 39 J/mg for common animal lipids, the latter being 2.4 times as high.[47,59] The relatively consistent differences between mean values of dry weight vs. energy utilization were reviewed by Slansky.[95] The same data as cited by Slansky[95] can be examined with respect to the variability in dry matter ECD (ECD(d)) vs. caloric ECD (ECD(c)) instead of differences between mean values. From data on effects of alternative host plants for three species of

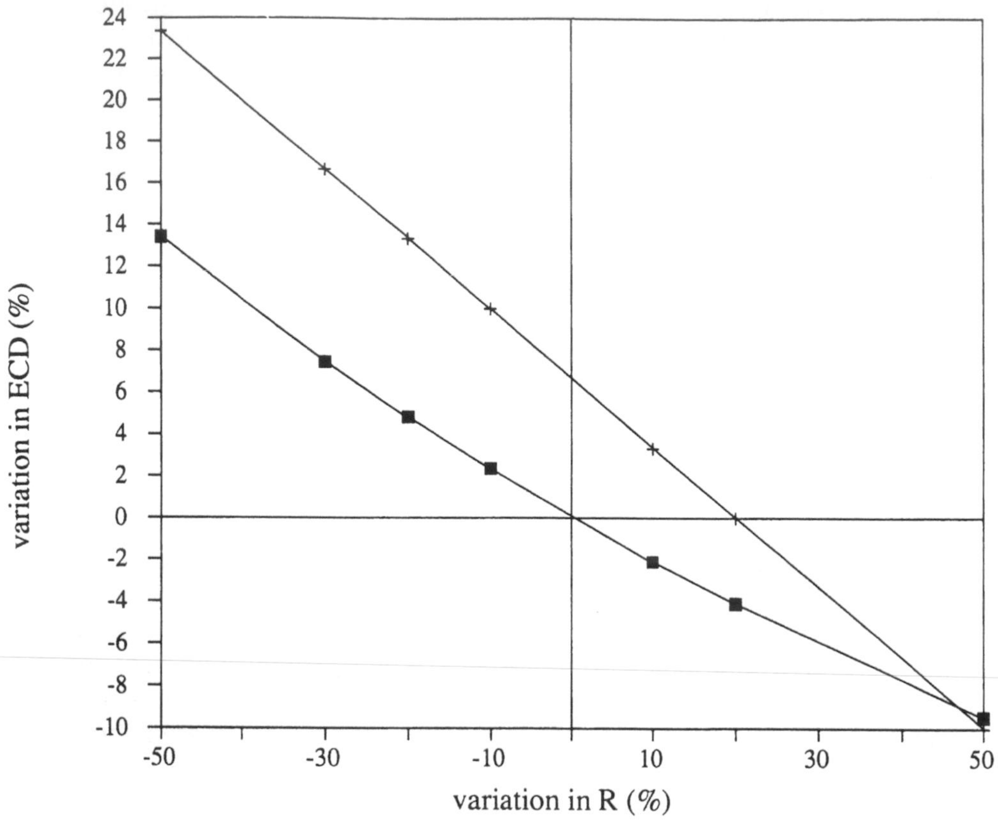

FIGURE 1. Relationship between variation in the amount of dry matter respired, R, and variation in efficiency of conversion of digested food, ECD. Values refer to an insect that has achieved a body growth of 100 mg dry matter and has respired 50 mg of dry matter (R) during the same period. Two situations are shown: ■: R varies while G is constant; +: R and G covary inversely, keeping the amount of assimilation A (= G + R) constant.

lepidopterous larvae that allow such a comparison (*Danaus plexippus*,[30] *Spodoptera eridania*,[82] and *Plutella xylostella*[102]), it can be calculated that variability in ECD(c) (expressed as coefficient of variation) is on the average 1.74 times as high as ECD(d), which supports the prediction made above. The same is also true for a comparison of variability in ECD(d) vs. ECD(c) over the consecutive larval stages of several species of lepidopterous larvae.[95]

IV. SOURCES OF ERROR IN GRAVIMETRY AND THEIR EFFECTS

This section discusses sources of error for each budget item and assesses the extent of their effects. The discussion is limited to the more substantial systematic and random errors and the methods by which they may be reduced or avoided altogether.

A. CONSUMPTION

1. Estimation of Percentage Dry Matter and Amount of Food Offered

Referring to the way in which C is found (Equation 5), it is seen that two parameters ($F_{i,f}$ and F_e) are determined by direct weighing, while %dm of food results from two additional direct weighings, i.e., fresh and dry weight, of the same piece of food material. By means of a sensitivity analysis it has been demonstrated that random errors in %dm produce an incorrect estimation of C, which has a substantial effect on the values of nutritional indices.[75] Moreover, this error is strongly dependent on a less conspicuous parameter, the fraction of food remaining at the termination of the experiment relative to the amount of food offered initially (both on a dry matter basis). The %dm used in the calculation of C usually is the mean of a series of dry:fresh weight ratios. The concomitant variance is given only in a minority of papers. Nevertheless, an error of 0.2% in a dry matter content of the food of 15% is well within the normal variability in such determinations with artificial diets. As Schmidt and Reese[75] showed, this resulted in errors of 20 to 25% in ECD (i.e., an actual value of 60% is erroneously estimated to be 35 to 40%) under the set of realistic assumptions they made for utilization experiments with *Agrotis ipsilon*. The magnitude of errors in derived parameters can be expected, as all matter and energy is furnished from C (Equation 1). Random errors in C on different test diets or host plants may thus produce either exaggerated differences in AD and ECD between such diets or disguise actual differences. Such errors necessarily arise from the use of aliquots, for which no alternative is available simply because the drying of the actual food offered would make it unsuitable as such. It is evident that in order to obtain a reliable estimation of C, the differences (in %dm and other properties) between aliquot and actual food offered must be minimized as much as possible. To achieve this, fresh leaves are split along the midrib and one half is used as food offered, the other half as aliquot.[110] This type of within-leaf matching is the best method available and will result in minimal differences between aliquot and experimental food, although the magnitude of these differences will depend on the plant species (ranging from 0.14 to 0.56% for four plant species[110]), leaf morphology and symmetry, leaf size, leaf light exposure, leaf age, etc. In very few publications that refer to Waldbauer[110] for their methods is it explicitly stated whether within-leaf matching was applied and what the extent of difference was in %dm between matched leaf halves. With plant food, that an error of 0.5% in the estimation of %dm readily occurs seems a modest assumption. The undesired effects this error may have on the estimation of utilization efficiency is elaborated in an example below. To illustrate how variable dry matter content of leaves may be, an example is given of variability in %dm of leaves of *Brassica oleracea* (Figure 2).

Additional problems of a different kind are attached to the within-leaf matching, that result from the effect of separating the two halves. This brings about wounding of the half offered to the insect. Wounding in return may be accompanied by an increased evaporation and respiration that will reduce the actual %dm of leaf material ingested by the insect (see below). It is important,

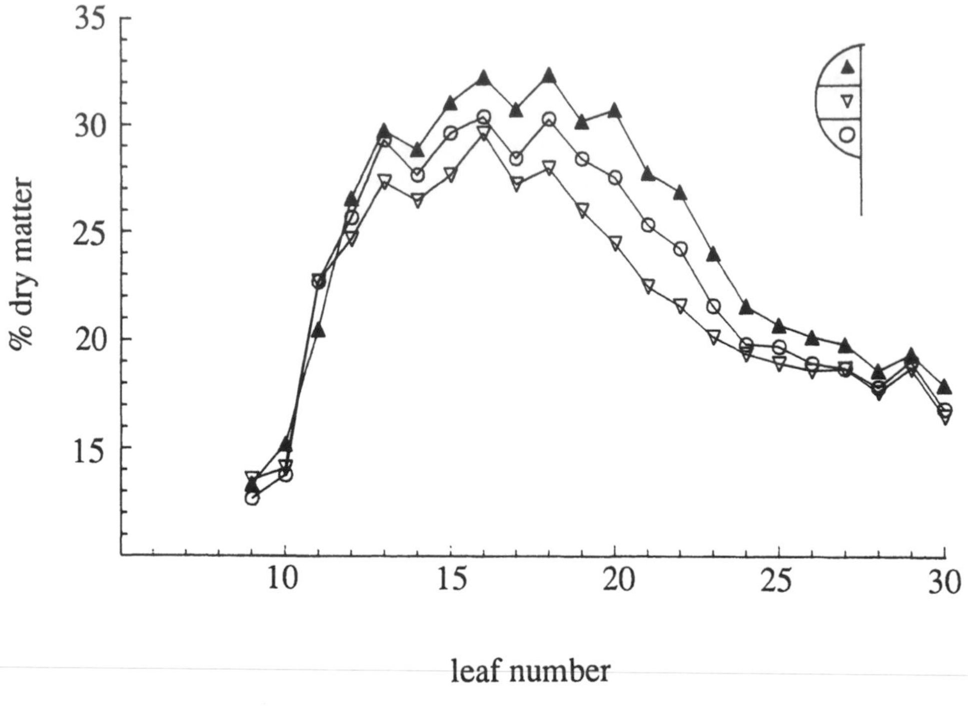

A

FIGURE 2. Dry matter content of leaves (as a percentage of fresh weight) of *Brassica oleracea* var. *gemmifera* as a function of leaf number counted from stem base to plant top of a greenhouse-reared plant (age 87 d); a higher number represents a younger leaf. Each leaf was split along the mid rib and each half immediately divided into 3 segments as shown in the drawing: a top, middle, and base segment. All 6 segments were then oven dried at 70°C to constant weight. Graph A gives the resulting values of %dm for the 3 segments of the left leaf half and graph B for the right leaf half. Differences of up to 0.5% were commonly observed between these symmetrically matched leaf halves, especially for the middle segments.

therefore, to minimize the interval between preparing the food and the onset of actual feeding by the insect.

Errors in the determination of %dm may also occur during the actual weighing. When a large series of weighings on dried leaf material is performed, this material will gradually increase in apparent dry weight with time, because it is highly hygroscopic.[48] Serial trends can be removed from such measurements by randomizing the sequence of weighings with respect to treatments. Also, during weighings the material can be stored in a desiccator instead of exposing it to the room environment.

2. Excess of Food Offered

A rather unexpected source of error in gravimetric studies (see above) is an excess of food offered. Food utilization experiments are preferably carried out under *ad libitum* conditions. Because the amount of food intake is often only approximately known in advance, one tends to offer an excess of food to

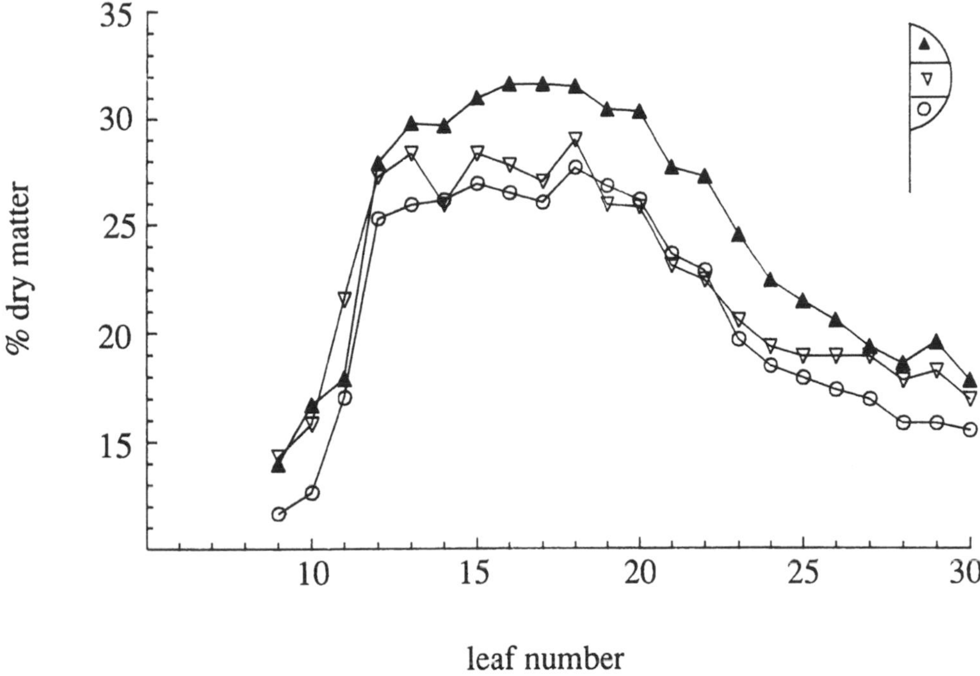

FIGURE 2B.

prevent temporary deprivation. This excess causes serious errors in the estimation of C, both with respiring plant food[5] and with the nonrespiring (or at least minimally respiring) artificial diets.[75] It is not feasible, however, to assess the possible role that excesses of food have played in the studies to date, as these data are reported only rarely in publications. It is highly likely that some differences among food types on insect utilization indices reported in the literature resulted from systematic differences in proportions of the different foods eaten.

To reduce this error, pilot experiments may be carried out to produce an estimate of food consumption in response to the different treatments, the effects of which should be investigated in the actual experiment.[75] The results of such a pilot study provide a basis for presenting the treatment groups with varying initial amounts of food offered in order to minimize the excess of food. It should be realized, however, that the experimental circumstances are no longer identical for all treatment groups. This implies the risk that the varying amounts of food offered as such will influence the behavior of the insects, which should be checked separately.

3. Food Respiration

Contrary to the suggestion of Waldbauer,[110] Axelsson and Ågren[5] demonstrated that neglecting respiration of the leaf material can result in a serious overestimation of C, especially in combination with an excess of food. They constructed a correction factor to account for the resulting reduction in dry weight of food during the experiment. Unfortunately, this correction factor has not been applied very often. This may have to do with the tacit assumption that the error involved is of a systematic nature and of no real importance in comparative studies into differential utilization of host plant species. This assumption, however, is incorrect as plant species may differ considerably in their rates of maintenance respiration.[64] Thus, when different plant species are compared, differences in the rates of maintenance respiration that are not accounted for will introduce errors in the estimation of food consumption. Examples of the extent of such differences are presented later in the chapter.

The investigation of the effects of leaf respiration on nutritional indices can be undertaken in two ways. In theory, it is possible to observe in separate control series, by leaving out the insects, a decrease of dry matter content caused by respiration. This method is subject to the same errors as mentioned above with respect to the measurement of %dm. A second method is to determine the respiration of the leaves by measuring the oxygen consumption, e.g., with a Gilson apparatus.[34] The latter method is considerably more precise and can be repeated several times during the experiment. This gives an indication about the time course of the respiratory rates of excised leaves, which may not be constant.

Wound respiration may exert confounding effects on the estimation of plant respiration when control leaves are used, i.e., leaves without insects but otherwise treated identically to the experimental leaves. Although wound respiration is known to occur in plant tissues, its magnitude in different plant species and in response to several types of mechanical damage is largely unknown.[50,103] It is clear that respirometry is the proper method to assess the importance of wound respiration.

The factors discussed thus far all tend to produce an overestimation of C, and ways of reducing the effect of the respiratory losses of dry matter from the plant have been indicated; however, as Wightman[116] rightly stated, in the natural feeding situation of a herbivore the plant will photosynthesize and thus one would expect corrections for an increase in food dry matter instead of a decrease. This is a relevant but often neglected point and is discussed in a separate section.

The three sources of error discussed in the preceding paragraphs are relevant to the utilization of dry matter. When utilization of energy or nitrogen is studied, changes in the biochemical composition of the plant food during the experiment, other than those caused by respiration, also come into play.[15,95] Systematic errors in the estimation of C may contribute to the discrepancies found in checking dry matter budgets with respirometric methods (see [1] in the list in Section III). Their main cause probably is the use of inadequate drying methods, yielding incorrect estimations of %dm. Indeed, a compilation of data from the literature reveals that drying has been performed at a wide range of temperatures (50 to 110°C) and with variable durations (a few hours to several days). A common practice has been the drying to constant weight. This yields no guarantee that all water has been removed, however. Conversely, drying at high temperatures will result in breakdown and volatilization of organic compounds, and consequently an underestimate of the dry matter content. These processes may affect the different types of nutrients differently. More recently, lyophilization (freeze drying) has been applied in a growing number of studies, which eliminates the risks of thermal degradation. In several studies, drying methods of leaf material and insect material have been different for reasons that were not stated. A comparison of drying of either cabbage leaves or an agar-based artificial diet at 60 and 80°C and freeze drying at –35°C at 0.02 bar resulted in significant differences in mean %dm, with freeze drying giving values between the two heating temperatures. Moreover, with the artificial diet an increase in the size of the piece of diet to be dried was significantly positively correlated to the %dm for each of these three drying regimes, indicating retention of water in the larger sized pieces.[107] These findings indicate the difficulty in estimating the actual %dm. When the utilization of nutritive components of the dry matter, such as nitrogen or carbohydrates, is studied separately, an incorrect estimation of the dry matter content carries over to the content of components. Admittedly, while it is not particularly difficult to put forward such possible sources of error, it is not feasible to assess the extent to which they may have played a role in utilization studies. Leaf thickness and fiber content are a few of the more apparent parameters that will affect the release of water during drying. When plant species differ in these characteristics, errors in estimation of %dm will likely play a role. This was found to be true even for different cultivars of maize.[101] The adequate drying regimes for particular insect, fecal, and plant materials deserve more attention than they commonly receive and the optimal procedures may be different for different plant or insect species, depending mainly on morphological and physical characteristics. Lyophilization seems the preferred method for the future.

B. FECES

In estimating FU, sources of error different from those involved in estimating C play their part. Several possible sources deserve mention.

Where separation of food and feces is difficult, errors may easily occur in the measures of amount of feces produced, as well as in the amount of food consumed.[75] Coprophagy is another undesired phenomenon and likewise complicates a reliable measurement of both FU and C. A countermeasure could be the removal of feces at short intervals, but this will often lead to disturbance of the insects. An alternative method is to suspend the food in the top of the experimental container or at least on a certain level above its bottom; feces will then drop onto the bottom, which may be a grid through which feces can pass. Depending on the relative humidity prevailing under the experimental circumstances, feces may or may not dry quickly. Substances may escape from the feces, leading to an overestimation of AD. Again, plant tissue respiration may be involved. One gram of freshly defecated frass pellets of *P. brassicae* had a carbon dioxide release rate (per unit fresh weight) that was approximately 35 to 40% of the rate of the intact leaf tissue.[107] Comparable data on the dry matter loss of feces are not available. It indicates, however, that this factor should not be neglected in experiments with leaf material. Sustained respiration of leaf material that has passed through the alimentary canal of the insect is not improbable, as the passage time of food through the gut between intake and defecation is not more than 2 to 4 h in some herbivorous insects[19,107] and digestive lysis of cellular structures or extraction of substrates for plant respiration will be incomplete. Dry matter loss from feces may not only be due to endogenous respiration of the leaf material but also due to the activity of microbial decomposers present on the leaf prior to intake or inhabitants of the insect gut egested with the feces. Dry matter losses from feces due to respiratory processes will effectively be reduced by quick drying of the feces, keeping in mind what has been said previously about drying techniques; however, a dry environment is unusual and often unfavorable for the herbivores, so that regular collection of feces, taking care to minimize the disturbance of the insects, is desirable.

Retention of an unknown amount of food in the gut at the conclusion of the experiment causes an overestimation of the amount of digested material. The retained undigested material will usually be mainly egested as feces, if the insects are given the time to do so within the experimental period. Waldbauer[110] discussed this point and it was briefly mentioned above in connection with the importance of physiological timing of experiments. Clearly, the retained material will also cause an overestimation of G (see below). An obvious procedure to avoid this error in routine experiments is to allow the natural emptying of the gut, generally occurring prior to moult and pupation, to proceed until it ceases and to consider the time of cessation as the endpoint of the experiment. When feeding experiments are not physiologically timed in this way, gut emptying may require periods that are long relative to the intended experimental duration. This is caused by the fact that under such forced deprivation insects will tend to retain food considerably longer than under *ad libitum* food availability and subsequent voluntary cessation of feeding. Still, the gut is rarely completely devoid of food during a moult or the pupal stage and this should be checked separately for each species and treatment via dissection of guts of an aliquot group, a practice reported only rarely.[72]

In contrast to the overestimation of AD associated with the sources of error in fecal dry matter thus far discussed, a possible cause of underestimation comes from those substances egested as components of the feces that have a metabolic or endogenous origin. Examples of these are uric, allantoic, and excreted amino acids and other waste products of nitrogen metabolism and turnover material from the peritrophic membrane, the intestinal epithelium, and digestive enzyme secretions. The presence of this type of material as an integral part of the feces has led to the term "approximate" digestibility.[110,118] While this error may be considered as systematic within one type of diet, it has been shown that it cannot be neglected when different diets are compared[12,39,43,54] Evidently, when the measurement of nitrogen or protein utilization is the objective, the presence of metabolic or endogenous nitrogen-containing material becomes a factor of major importance for measurements to be correct.[110]

C. GROWTH

The amount of growth of the insect body is the budget item that can be determined with greatest accuracy. Yet, in this case, actual biomass produced may also be underestimated by the failure to collect exuviae and secretions such as silk. For four species of lepidopterous larvae, exuviae constituted from

1.3 to 8.0% of total body dry weight[10] and from 2.6 to 10.2% of the total amount of energy accumulated during the instar.[98] The biomass production invested in cuticle is far more important in hemimetabolous insects such as Orthoptera, ranging from 40 to 56% of total dry weight accumulated for five species.[10] Depending on the species, silk may consist of a considerable quantity of *de novo* synthesized protein-rich material, ranging from 3 to 4% of the total biomass produced in the moderately spinning final instar of *Pieris brassicae*[107] to 20% in the silkworm *Bombyx mori.*[110] Net dry matter body growth, G, taken at the end of the experimental period (Equation 8), can be more accurately determined than gross body growth. This growth often has been temporarily higher than the net value, but is confounded by the presence of an unknown amount of food in the gut. The amount of food in the gut at the start and end of the experiment has not been reported much in utilization studies, but the few values available indicate that this parameter varies widely over species, ranging from 10 to 50% of the dry body weight of lepidopterous, coleopterous, and orthopterous herbivorous insects.[11,15,28,72,90,110] Thus it seems important to minimize the contribution of the amount of food retained in the gut to the "apparent" net body growth.

Another error in the measurement of G may arise in the estimation of the initial dry weight of the insect body, which must rely on the measurement of the dry matter content of an aliquot of insects. Although errors inherent in using aliquots may occur, their magnitude will in most cases be of minor importance relative to errors in the final weight, due to the fast growth rates of many herbivorous insects.

V. QUANTIFYING THE EFFECTS OF VARIOUS ERRORS: EXAMPLES

The main types of error discussed above are quantified here in their effect on ECD by means of a sensitivity analysis that integrates the approaches of Schmidt and Reese,[75] limited to artificial diet studies, and that of Axelsson and Ågren.[5] The author chose to elaborate an example, rather than attempting a generalized analysis.

Imagine a medium-sized herbivorous caterpillar that grows from 100 to 450 mg weight (from 12 to 110 mg in dry matter units) in its final instar (duration of around 90 h at 25°C), feeding on excised, water-supplemented leaves of a suitable food plant. Representatives of such a situation would be final instar noctuid lepidopterans such as *Spodoptera* spp., *Heliothis* spp., *Agrotis* spp., or *Mamestra* spp; actual values are representative for *Pieris brassicae*[106] and are given in Table 1.

Errors were quantified for a situation of a typical gravimetric food utilization experiment. The starting point for the sensitivity analysis was the assumption of a set of realistic parameters needed to calculate ECD. The calculated value, 62.5% in this case, is taken as a reference value and errors are expressed as the number of percentage points by which they differ by subtracting this value (i.e., an error of +10% on the vertical axis means that the calculated value of 62.5% should in fact be 52.5%). Subsequently, a number of alternative values

Table 1
DEFAULT SETTINGS FOR PARAMETER VALUES USED IN THE SENSITIVITY ANALYSES QUANTIFYING THE EFFECT OF VARIOUS ERRORS ON ECD (FIGURE 3, *PIERIS BRASSICAE*) AND RMR (FIGURE 5, *SPODOPTERA ERIDANIA*)

		Species	
Parameter	Unit	*Pieris brassicae*	*Spodoptera eridania*
C	mg	400	40
G	mg	100	10
FU	mg	240	10
R	mg	60	20
T[a]	h	72	74
W[b]	mg	55	10
ECD	%	62.5	66.6
RMR	mg/mg/d	0.54	0.65
%dm	%	15	12.5
α[c]	mg/mg/d	0.65	0.45
r[d]	mg/mg/day	0.03	0.03

Note: mg are dry matter.

[a] Duration of experiment, considered to be equivalent to duration of active feeding period.
[b] Mean body mass during the experiment, for which the arithmetic mean of initial and final larval weights was used.
[c] Relative growth rate of larva.[5]
[d] Respiration rate of plant tissue, taken to be constant over the experimental period. For the analysis depicted in Figure 3B, r was set at 0.

of the actual %dm present in the leaf (both at the start and during the experiment) is introduced and the extent to which the originally calculated value of 62.5% is in error, is quantified. The alternatives are focused on quantifying the effect of errors on the estimation of dry matter of food consumption, which has been shown to be especially sensitive to incorrect estimations of three parameters: %dm (Equation 4), the amount of food offered in excess of the quantity consumed at the start of the experiment, and the respiration rate of the plant tissue.[5,75] For this reason the curves describing the magnitude of errors in ECD have been plotted as a function of the fraction of food remaining relative to the food offered at the start of the experiment (abbreviated as %FE in the following), which was varied between approximately 20 and 100%. The apparently correct value for ECD (62.5%) was set to be independent of %FE. The quantitative consequences of two main

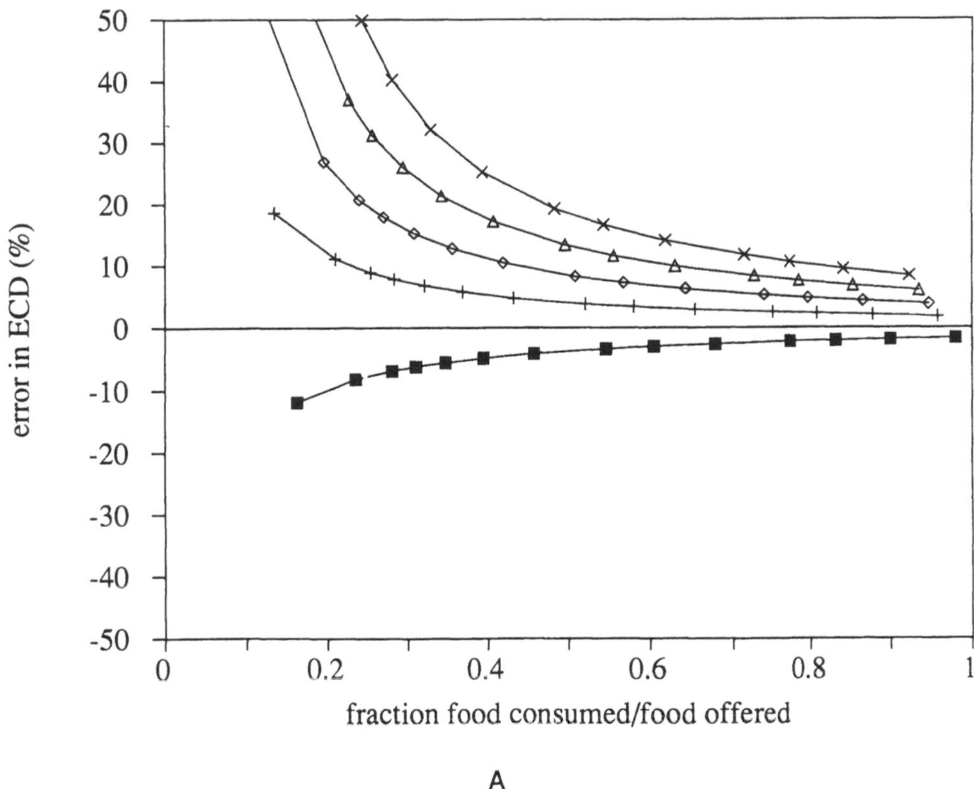

A

FIGURE 3. Results of a sensitivity analysis that quantifies the error in ECD as a function of the fraction of food consumed relative to the amount offered at the start of a feeding experiment. Original settings of parameter values are given in Table 1, headed under *Pieris brassicae*. Ordinate scale: actual ECD-percentage points. For further details see text. (A) Errors in ECD resulting from differences in experimental duration when no correction is made for leaf respiration (r = 0.03). The curves belong to the following durations: ■: T = 60 h; +: T = 84 h; ◊: T = 96 h; Δ: T = 108 h; x: T = 120 h. (B) Errors in ECD resulting from deviations in dry matter content (%dm) and respiration rate (r) relative to the values assumed for these parameters in the original calculation of ECD, in which leaf respiration was not accounted for (r = 0). ■: -0.5% error in %dm; +: +0.5% error in %dm (these first two cases are comparable to those presented by Schmidt and Reese,[75] who show examples for ±0.2% error in %dm); ◊: r = 0.02; Δ: r = 0.03; ∇: r = 0.04. (C) Errors in ECD resulting from deviations in actual respiration rates relative to an assumed rate (r = 0.03) and combinations of errors in r and %dm. ■: r = 0.02; +: r = 0.04; ◊: r = 0.02 and error in %dm +0.5%; Δ: r = 0.02 and error in %dm -0.5%; x: r = 0.04 and error in %dm +0.5%; ∇: r = 0.04 and error in %dm -0.5%.

sources of error, %dm and plant respiration rate have been examined. Figure 3A visualizes the effect of neglecting leaf respiration (i.e., the false assumption that leaf respiration is nil) when durations of the active feeding period in the instar differ. This has often been observed when different host plant species are compared for their nutritive suitability. Two simplifying assumptions underlie this sensitivity analysis: the first implies that leaf respiration is constant over

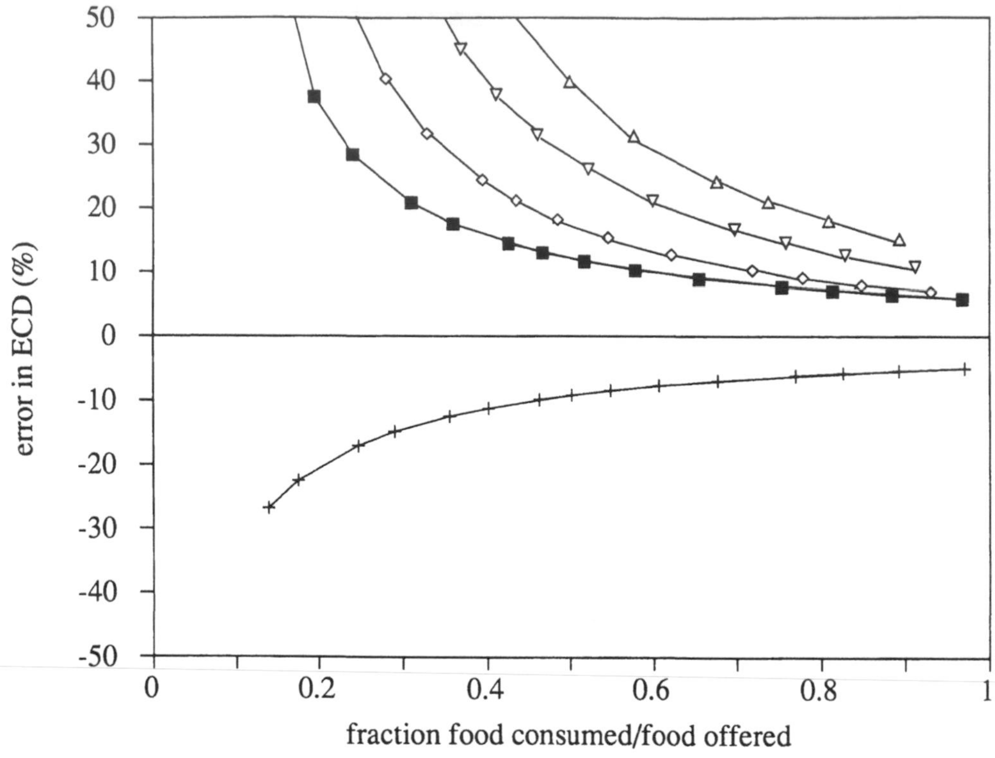

FIGURE 3B.

time, which is a premise also in the next two analyses; the second is that the different host plant species under study have identical respiration rates. It appears from Figure 3A that a 12-h difference (either shorter or longer) relative to the standard duration of 72 h results in errors of <10% when %FE varies between 20 and 50% and <5% when %FE is >50%. Larger differences in duration of the active feeding period (prolongations of 24 to 48 h) lead to more substantial errors in the estimation of ECD. This means that differences in ECD associated with different host plants may be explained solely by the failure to correct for leaf respiration.

The effects of errors in the estimation of %dm in the leaf material either at the start of the experiment or due to leaf respiration during the course of the experiment are shown in Figure 3B. An error of 0.5% in %dm (e.g., 14.5 instead of 15%) is realistic in view of the variation found between symmetrical halves of leaves, as discussed above[110] (Figure 2). Errors in ECD between 5 and 10% easily result, even when %FE is >70%, and increase rather drastically when %FE is <50%. When leaf respiration is not accounted for, systematic underestimations of ECD will be the result, the extent of which depends quite strongly on the actual rate of leaf respiration. This confounds straightforward comparisons of nutritive suitability of plant species when it has not been assured that these species have almost identical respiration rates.

The third analysis combines errors in the estimation of %dm with incorrect values for plant respiration (Figure 3C). The situation here is that the correction

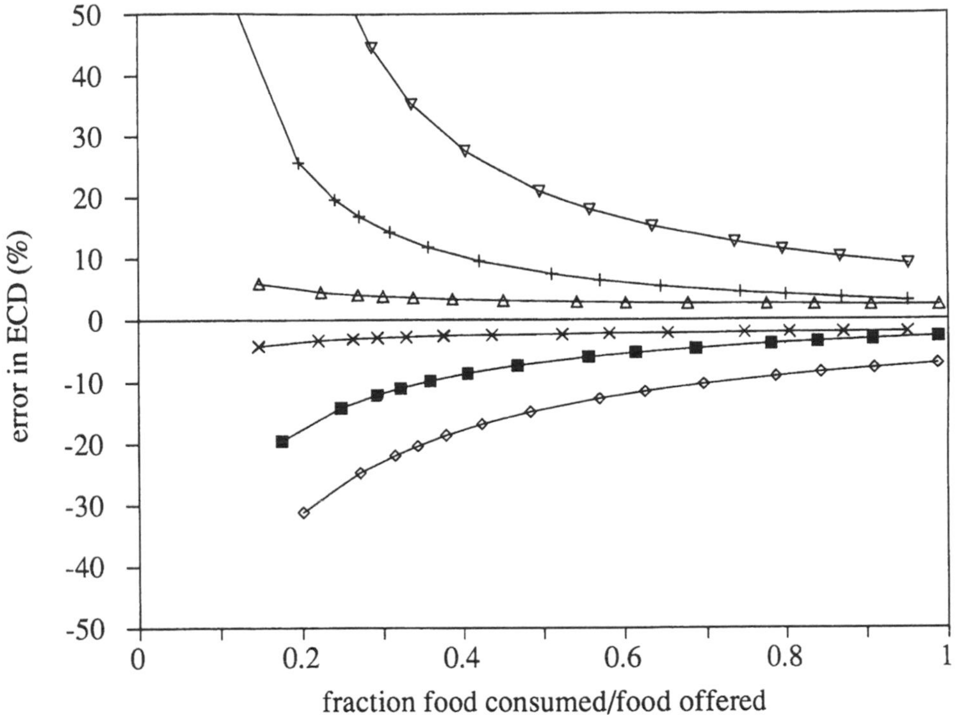

FIGURE 3C.

factor of Axelsson and Ågren[5] accounts for a respiration rate of 3% of the dry leaf matter per 24 h but that the host plant species to be compared have either lower or higher respiration rates. These different rates are then combined with either an under- or overestimation of leaf dry matter content of 0.5%. While an error in the actual respiration rate used to correct for a gradual decline in dry matter content yields errors in ECD <5% when %FE has values >80%, at values of %FE between 40 and 50% errors in ECD amount to 6.5 and 9%. When both types of error are combined, two different situations occur: they either reinforce or compensate each other, leading to either errors of >10% also at high values of %FE or to relatively small errors (2 to 5%) even when there would have been an excess of food (Figure 3C).

VI. VARIATION IN FOOD UTILIZATION EFFICIENCY: PHYSIOLOGICAL vs. METHODOLOGICAL CAUSES

Can one assess to what extent the foregoing errors have played a role in published studies? A straightforward answer to this question cannot be given. Data on experimental parameters that apparently can confound the accurate measurement of food utilization, such as temporal or spatial variability of dry matter content of leaves, the fraction of food consumed relative to the amount of food offered, and respiration rates of plant species under study, are lacking in all but a few papers on food utilization of herbivorous insects. This paucity of relevant experimental details prevents a proper evaluation of the causes of the relatively high variation in ECD; however, on the basis of the sensitivity analyses presented here, it can be concluded that only part of this variation is of biological origin. Which part is the interesting question, answered perhaps by the following section.

Mention has already been made of the problem of straightforward comparisons of ECD values when considerable differences were observed in the duration of the active feeding period (see also Figure 3A). A possible accounting of the effect of different active feeding or developmental durations per se is to base the comparison on rates, thus incorporating the time dimension. Due to the way in which they are calculated, rates and efficiencies can be interrelated. The causation of observed relationships between rates and efficiencies have been discussed in several reviews [86,87,93,98] and from these it has become evident that the causal factors are manifold and interact in complicated ways (see below). The relevant rate parameters are those of the budget items used to calculate ECD, growth rate, and metabolic rate (Equation 10). This section puts forward another way of looking at variation in ECD by interrelating growth and metabolic rates. A comparative compilation of these parameters for a range of insect species was composed by Slansky and Scriber,[98] but provides both data only for three species. Relating growth rates to metabolic rates offers the possibility of comparing the variation in both components of ECD while accounting for putative confounding effects of variable durations per se. Moreover, using relative growth rate (RGR, i.e., milligrams of dry matter of growth per milligrams dry body weight per day) and relative metabolic rate (RMR, expressed as milligrams of dry matter respired per milligram of dry body weight per day; by convention, the symbol M instead of R is used in this expression[98]), correction is obtained for absolute differences in growth rate due, for example, to differences between cohorts over time or between different laboratory strains. A reevaluation is presented of published data on a series of utilization experiments under controlled conditions using penultimate instar caterpillars of *Spodoptera eridania*.[43,53,82,84,85,99] From these papers it was possible to calculate RMR and RGR when they were not already presented in the papers themselves. Of the data presented by Soo Hoo and Fraenkel[99] those were examined that met the criteria of initial %dm of leaf material being <20% and overall larval survival being >90%. Although experimental temperatures differed between studies (27°C,[99] 24°C,[43] and 22°C[53,82,84,85]) data were pooled as no differences in RMR or RGR were apparent between results at these temperatures. Summarized statistics for the 35 remaining experiments are given in Table 2 for a case that includes all the studies and another that excludes one of them that contributes disproportionally to the calculated variation.[53] Over all studies a coefficient of variation (CV) for RMR is calculated that is 3.9 times higher than the CV for RGR. When the study mentioned above is excluded (three experiments on maize cultivars), the CV for RMR is 1.77 times higher than the CV for RGR. The ranges found for RMR are also considerably wider. Comparable data for final instars from 12 experiments are also included in Table 2. The trends observed for final instars follow those of penultimate instars; the CV for RMR is 1.65 times higher than the CV for RGR, but the range for RMR is narrower. Data on 32 experiments

Table 2
SUMMARY OF REPORTED VALUES FOR RELATIVE GROWTH RATE (RGR) AND RELATIVE METABOLIC RATE (RMR) AND THEIR VARIABILITY FROM SIX STUDIES ON *SPODOPTERA ERIDANIA*[43,53,82,84,85,99]

	Penultimate instar				Final instar	
	RGR	RMR	RGR[a]	RMR[a]	RGR	RMR
Mean value	0.444	0.767	0.437	0.652	0.219	0.707
SD	0.082	0.544	0.079	0.208	0.057	0.302
CV[b] (%)	18	71	18	32	26	43
Minimum	0.31	0.153	0.31	0.153	0.11	0.35
Maximum	0.62	3.36	0.55	1.15	0.31	1.38
Ratio max/min	2	22	1.8	7.5	2.8	3.9
n[c]	35	35	32	32	12	12

Note: Both RGR and RMR are expressed in mg/mg/d.

[a] Values in these columns originate from the same experiments as in the first two columns, but the three experiments reported by Manuwoto and Scriber[53] were excluded.
[b] CV is coefficient of variation (SD divided by mean).
[c] Number of experiments.

on penultimate instars are represented graphically in the scatter diagram of Figure 4, which shows a lack of correlation between RGR and RMR (Spearman's rho is –0.13, $P = 0.47$). It is noted that values for RMR from an artificial diet study[43] lie at the low end of the range reported, indicating that neglect of plant respiration may indeed have led to overestimation of RMR. In the latter study,[43] care was taken to maintain %FE above 70%. As a means for comparison, three lines describing particular values of ECD are plotted in Figure 4. The course of these lines shows that the considerable variation in RMR is largely contained within an ECD range of approximately 30 to 60% (see Figure 1). This reevaluation of published data leads to the conclusion that variation in the metabolic rate for these particular data is substantially higher than the variation in growth rate and thus will make an important contribution to the reported variations in ECD.

As a next step in the reevaluation of these data, a sensitivity analysis of the kind described above was applied to the data. Parameter values used in the analysis are given in Table 1 and the results are given in Figure 5. In this case the error in RMR plotted on the ordinate was calculated as a percentage (error in RMR [%] equals the absolute error in RMR divided by the original value, which was set at 0.65 mg/mg/d, the average of 32 experiments). In this way a measure of deviation from the mean was obtained that can be compared to the CV. Depending on the assumptions made about the degree to which the actual

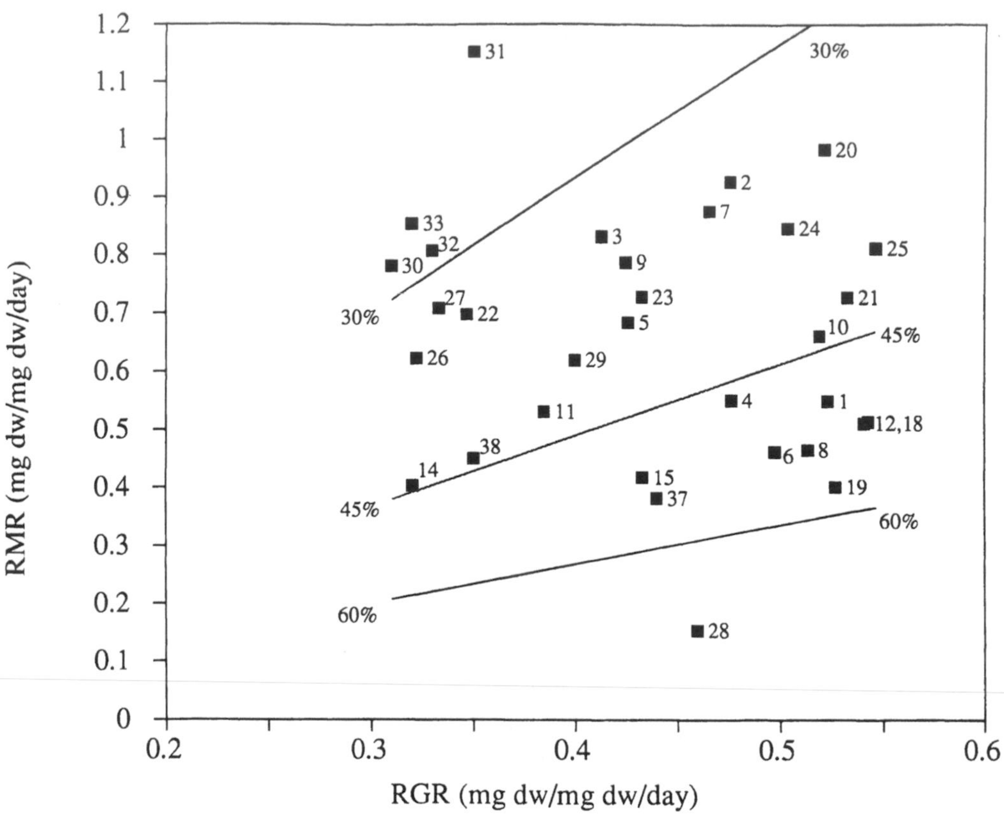

FIGURE 4. Scatter diagram of mean relative growth rates (RGR) and mean relative metabolic rates (RMR) of penultimate instars of *Spodoptera eridania* in 32 feeding experiments from 5 studies, involving 16 host plants and 2 artificial diets. Data points 1—21 were taken from Soo Hoo and Fraenkel:[99] 1—10: *Phaseolus lunatus,* exp. II—XI; 11: *Morus alba;* 12: *Taraxacum officinale*; 14: *Ipomea batatas;* 15: *Chenopodium album;* 18: *Phytolacca americana;* 19: *Solanum tuberosum;* 20: *Lycopersicon esculentum*; 21: *Spinacia oleracea*; points 22—27:[82] 22 and 23: *Phaseolus vulgaris*; 24—27: *Lotus corniculatus* CN+, 4 weeks old; 25: CN–, 4 weeks old; 26: CN+, 20 weeks old; 27: CN–, 20 weeks old; points 28—30:[84] 28: *Anethum graveolens*; 29: *Phaseolus lunatus*; 30: *Brassica oleracea*; points 31—33:[85] 31: *Sorbus americana*; 32: *Betula papyrifera*; 33: *Prunus serotina*; 34:[43] 7.5% casein; 35:[43] 19.5% casein.

values of leaf dry matter content and leaf respiration rate differed from the values in the original calculation (not corrected for leaf respiration; Table 2; Figure 5), the resulting curves show that an important amount of variation could have been caused by relatively small errors in %dm (±0.5% relative to an assumed value of 12.5% dry matter content) and small differences in respiration rates between plant species. For a situation where %dm was 12% instead of 12.5%, the leaf respiration rate of two host plants differed by 0.5% of dry matter per day and %FE is either 80 or 50%, RMR will be in error by 14 and 19% respectively, which are 44 and 60% of the overall CV of 31.9% (Table 2).

The foregoing discussion addresses questions of accuracy. Accuracy of measurement is especially relevant when the systematic nature of errors has not been ascertained and the experiment is aimed at comparing treatment

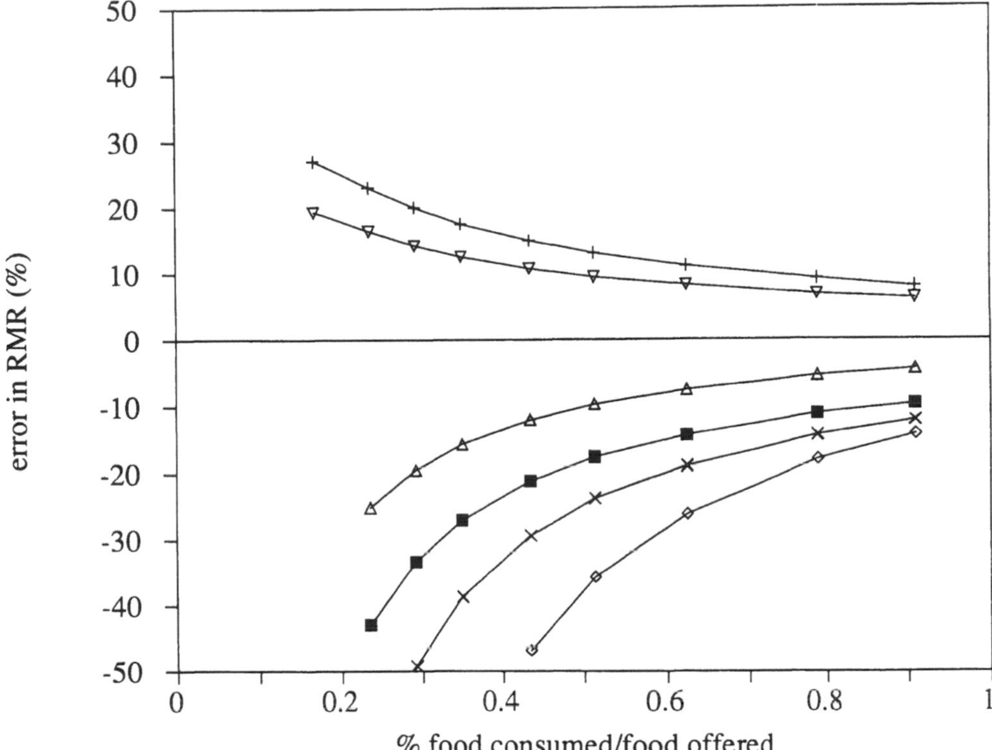

FIGURE 5. Results of a sensitivity analysis comparable to that presented in Figure 3. Parameter values used in the calculations are given in Table 1, headed under *Spodoptera eridania*. Errors in RMR (relative metabolic rate) as a percentage deviation from a mean value of 0.65. ■: Error in %dm -0.5%; ◊: r = 0.03; x: a difference in r between plant species of 0.005% combined with an error of–0.5% in %dm; +: error in %dm 0.5%; Δ: r = 0.01; ∇: a difference in r between plant species of 0.005% combined with an error in %dm of 0.5%.

effects, e.g., different host plants. Several arguments have already been posited here supporting the notion that errors are not all systematic. Yet another source of variation is measurement precision. In a comparative methodological study using artificial media,[63] precision in gravimetric ECI and ECD was calculated to be 80 to 85%. Schroeder[80] demonstrated that a minimum of 20% (but more probably 40%) of the observed variability in ECI could be attributed to gravimetric instrument precision in a study on three caterpillar species fed leaf material. Very few comparable studies aimed at quantifying variation due to measurement are available, but the latter study shows the importance of the contribution of methodological sources of variation.

A sensitivity analysis as presented here provides a way of determining to what extent measuring errors may have contributed to observed variation in food utilization efficiency; however, the execution of this analysis is not meant to imply that the putative errors should lead to doubt surrounding the particular data analyzed. The magnitude of variation reported for RMR and ECD in

Spodoptera also can be detected from published data on other herbivorous insects.[55,106] This is in accordance with a theoretical expectation based on Equation 2, which implies that all errors in the measured parameters C, G, and FU will be reflected in the calculated parameter, R. Thus, the view can be defended that due to the fact that the R-term is not measured directly in the gravimetric rationale, the foregoing analysis raises at least some question as to the accuracy with which gravimetry determines this term. It may be concluded that methodological variation, although difficult to separate from biological variation, may contribute significantly to the total variation observed when there is an excess of food and when plant respiration is neglected.

VII. ALTERNATIVE METHODS TO MEASURE FOOD UTILIZATION

In view of the methodological and technical problems inherent to the gravimetric measurement of food utilization, the logical question arises: are there alternative methods that circumvent these problems? An explicit point of view defended in this review is that sound alternatives must include the actual measurement of metabolism. Methods to achieve this are dealt with in more detail below. In the following three sections, other alternatives are dicussed briefly.

A. INDIRECT TECHNIQUES

1. Markers

At least five alternative indirect techniques have been used to measure food consumption and absorption efficiency in herbivores, and are reviewed elsewhere;[48,108,110] however, Kogan[48] stated that all of these alternatives involve a combination of gravimetry with an indirect indicator of consumption. Thus, they are subject to the same problems as the gravimetric method, although to a variable degree. Relatively few comparative studies have been performed that allow a direct comparison of the accuracy and precision of different techniques. From one such study, Parra and Kogan[63] concluded that gravimetry was more reliable than three other indirect methods studied. One of the indirect approaches that holds some promise employs indigestible markers such as chromium.[48] Such markers should meet a number of biological requirements to be suitable for the quantification of food utilization, the most important being their lack of deterrency, toxicity, or nutritional value.[48,110] Additionally, the marker should not be preferentially adsorbed to the gut wall. In theory, the use of such a suitable marker may prevent the necessity of measuring C in terms of dry matter. Instead, concentrations of the element (as percentage of dry matter) in food and feces are measured. When total dry matter of feces produced is also determined, C can be calculated according to:

$$C = \frac{FU \cdot (\%m(FU))}{(\%M(C))} \qquad (12)$$

where FU is the total amount of fecal dry matter produced, %m(FU) the concentration of the marker (percentage of fecal dry matter), and %m(C) the concentration of the marker as %dm in the food.

Several additional problems are associated with the use of markers. When an exogenous marker is employed, it should be checked after application for homogeneous distribution in or on the food. The same requirement should be met by endogenous markers, which are naturally present in the food, such as silicon.[8] It is difficult to prove that this is the case and the magnitude of the error that is introduced depends on the magnitude of within-tissue differences, the extent to which the unit of plant material subjected to analysis is representative for the tissue that was actually ingested, and the degree of selectivity of feeding behavior toward different tissues. It is noted that these considerations, in fact, also pertain to the %dm. Moreover, concentrations of the marker in any food tissue may change with time, as may the concentration in feces when there is a time lag between production of fecal pellets and their fixation by drying. Concentration changes in food or feces may occur both by respiratory losses of dry matter over time or by gaseous escape of the marker or its reaction products from feces. In view of these complications indirect methods may not be viewed as an improvement relative to gravimetry, especially so because they are still dependent on it.

2. Elemental Budgets

Food consumption may be estimated from the budget of a particular element that is a nutrient or is contained in nutritive compounds. A relevant example is nitrogen. By determining the total amount of nitrogen accumulated in the body during growth (N(G)), the total amount egested in the feces [N(FU), i.e., including uric acid and other products of nitrogen metabolism], and the concentration of nitrogen in the leaves (%N(C)), food consumption can be calculated according to:

$$C = \frac{(N(G) + N(FU))}{(\%N(C))} \tag{13}$$

This method is subject to comparable sources of error as discussed for other indirect methods. Indeed, in several studies that measured consumption, growth, and fecal production gravimetrically and also determined nitrogen concentrations in these three budget items, the nitrogen budget did not balance.[22,39,43,106,118] One explanation could be that nitrogen may escape in a gaseous form from feces as a result of microbial activity or even more so as a result of high temperature drying. While these studies were performed using artificial diets, in leaves temporal changes of nitrogen concentration are to be expected due to leaf metabolism and translocation.[15,112] The initial concentration is tacitly assumed to be fixed over time and the dynamics of elements or nutrients in leaves during utilization experiments has received little attention.[83]

B. PLANIMETRY

Measurement of the surface area of plant tissue consumed by a herbivore has some important potential advantages over gravimetric measurement of consumption.[80] The most important advantage is of a biological nature, being that planimetry can be performed on intact plants.[37] The relevance of studying the intact plant situation as compared to excised food tissues is discussed later in the chapter. A second advantage is the fact that the amount of uneaten tissue is not used in the calculation, thus canceling out the errors in the percentage of food eaten shown in the sensitivity analysis; however, gravimetry comes into play when the surface is converted to dry weight. Here again, problems emerge with respect to homogeneity and dynamics of specific leaf weight, i.e., amount of dry matter per unit of leaf surface area.[111] An additional problem is the change in leaf area caused by leaf growth or shrinkage during the experiment. In the intact situation rather than in an excised condition, wound healing may occur that introduces errors in the planimetric estimation of consumption.[37] Few studies have been devoted to a methodological comparison of gravimetry and planimetry. Schroeder[80] performed such a direct comparison on three caterpillar species feeding on one host plant under laboratory conditions. Although the planimetric method was less precise, observed variability was similar for both methods. Absolute values of budget items were the same for two species, but for a third, growth efficiencies differed between the situations. In view of the advantages mentioned, more information confirming the usefulness of planimetry for utilization studies would be desirable.

C. RESPIROMETRY: THE METABOLIC ALTERNATIVE

It has been pointed out in this chapter that the gravimetric rationale relies on the accurate measurement of consumption and subsequently derives the amount of dry weight that was lost in respiration (Equations 1 and 2) without an independent check. This section advocates that dry matter or energy budgets should be instead established by the reverse, i.e., measuring respiration, growth, and fecal production and deriving consumption. In fact, this procedure has been the default method in comparative physiological studies on aphids, relatively small sapsucking herbivores, for which it is technically difficult to measure consumption directly.[52] It has been used also for caterpillars and locusts in detailed studies of assimilation efficiency of different nutrients, in order to obtain an accurate estimate of consumption independent from possible confounding errors in %dm.[90,93,106] In contrast to studies in nutritional ecology,[98] in the field of ecological energetics measurement of respiration has long been common practice;[17,65,100] however, as has been noted previously, the comparison of R determined by respirometry and oxycalorific conversion factors (R(m)) with R calculated by difference after gravimetric determination of C, FU, and G (R(c); see Equation 2) led to considerable discrepancies or imbalances.[58,116]

Explanations for these discrepancies have been the subject of debate in which either the overestimation of consumption[1,4] or the underestimation of respiration[115,116] was emphasized. The analysis of published studies by McEvoy[58] revealed that both these factors contribute to an explanation; indeed, they are not mutually exclusive. Wightman[115,116] provided evidence for the reality of what he called "flask effects", effects associated with confinement in closed flasks leading to a decrease in respiratory rate.

1. Closed-Vessel Respirometry

Respirometry on insect herbivores has been commonly performed using volumetric methods such as Gilson's, a modification of the classical Warburg constant volume respirometer.[34,45,59] These require a closed system in which oxygen concentration gradually decreases and carbon dioxide concentration is assumed to be zero. Although several effects of carbon dioxide concentrations higher than normal have been documented,[61] the effects of a very low partial pressure of carbon dioxide have been sparsely investigated, but do seem to affect respiration.[115] Strong alkaline solutions have been commonly used for the absorption of carbon dioxide. These absorbing agents will also cause a low relative humidity in the flask atmosphere that may force the insects to limit their tracheal ventilation in order to prevent desiccation. Confinement in small vessels may also affect behavior by the restriction of movement for larger species. Underwater agitation of the flasks intended to maintain a constant temperature and to ensure a rapid diffusion of exchanged gases may be expected to inhibit normal behavior.[115] Prior to respirometry, insects have commonly been deprived of food, which has been argued to bring about a standardized physiological condition.[45] No food was offered during respirometry.

An important improvement in closed vessel respirometry was introduced by Aidley[2] in an original study on *Spodoptera exempta*, where he processed leaves of maize or sorghum in a Warburg respirometer and found that the respiration rate became twice that at rest. In a comparable study by McEvoy,[57] feeding by caterpillars of *Tyria jacobaeae* resulted in an increase of oxygen consumption by a factor of 1.6. A considerable increase in respiration rate associated with feeding was noted for other lepidopterans and a locust,[10,105] suggesting that the calorigenic effect of feeding is as widespread in insects as in vertebrates.[47] Thus, food deprivation will certainly be partly responsible for the discrepancy between R(m) and R(c). On the other hand, deprivation of food prior to and during respirometry will reduce the underestimation of the weight-specific respiration rate caused by the contribution of gut contents to the apparent live weight. This contribution may be considerable, as indicated elsewhere in the chapter, and may also vary over the instar.[72] Thus, when food is available, weight-specific respiration rates should at least be corrected for gut contents to justify extrapolation from short- to long-term measurements.[105]

A lingering difficulty with the standard Gilson method, however, is that it does not permit measurements longer than 1 to a few hours. This is a short period relative to the duration of budget experiments that usually are run for the duration of a developmental phase, commonly 2 or more days. The validity of extrapolating short-term measurements to the duration of an entire instar is doubtful. To check this, the pattern of respiration rate over time should be known to judge whether measurements during a certain period are representative for the entire phase under study. Time patterns in respiration rates have been established in a number of instances as an integral part of growth and utilization studies, although the degree of continuity[21,73,105,106,109,117,118] varied, and in these cases they showed profound and rapid changes over time.

2. Flow-Through Respirometers

Recently, respirometers that operate with a flowing gas stream have been used with insects.[4,25,88,105] Wightman[115] compared five types of respirometer for their suitability for continuous monitoring of the respiration of larvae of the scarabaeid, *Pachnoda ephippiata*. He rediscovered the older work of Punt,[67,68] who used a sensitive thermal conductivity detector (diaferometer) to monitor the carbon dioxide release of beetles and judged this principle as approaching the ideal. Van Loon[105] developed a flow-through respirometer that makes use of a paramagnetic oxygen analyzer and a diaferometer in series. The diaferometer has the advantage of functioning reliably at low gas flow rates, thus increasing measurement sensitivity. A disadvantage, however, is that it is not specific for carbon dioxide. This is not an important drawback if it is analyzed regularly via gas chromatography (GC) for improper exchange of gases by the experimental insects (no gases other than carbon dioxide and oxygen should be exchanged). Hydrogen and methane gases, in particular, may disturb diaferometer analyses. If, for example, microbes, present in the gut or in the feces, produce such combustible gases, these may be removed prior to diaferometer detection by combustion over a red-hot platinum wire. Recent improvements in the technology of infrared carbon dioxide analyzers renders a sensitivity approaching that of diaferometers, including the advantage of specificity. Construction, operation, and calibration of flow-through respirometers has been extensively reviewed by McLean and Tobin[59] and are beyond the scope of this review. Application of these methods to studies with herbivorous insects is mainly a matter of scaling.

Respirometry is a form of indirect calorimetry; thus, a conversion of units of gas exchange to caloric units (Joules) is needed. In this conversion, different chemical forms and quantity of products of nitrogen metabolism that are excreted as part of the feces or stored in the body must be taken into account, as they may be quantitatively important in insects, depending on the nutritional quality of food protein.[12,39,43] Both oxygen consumption and carbon dioxide release are needed to allow an accurate calculation of heat loss as well as the determination of the respiratory quotient. Extensive experimental work in the field of animal energetics has provided a solid basis for the validity of such conversions, as reviewed by McLean and Tobin.[59]

D. RESPIROMETRY vs. GRAVIMETRY: PROS AND CONS

The gravimetric method yields one static figure of the magnitude of R, which is rather sensitive to a number of errors in the three basic budget items and is expressed in dry matter units. In contrast, respirometry yields an instantaneous and, if desired, continuous quantification of metabolic activity in the insect in a manner independent of errors in food intake and fecal production.[105,107] A study comparing both methods indicated that respirometry yielded a significantly smaller degree of variation in metabolic efficiency than gravimetry when excised plant tissues were fed, despite the application of correction factors.[105,107] Respirometry makes it possible to determine when and in what way treatments, e.g., an alternative host plant or application of an allelochemical, affect the metabolic rates of both the insect and the plant. These dynamic properties are important advantages over the gravimetric approach, especially when they are considered in connection with the possible metabolic responses of the plant to herbivore feeding. By now it has become evident that plants do not passively undergo herbivore damage. This point is addressed in more detail in the next section.

A disadvantage of the respirometric approach is the requirement of continuity of measurement when total metabolic investment is to be measured over a developmental phase. This demands a separate set of gas analyzers for each experimental unit, either an individual or a group. Although this is only a technical and financial limitation, it decreases the feasibility of large-scale experiments involving many replications. A compromise involves performing continuous pilot experiments that focus on the establishment of time pattern of the respiration rate and on the amplitude of fluctuations that occur for the different experimental situations to be studied. Based on the results, the accuracy of discontinuous measurements can be estimated, which in turn will depend on the duration and interval of sampling of the emerging gas stream.[88] In this way, a larger number of situations or replications can be monitored with a limited number of gas analyzers. The validity and precision of such a discontinuous procedure will depend on the predictability of the time pattern of respiration and interindividual variation in respiration rate. CV of the rate of carbon dioxide production ranged between 1.8 and 6.3% in final instar caterpillars of *Pieris brassicae* feeding on an artificial diet, which is remarkably low compared to variability in metabolic rates derived from gravimetric data for this species.[106]

The maximum sensitivity of flow-through respirometry currently available may still prevent its application with smaller herbivores such as aphids, although measurements on groups or clones may be feasible.

A most interesting optional technique for utilization studies under field conditions may be the doubly labeled water method.[76] This is a form of indirect calorimetry that measures total carbon dioxide production by establishing the elimination rates of the radioactive isotopes of oxygen and hydrogen from body water. This method holds considerable promise as it can be applied on free-living herbivores under field conditions. Its validation with locusts was quite satisfactory,[16] but some problems were encountered with beetles.[24] A very accurate mass spectrometer is needed for sample analysis, which involves fairly high costs. Nevertheless, research on the accuracy and usefulness of this method definitely deserves more attention.

VIII. THE ENERGETIC COST OF GROWTH IN HERBIVOROUS INSECTS: PHYSIOLOGICAL CONSIDERATIONS

Respirometry allows an accurate determination of the energetic cost of growth as well as the monitoring of the time pattern in metabolic rate. This offers the possibility of dissecting the energy budget of the insect into a number of basic components and evaluating their relative contribution. When this approach is combined with a theoretical approach that is based on the energetics of general biochemical pathways of anabolism, increased insight into the relative contribution of biological factors that cause variation in food utilization efficiency is gained. Relevant questions in this respect are which part of the energetic costs of food processing and growth are fixed, which are variable, and how does the sum of both relate to costs actually observed? In the biochemical energetics approach,[3,35,79] the costs of synthesis of the different main body components (i.e., proteins, lipids and carbohydrates) are estimated using the relatively universal energetic efficiencies of the biochemical conversions. A measure of the energetic cost of synthesis for each of these main body components can then be obtained by the formula:

$$SC = ((100/Ef_S) \times D_C) - D_C$$

in which SC is cost of synthesis of the component (J/mg), Ef_s is the efficiency of synthesis (%), and D_c is the caloric density of the component (J/mg). A number of assumptions must be made to make such a calculation possible. The effect of alternative assumptions can then be evaluated. Table 3 gives a calculation of the energetic cost of growth for the case of *Pieris brassicae*. First, values for the energetic efficiencies of synthesis processes must be chosen. In Table 3, efficiency of protein synthesis was set at 80%, assuming that no turnover of the newly synthesized protein occurs. Synthesis of lipids from carbohydrate was assumed to have an efficiency of 78%. Synthesis of glycogen from glucose was assumed to be 90% efficient; chitin synthesis was taken to be 80% efficient. Second, values for D_c must be chosen. For the calculation in Table 3, the values used were: protein, 23 J/mg; lipid, 39 J/mg; and glycogen and chitin, 17.5 J/mg. Third, body composition must be known from chemical analysis. The respective values were taken from the literature and are given in Table 3. In this way, a reasonable estimate of synthesis costs during growth can be obtained.

A. MAINTENANCE EXPENDITURE

The contribution of maintenance to the total energetic cost in this calculation poses problems, as surprisingly few experimental data are available on caterpillars or any other herbivore group. This may be explained by the fact that many herbivores grow quickly and it is difficult to discern when they are in a state of pure maintenance metabolism.[45] In a number of cases, prolonged food deprivation has been employed to obtain an estimate of the respiration rate associated with maintenance metabolism. In caterpillars of *Manduca sexta* in the middle of their final instars, respiration rate declined to 30 to 38% of the initial value after 10 to 15 h of food deprivation.[119] For food-deprived final instars of *Bombyx mori* oxygen consumption dropped to 40% of the values for fed insects within 24 h.[14] In the case of *Manduca sexta*, moulting from the penultimate to the final instar decreased the oxygen uptake to 55% of that in the prior feeding phase.[119] Second instars of *Locusta migratoria* decreased

Table 3
QUANTIFICATION OF THE COST OF GROWTH OF A CATERPILLAR (*PIERIS BRASSICAE* IN ITS FINAL INSTAR) AS COMPRISED OF THE MAIN ENERGY REQUIRING ITEMS

Budget item	Content (mg/100 mg)	Cost[a] (J/mg)	Cost[b] (J)	Ref.
Synthesis				
Proteins	50			
Minus free amino acids[c]	45	5.75	259	42, 49, 106
Lipids	15			
Minus absorbed[d]	7.5	11	83	44
Carbohydrates	30			
Glycogen	10	2	20	20
Chitin[e]	5	4	20	33
Digestion			55	79
Maintenance			300	106, 107
Work (no estimate available)			(-)	
Total			737	
Observed			900[f]	106, 107
Percentage of observed costs unaccounted for			18%	

Note: Total cost of growth of 100 mg of dry matter with a calorific content of 23 J/mg dm was calculated on the basis of the efficiencies of synthesis of the main biochemical components of the body, according to Equation 14. See text for further explanation.

[a] Cost in Joules per milligram of component calculated according to Equation 14.
[b] Total cost per component, found by multiplying the content of the component with the cost in Joules per milligram.
[c] 10% of the total amino acid content of the body was assumed to occur in free form. This was based on the free amino acid level in the hemolymph.[42]
[d] 50% of the lipids accumulated at the end of the feeding phase on an artificial diet have been estimated to originate from synthesis.[44]
[e] Exact synthesis costs for chitin are unknown but were assumed to be twice those for glycogen.
[f] Mean from 7 experiments, each carried out on a group of 6—7 caterpillars feeding on an artificial diet.

their rate of oxygen uptake to approximately 60% after 12 h of fasting.[21] Basal metabolism of the final instar of the omnivorous *Acheta domesticus* was considered to be reached after 48 h of food deprivation and found to be about 50% of the peak values in the active feeding and growing phase of the instar.[73]

In the latter case, the level of respiration reached after food deprivation was identical to the level found in crickets that ceased feeding voluntarily at the end of the instar. These decreases occur notwithstanding an oft-observed increased proportion of time spent on movement with an increasing speed in nonrestricted insects deprived of food.[89,106] This inevitably leads to an overestimation of basal metabolic rate, which by definition requires the absence of movement. To prevent movement, confinement of the body has been applied but is no guarantee that muscular activity is absent, as a considerable degree of struggling will often occur.[45] Data on metabolism of voluntarily inactive herbivores are scarce.[18] Moreover, commonly used respirometric methods tend to underestimate undisturbed metabolic rates, an aspect which is discoursed on below. Another complication with food deprivation as a method to measure maintenance metabolism is that it also decreases body weight and changes body composition. In the data cited above, respiration rates were expressed on a fresh weight basis, but it was not stated if either the initial weight or the weight after the deprivation period was corrected for gut contents or loss of dry matter.

A second estimate of maintenance metabolism has been derived from respiration rates of young nondiapausing pupae that perform no external work. The time course of pupal respiration has often been described by a U-shaped curve.[45,62] Respiration rates for several lepidopterous species with fresh weights ranging from 100 to 700 mg at 23 to 25°C were at maximum just after the larval-pupal ecdysis, with a range between 0.4 to 1.0 μl/mg fw/h, while steady state levels ranged between 0.2 to 0.4 μl/mg fw/h.[6,31,45] For *Pieris brassicae*, the oxygen consumption rate just after pupation was 0.40 μl/mg fw/h,[31,105] which is 15 to 18% of values found for actively feeding caterpillars in the middle of their final instars.[105] It is obvious that values for newly formed pupae lead to lower estimates of maintenance costs than those for food-deprived larvae that are actively moving in most cases. On the basis of the data on respiration rate as such, it is not really possible to make a very precise estimate of maintenance expenditure due to the rapid changes of respiration rate during food deprivation or different developmental phases, which in turn are associated with changes in body composition such as lipid and protein content. Accumulation of lipids during the instar will cause a lower weight-specific respiration rate as compared to its start. It must be noted that an estimate of maintenance, as derived from a rate value per unit body weight, also depends quite strongly on an estimate of the mean dry body weight, the absolute value of which varies considerably with the calculation method.[36,48,110] The discrepancy between the estimation based on food deprivation vs. that based on prepupal respiration is large and does not yield an estimate that reflects actual maintenance metabolism in a representative manner, due to the fast and profound changes in body composition.

For these reasons a different procedure was employed for the purpose of the calculation in Table 3. In this procedure, maintenance metabolism is estimated from a linear regression analysis of heat production as a function of growth rate.[104] The intercept of the regression line (heat production at zero growth) is an estimate of heat produced due to maintenance. For *Pieris brassicae,* a significant linear relationship was obtained between both parameters (y = 300 + 6.1x, r^2 = 0.847, p = 0.003; heat production (J) is regressed on growth (milligrams of dry matter); mean results of seven groups of six caterpillars).[106]

The costs of digestion have been conservatively estimated by using a value of 0.33 J/mg of digested matter,[79] with 160 mg of dry matter digested. The latter value and the 100 mg dry matter of growth achieved together yield an ECD value of 62.5%. Part of the digestion costs, such as synthesis and secretion of digestive enzymes, may be listed under "maintenance costs".[104]

The item "work" comprises all external work done on the environment — transport of the body by muscular movement and the work done by muscles of the feeding apparatus. For caterpillars, no data are available on this component, making a realistic estimate difficult.

B. COMPONENT ANALYSIS OF ENERGY EXPENDITURE

Several conclusions can be drawn from the component analysis given in Table 3. The observed cost of growth of *Pieris brassicae* adds up to 82% of the cost of growth predicted on the basis of theoretical considerations. This means that a large proportion of the observed costs is a rather fixed amount, although its exact magnitude depends on several assumptions. It sets the lower limit of the cost of growth. The amount of 18% not accounted for (Table 3) must in part have been spent as muscular activity during movement and food intake and also covers all additional costs, e.g., when estimates would actually be too low. The remaining 18% in this particular example can be interpreted as the budget item causing actual variation in the cost of growth and will determine the upper limit. Its magnitude is expected to increase by suboptimal nutritional quality of the food or any other biotic or abiotic stress factors. When the nutritional quality of the food is poor, the insect will have to spend more energy to achieve the same growth rate as on an ideal diet, assuming it is capable of doing so, and there is *ad libitum* food available. This may be achieved by a compensatory increase of the rate of food intake,[93] which will cost more energy per unit time and will increase the absolute amount of the variable portion of the budget; however, even when the rate of food intake is twice as high, the increase in the total cost of growth is expected to be smaller than 20%, because it would maximally double the unaccounted for amount of 18% (Table 3).

The cost of protein synthesis (29% of the observed costs) constitutes a major budget item. In the calculation presented in Table 3, turnover of the newly synthesized protein was assumed to be absent. Insect larvae are known to synthesize and accumulate large amounts of storage proteins in their hemolymph and fat body;[51] however, data on turnover rates of these proteins are lacking in the literature. Assuming that all newly synthesized protein is turned over once during the final instar (decreasing the protein synthesis efficiency to 75%), it can be calculated that the percentage of observed costs unaccounted for decreases to <10%.

Maintenance expenditure contributes considerably to the observed apparent cost of growth (33%). Indeed, it is much greater than the value of 0.92 kJ needed to achieve a growth from 0.005 to 1 g dry matter in 20 d, which has been adopted as a minimum rate in a calculation of the maximum efficiency of growth of 82% in insect herbivores.[79] It follows that an important part of higher energetic costs associated with lower growth rates may be explained by an extended duration of development per se, which is expected to result in a greater contribution of maintenance energy to total energy expenditure. Another important aspect of maintenance expenditure, that seems self-evident but tends to be neglected, is the fact that insects are poikilotherms and are largely dependent on ambient temperature as the main determinant of their maximum metabolic rate. The usual experimental circumstances of constant temperature will thus determine an upper limit of metabolic rate. Since total metabolic expenditure is the product of metabolic rate and time, a major cause of decreased efficiency is predicted to be an extended duration of development, in turn caused by a reduced growth rate that results in a relative increase in maintenance expenditure.

Data from the few studies that combined repeated respirometric measurements with determinations of dry matter growth support the quantitative importance of maintenance expenditure as an item of the energy budget (Table 4). Thus, it appears that *Acheta domesticus* and *Locusta migratoria*, both hemimetabolous species, have distinctly lower growth rates than the three caterpillar species and clearly seem to invest more energy per unit of dry matter growth, which confirms a direct comparison that has been made in this respect.[10] Moreover, a comparative evaluation of these data shows that the values reflecting the cost of growth, here defined as the ratio between heat production and growth, are restricted to a surprisingly narrow range (7.7 to 19.9 J/mg). This is certainly remarkable in view of the considerable differences in total amount of growth achieved, growth rate, body size, and lifestyle for these five species. Within any one species, the phase of fastest growth coincided with the highest respiration rates, demonstrating the energy demand of anabolism.[73,105,106,117,118] This is why one would expect a positive correlation between RGR and RMR, which is absent in the *Spodoptera* data (Figure 4).

Emerging from these physiological and biochemical considerations is the unexpected large degree of variation in energetic efficiency reported for herbivorous insects based on gravimetric measurements, especially when growth rates were not particularly (if at all) negatively affected.

C. METABOLIC EFFICIENCY AND GROWTH RATE: CAUSE - CONSEQUENCE PROBLEMS

The foregoing conclusions are relevant with regard to a common explanation of decreased utilization efficiencies, which could be called the "metabolic load" hypothesis: so-called increased processing costs (to be distinguished from maintenance costs sensu stricto) are a direct cause of lower growth rates, suggesting a trade-off between the two. This idea has been repeated without proper experimental evidence.[70,78,84,86,87,98] Induction of the mixed function oxidase (MFO) enzyme system in response to allelochemicals in the food has been one of the supposedly more important energy requiring processes in the "metabolic load" hypothesis; however, an actual experiment established in order to quantify the cost of this induction gravimetrically led to the conclusion that no increased costs could be demonstrated, and indeed would be unexpected in view of the amount of enzymic protein synthesis involved.[60]

Very few data are available on direct measurements of metabolism as affected by dietary quality; however, the available data cited in Table 4 suggest that the cost of growth in the presence of the flavonoid allelochemical quercetin (*Pieris* spp.) or under conditions of water stress (*Manduca sexta*) determined by respirometry differ little from the control situation.

In relating consumption or growth rates to gravimetrically determined nutritional indices, inverse relationships have commonly been reported.[86,87,98] Such relationships do not allow a distinction of cause and consequence, however; i.e., is growth rate reduced because of a lower metabolic efficiency or is metabolic efficiency reduced due to a lower growth rate? The growth rate of herbivorous insects is assumed to be nutrient-limited rather than limited by energy constraints;[81,96,109,114] thus, a suboptimal availability of a limiting nutrient, often nitrogen or water, reduces growth rate, increases maintenance costs, and initiates a lower metabolic efficiency. The availability of the limiting nutrient may be increased by a compensatory feeding response that by now is known to be well developed in several phytophagous species (reviewed recently in this series[93]). Such compensatory behavior counteracts a reduced growth rate when the added costs of a higher feeding rate are smaller than increased maintenance costs.[17,57] Unfortunately, very few data are available about the energetic demands of feeding activity, but those data suggest that the costs are small.[2,57] This is also supported by the component analysis in Table 3.

Another problem associated with the inverse relationships between consumption rate and ECI was alluded to in the foregoing discussion on sources of error in the measurement of budget parameters. Recalling the example of *Spodoptera eridania*, growth often varies within rather narrow limits, while consumption shows a much greater variability. Evidently, as the calculation of ECI contains consumption as the quantitatively larger term (Equation 11), errors in the estimation of consumption will create a variability range in which a negative correlation with ECI is not at all surprising. The commonly observed inverse relationship between AD and ECD likewise is probably a function of methodological problems.[93]

IX. THE PHYSIOLOGICAL CONDITION OF THE PLANT IN UTILIZATION STUDIES

The gravimetric approach necessitates the excision of leaf material to ensure an acceptable level of measurement precision. The few data that are available, however, point to significant differences in feeding and growth rates of insects fed on intact or excised plant tissue for locusts,[56] beetles,[69] and caterpillars.[80] A great advantage of flow-through respirometry is the possibility of studying herbivore food utilization by insects feeding on intact plants. This is a crucial point when the ecological relevance of the utilization data obtained is an issue,[28,116] but has received very little attention in the literature. Excision of leaves has often been performed without subsequent supply of water to the leaves, leading to a loss of leaf turgor,[83] and this will have negatively influenced or prevented any photosynthesis. Moreover, light intensity is hardly (if ever) specified; therefore it is no exaggeration to state that no single utilization study

Table 4
COSTS OF GROWTH AS CALCULATED FOR FIVE INSECT SPECIES IN THEIR FINAL LARVAL STAGES

Species	Diet	Duration (h)	G (mg)	RGR (mg/mg/d)	H (J)	H/G (J/mg)	Ref.
Pieris brassicae	Control	90	96	0.713	864	9.00	106
	Quercetin, 1.0 m*M*	90	67	0.582	515	7.69	
Pieris rapae	Control	90	34	0.652	291	8.56	106
	Quercetin, 2.5 m*M*	90	31	0.617	232	7.49	
Manduca sexta	Water, 82%	67	14.5[a]	0.534	188	13.00	109
	Water, 65%	83	23.7[a]	0.577	311	13.10	
Acheta domestica	Artificial diet	144[b]	58	0.176	1155	19.9	73
Locusta migratoria	–[c]	240[d]	258	0.124	4536	17.6	21

Note: Continuous or repeated respirometric data as well as gravimetric growth data are available. For the Caterpillar species, the effects of an alternative diet that affected growth rate is also given. G: growth(mg dry matter) R: GR: relative growth rate, H: heat production. For both *Pieris* species, heat production was calculated according to Brouwer's formula,[59] for the other species it was obtained by multiplying oxygen consumption (ml/larva/instar) with a caloric equivalent of 21 J/ml. H/G signifies the amount of heat produced per unit of growth. Experiments were performed at 25°C, except with *Locusta migratoria*, which was studied at 28°C.

[a] Data on dry weight growth were not given by the authors in the same paper with the respirometric data, but could be obtained by combining data from van't Hof and Martin.[54,109]

[b] Data for the active feeding phase (days 2 to 7) were taken.

[c] Diet not stated, but most probably wheat.[90]

[d] Duration taken from Simpson.[90]

has been documented that is representative for a herbivorous insect feeding on a plant under field conditions. It is unknown whether this leads to an underestimation of actual food utilization under natural foraging conditions or the reverse for a particular insect-plant combination. The suboptimal physiological condition of the plant may influence its nutritional quality negatively or positively because of the many and varied changes that take place in foliage after excision.

The accumulated evidence on changes in plant gas exchange in response to herbivore feeding was reviewed earlier in this series.[113] The time course of induced defenses[7,29,38] may be followed both phytochemically and by respirometry, if putative metabolic consequences of these induced responses occur, and these can be related to associated changes in herbivore metabolism over time. The same applies to plant responses to wounding, which have been estimated to increase plant maintenance respiration by a factor of 2 to 4 (see References 50, 103, and 107), although some plant species seem to be little affected.[15] The usefulness of the metabolic approach in these largely unexploited but relevant areas seems evident. The fact that very few utilization values, if any, have been determined under circumstances representative of field situations, has not been adequately recognized in the field of quantitative ecological energetics. The extent to which realistic values would change calculated flows at an ecosystem level can only be speculated upon.[79,114]

X. CONCLUSIONS AND RECOMMENDATIONS

Since Waldbauer[110] formalized the foundations of the gravimetric technique for studies on insects, both ecologists and physiologists studying nutritional aspects of insect-plant relationships have been using this technique as a tool for the quantification of the utilization efficiency of matter and energy of herbivorous insects. Although Waldbauer also pointed to several sources of error inherent to gravimetry, the quantification of these errors by means of sensitivity analyses has been neglected by most workers in the field, with a few exceptions.[5,75] A basic methodological requirement, i.e., a check of the gravimetric budget via an independent method has not been met and was only attempted recently.[105,106] Relevant work concerning discrepancies found in the attempt to perform such an independent check in the related field of ecological energetics[1,58,115,116] has taken place without receiving recognition from students of insect-plant relationships. While the considerable differences in gravimetric utilization efficiencies found for different feeding guilds, such as tree vs. herb feeding herbivores, may be assumed to represent actual differences in view of the major differences in nutritional quality of these plant categories, the reliability of gravimetry to measure more subtle differences in utilization efficiency is questionable.[75] The accurate measurement of the largest budget item, consumption, is especially problematic because the food plant is a living organism, characterized by temporal and spatial changes. This is reflected in a large calculated variability in energy expenditure, which is unrealistic from a physiological point of view. Errors in the measurement of consumption depend on a number of parameters that have been reported in few utilization studies, making it difficult to evaluate the reliability of published data; therefore, it is recommended that in the future, these parameters should be documented. Important examples of such parameters are: percentage of food consumed relative to the amount offered, duration of feeding periods and total duration of experiment, respiration rates of plant tissues, homogeneity and dynamics of contents of plant dry matter, and of nutritive elements or compounds, where applicable. Possible measures that have been employed to avoid errors should be mentioned explicitly. The variability in dry matter content should be given and its consequences on the variability of nutritional indices should be assessed by means of a sensitivity analysis.

Apart from these methodological considerations, the excised condition of plant tissues in gravimetric studies is an inherent biological drawback. The growing body of data on plant responses, induced by insect feeding (such as induced defenses), wound responses or considerable changes in photosynthesis and respiration must lead one to reconsider the appropriateness of the gravimetric technique in accounting for the dynamics of the insect-plant interaction.

This chapter explicitly advocates that respirometry is a real and necessary alternative for gravimetry as a method to properly measure the efficiency of food utilization and growth. Continuous flow-through respirometry has turned out to be feasible on individual insects[105] as well as on groups and allows a theoretically well-founded estimation of heat loss by the herbivore and also the calculation of the heat production due to growth independent of an estimate of food consumption. The data obtained using respirometry indicate a distinctly smaller variability in the cost of growth than is suggested by gravimetric data. This furnishes a reason to reevaluate current hypotheses concerning trade-offs between growth and energy expenditure. Respirometry allows the monitoring of time patterns in the metabolic rate, allowing for the dynamics of the interaction between herbivore and host plant.[113] The application of respirometric methods in the study of herbivore performance on intact plants will allow important improvements and may open up interesting perspectives in the field of insect-plant interactions.

ACKNOWLEDGMENTS

I would like to thank Aren van Es, my guide in the field of animal energetics, for his invaluable advice and encouragement which he gave so willingly since my interest in the subject emerged. His suggestions and lucid comments have been stimulating and supportive. I thank Liz Bernays for challenging me to put respirometry into perspective. I also thank her and a reviewer for their helpful comments on earlier drafts of the manuscript. Thanks are due Ray Barbehenn for allowing me to cite unpublished data and Steve Simpson for giving me access to an unpublished review.

REFERENCES

1. **Ågren, G. I. and Axelsson, B.**, Energy budgets do balance - a comment on a paper by Wightman and Rogers, *Oecologia,* 42, 375, 1979.
2. **Aidley, D. J.**, Increase in respiratory rate during feeding in larvae of the armyworm, *Spodoptera exempta, Physiol. Entomol.,* 1, 73, 1976.
3. **Armstrong, D. G.**, Cell bioenergetics and energy metabolism, in *Handbuch der Tierernährung,* Vol. 1, Lenkeit, W., Breirem, K., and Grasemann, E., Eds., Parey, Hamburg, 1969, 385.
4. **Armstrong, G. and Mordue, W.**, Oxygen consumption of flying locusts, *Physiol. Entomol.,* 10, 353, 1985.
5. **Axelsson, B. and Ågren, G. I.**, A correction for food respiration in balancing energy budgets, *Entomol. Exp. Appl.,* 25, 260, 1979.
6. **Bailey, C. G. and Singh, N. B.**, An energy budget for *Mamestra configurata* (Lepidoptera, Noctuidae), *Can. Entomol.*, 109, 687, 1977.
7. **Baldwin, I. T. and Schultz, J. C.,** Rapid changes in leaf chemistry induced by damage: evidence for communication between plants, *Science,* 221, 277, 1983.
8. **Barbehenn, R.,** The use of silicon as an indigestible marker for measuring consumption and digestion by insects, unpublished results.
9. **Berenbaum, M.,** Postingestive effects of phytochemicals on insects. On Paracelsus and plant products, in *Insect-Plant Interactions,* Miller, J. R. and Miller, T. A., Eds., Springer-Verlag, New York, 1986, 121.
10. **Bernays, E. A.,** Evolutionary contrasts in insects: nutritional advantages of holometabolous development, *Physiol. Entomol.,* 11, 377, 1986.
11. **Bernays, E. A.,** personal communication, 1990.
12. **Bhattacharya, A. K. and Waldbauer, G. P.**, The effect of diet on the nitrogenous end products excreted by larval *Tribolium confusum*, with notes on the correction of A.D. and E.C.D. for fecal urine, *Entomol. Exp. Appl.,* 15, 238, 1972.
13. **Blaney, W. M. and Chapman, R. F.**, How animals perceive secondary compounds, in *Herbivores. Their Interactions with Secondary Plant Metabolites*, Rosenthal, G. A. and Janzen, D. H., Eds., Academic Press, New York, 1979, 161.
14. **Bosquet, G.**, Glycine incorporation during starvation in *Bombyx mori.* Relation to respiratory metabolism. *J. Insect Physiol.,* 22, 541, 1976.
15. **Bowers, M. D., Stamp, N. E., and Fajer, E. D.,** Factors affecting calculation of nutritional indices for foliage-feeding insects, an experimental approach, in press.
16. **Buscarlet, L. A., Proux, J., and Gester, R.,** Utilization du double marguage $HT^{18}O$ dans un étude du bilan métabolique chez *Locusta migrandia migratoriodes, J. Insect Physiol.,* 24, 225, 1978.
17. **Calow, P.**, Ecology, evolution and energetics, a study in metabolic adaptation, *Adv. Ecol. Res.,* 10, 1, 1977.
18. **Casey, T. M. and Knapp, R.,** Caterpillar thermal adaptation: behavioral differences reflect metabolic thermal sensitivities. *Comp. Biochem. Physiol.,* 86A, 679, 1987.
19. **Chapman, R. F.,** Coordination of digestion, in *Comprehensive Insect Physiology, Biochemistry and Pharmacology,* Vol. 4, Kerkut, G. A. and Gilbert, L. I., Eds., Pergamon Press, New York, 1985, 213.

20. **Chippendale, G. M. and Kilby, B. A.**, Relationship between the proteins of the haemolymph and fat body during development of *Pieris brassicae*, *J. Insect Physiol.*, 15, 905, 1969.
21. **Clarke, K. U.**, The relationship of oxygen consumption to age and weight during the post-embryonic growth of *Locusta migratoria* L., *J. Exp. Biol.*, 34, 29, 1957.
22. **Cohen, A. C. and Patana, R.**, Efficiency of food utilization by *Heliothis zea* (Lepidoptera, Noctuidae) fed artificial diets or green beans, *Can. Entomol.*, 139, 1984.
23. **Cohen, R. W., Waldbauer, G. P., and Friedman, S.**, Natural diets and self-selection: *Heliothis zea* larvae and maize, *Entomol. Exp. Appl.*, 46, 161, 1988.
24. **Cooper, P. D.**, Validation of doubly labeled water ($H^3H^{18}O$) method for measuring water flux and energy metabolism in tenebrionid beetles, *Physiol. Zool.*, 56, 41, 1983.
25. **Culik, B. M. and McQueen, D. J.**, Monitoring respiration and activity in the spider *Geolycosa domifex* (Hancock) using time-lapse television and CO_2 gas analysis, *Can. J. Zool.*, 63, 843, 1985.
26. **Dadd, R. H.**, Nutrition: organisms, in *Comprehensive Insect Physiology, Biochemistry and Pharmacology*, Vol. 4, Kerkut, G. A. and Gilbert, L. A., Eds., Pergamon Press, New York, 1985, 313.
27. **Dethier, V. G.**, Mechanism of host-plant recognition, *Entomol. Exp. Appl.*, 31, 49, 1982.
28. **Edwards, P. B. and Wightman, J. A.**, Energy and nitrogen budgets for larval and adult *Paropsis charybdis* (Coleoptera, Chrysomelidae) feeding on *Eucalyptus viminalis*, *Oecologia*, 61, 302, 1984.
29. **Edwards, P. J., Wratten, S. D., and Cox, H.**, Wound-induced changes in the acceptability of tomato to larvae of *Spodoptera littoralis*, a laboratory bioassay, *Ecol. Entomol.*, 10, 155, 1986.
30. **Erickson, J. M.**, The utilization of various *Asclepias* species by larvae of the monarch butterfly, *Danaus plexippus*, *Psyche*, 80, 230, 1973.
31. **Fourche, J., Guillet, C., Calvez, B., and Bosquet, G.**, Le métabolisme énergétique des nymphes de *Pieris brassicae* (Lépidoptères) au cours de la métamorphose et de la diapause. Essai d'établissement d'un bilan, *Ann. Zool. Ecol. Anim.*, 9, 51, 1977.
32. **Fraenkel, G. S.**, The raison d'être of secondary plant substances, *Science*, 129, 1466, 1959.
33. **Gijswijt, M. J., Deul, D. H., and de Jong, B. J.**, Inhibition of chitin synthesis by benzoyl-phenylurea insecticides. III. Similarity in action in *Pieris brassicae* L. with polyoxin D, *Pest. Biochem. Physiol.*, 12, 87, 1979.
34. **Gilson, W. E.**, Differential respirometer of simplified and improved design, *Science*, 141, 531, 1963.
35. **Gordon, H. T.**, Interpretations of insect quantitative nutrition, in *Insect and Mite Nutrition*, Rodriguez, J. G., Ed., North-Holland, Amsterdam, 1972, 73.
36. **Gordon, H. T.**, Growth and development of insects, in *Ecological Entomology*, Huffaker, C. B. and Rabb, R. L., Eds., John Wiley & Sons, New York, 1984, 53.
37. **Hargrove, W. W.**, A photographic technique for tracking herbivory on individual leaves through time, *Ecol. Entomol.*, 13, 359, 1988.
38. **Haukioja, E. and Neuvonen, S.**, Insect population and induction of plant resistance: the testing of hypotheses, in *Insect Outbreaks*, Barbosa, P. and Schultz, J., Eds., Academic Press, New York, 1987, 411.
39. **Horie, Y. and Watanabe, K.**, Ef-

fect of various kinds of dietary protein and supplementation with limiting amino acids on growth, haemolymph components and uric acid excretion in the silkworm, *Bombyx mori*, *J. Insect Physiol.*, 29, 187, 1983.

40. **Ishaaya, I.,** Nutritional and allelochemic insect-plant interactions relating to digestion and food intake: some examples, in *Insect-Plant Interactions*, Miller J. R. and Miller, T. A., Eds., Springer-Verlag, New York, 1986, 191.
41. **Jermy, T., Hanson, F. E., and Dethier, V. G.,** Induction of specific preference in lepidopterous larvae, *Entomol. Exp. Appl.*, 11, 211, 1968.
42. **Junnikkala, E.**, Effect of a semisynthetic diet on the level of the main nitrogenous compounds in the haemolymph of larvae of *Pieris brassicae* L., *Ann. Acad. Sci. Fenn. A: IV,* 155, 1, 1969.
43. **Karowe, D. N. and Martin, M. M.,** The effects of quantity and quality of diet nitrogen on the growth, efficiency of food utilization, nitrogen budget and metabolic rate of fifth instar *Spodoptera eridania* larvae (Lepidoptera, Noctuidae), *J. Insect Physiol.*, 35, 699, 1989.
44. **Kastari, T. and Turunen, S.**, Lipid utilization in *Pieris brassicae* reared on meridic and natural diets: implications for dietary improvement, *Entomol. Exp. Appl.*, 22, 71, 1977.
45. **Keister, M. and Buck, J.,** Respiration. Some exogenous and endogenous effects on rate of respiration, in *The Physiology of Insecta*, Vol. 6, 2nd ed., Rockstein, M., Ed., Academic Press, New York, 1974, 469.
46. **Kennedy, J. S. and Booth, C. O.,** Host alternation in *Aphis fabae* Scop. I. Feeding preferences and fecundity in relation to the age and kind of leaves, *Ann. Appl. Biol.* 38, 25, 1951.
47. **Kleiber, M.,** *The Fire of Life*, Wiley, New York, 1961.
48. **Kogan, M.,** Bioassays for measuring quality of insect food, in *Insect-Plant Interactions*, Miller, J. R. and Miller, T. A., Springer-Verlag, New York, 1986, 155.
49. **Lafont, R., Mauchamp, B., Boulay, G., and Tarroux, P.**, Developmental studies in *Pieris brassicae* (Lepidoptera). I. Growth of various tissues during the last larval instar, *Comp. Biochem. Physiol.*, 51B, 439, 1975.
50. **Lambers, H.,** Respiration in intact plants and tissues: its regulation and dependence on environmental factors, metabolism and invaded organisms, in *Encyclopedia of Plant Physiology, New Series*, Vol. 18, Duce, R. and Day, D. A., Eds., Springer-Verlag, Berlin, 1985, 418.
51. **Levenbook, L.**, Insect storage proteins, in *Comprehensive Insect Physiology, Biochemistry and Pharmacology,* Vol. 10, Kerkut, G. A. and Gilbert, L. I., Eds., Pergamon Press, New York, 1985, 307.
52. **Llewellyn, M.,** Aphid energy budgets, in *Aphids. Their Biology, Natural Enemies and Control*, *World Crop Pests*, Vol. 2B, Minks, A. K. and Harrewijn, P., Eds., Elsevier, Amsterdam, 1988, 109.
53. **Manuwoto, S. and Scriber, J. M.,** Consumption and utilization of three maize genotypes by the southern armyworm, *J. Econ. Entomol.,* 75, 163, 1982.
54. **Martin, M. M. and Van't Hof, H. M.,** The cause of reduced growth of *Manduca sexta* larvae on a low-water diet: increased metabolic processing costs or nutrient limitation?, *J. Insect Physiol,.* 34, 515, 1988.
55. **Mattson, W. J., Jr.,** Herbivory in relation to plant nitrogen content, *Ann. Rev. Ecol. Syst.,* 11, 119, 1980.
56. **McCaffery, R.,** Difference in the acceptability of excised and grow-

ing cassava leaves to *Zonocerus variegatus*, *Entomol. Exp. Appl.*, 32, 111, 1982.

57. **McEvoy, P. B.,** Increase in respiratory rate during feeding in larvae of the cinnabar moth *Tyria jacobaeae*, *Physiol. Entomol.*, 9, 191, 1984.
58. **McEvoy, P. B.,** Balancing insect energy budgets, *Oecologia*, 66, 154, 1985.
59. **McLean, J. A. and Tobin, G.,** *Animal and Human Calorimetry*, Cambridge University Press, Cambridge, 1987.
60. **Neal, J. J.,** Metabolic costs of mixed-function oxidase induction in *Heliothis zea*, *Entomol. Exp. Appl.*, 43, 175, 1987.
61. **Nicolas, G. and Sillans, D.,** Immediate and latent effects of carbon dioxide on insects, *Ann. Rev. Entomol.*, 34, 97, 1989.
62. **Peakin, G. J.,** The measurements of the costs of maintenance in terrestrial poikilotherms, a comparison between respirometry and calorimetry, *Experientia*, 29, 801, 1973.
63. **Parra, J. R. P. and Kogan, M.,** Comparative analysis of methods for measurements of food intake and utilization using the soybean looper, *Pseudoplusia includens* and artificial media, *Entomol. Exp. Appl.*, 30, 45, 1981.
64. **Penning de Vries, F. W. T.,** The cost of maintenance processes in plant cells, *Ann. Bot.*, 39, 77, 1975.
65. **Petrusewicz, K. and MacFayden, A.,** *Productivity of Terrestrial Animals: Principles and Methods*. IBP Handb. 13, Blackwell Scientific, Oxford, 1970.
66. **Phillipson, J.,** A miniature bomb calorimeter for small biological samples, *Oikos*, 15, 130, 1964.
67. **Punt, A.,** The respiration of insects, *Physiol. Compar.*, 2, 59, 1950.
68. **Punt, A.,** Further investigations on the respiration of insects, *Physiol. Compar.*, 4, 212, 1956.
69. **Raina, A. K., Benepal, P. S., and Sheikh, A. Q.,** Effects of excised and intact leaf methods, leaf size, and plant age on Mexican bean beetle feeding, *Entomol. Exp. Appl.*, 27, 303, 1980.
70. **Rausher, M. D.,** Population differentiation in *Euphydras editha* butterflies, larval adaptation to different hosts, *Evolution*, 36, 581, 1982.
71. **Reese, J. C.,** Interactions of allelochemicals with nutrients in herbivore food, in *Herbivores. Their Interactions with Secondary Plant Metabolites*, Rosenthal, G. A. and Janzen, D. H., Eds., Academic Press, New York, 1979, 309.
72. **Reynolds, S. E., Nottingham, S. F., and Stephens, A. E.,** Food and water economy and its relation to growth in fifth-instar larvae of the tobacco hornworm, Manduca sexta, *J. Insect Physiol.*, 31, 119, 1985.
73. **Roe, R. M., Clifford, C. W., and Woodring, J. P.,** The effect of temperature on feeding, growth, and metabolism during the last larval stadium of the female house cricket, *Acheta domesticus*, *J. Insect Physiol.*, 26, 639, 1980.
74. **Schiff, N. M., Waldbauer, G. P., and Friedman, S.,** Dietary self-selection by *Heliothis zea* larvae: roles of metabolic feedback and chemosensory stimuli, *Entomol. Exp. Appl.*, 52, 261, 1989.
75. **Schmidt, D. J. and Reese, J. C.,** Sources of error in nutritional index studies of insects on artificial diet, *J. Insect Physiol.*, 32, 193, 1986.
76. **Schoeller, D. A.,** Measurement of energy expenditure in free-living humans by using doubly labeled water, *J. Nutr.*, 118, 1278, 1988.
77. **Schoonhoven, L. M.,** What makes a caterpillar eat? The sensory code underlying feeding behavior, in *Advances in Chemore-*

ception and Behaviour, Chapman, R. F., Bernays, E. A., and Stoffolano, J. G., Eds., Springer-Verlag, New York, 1987, 69.

78. **Schoonhoven, L. M. and Meerman, J.,** Metabolic costs of changes in diet and neutralization of allelochemics, *Entomol. Exp. Appl.,* 24, 689, 1978.
79. **Schroeder, L. A.,** Consumer growth efficiencies, their limits and relationships to ecological energetics, *J. Theor. Biol.,* 93, 805, 1981.
80. **Schroeder, L. A.,** Comparison of gravimetry and planimetry in determining dry matter budgets for three species of phytophagous lepidopteran larvae, *Entomol. Exp. Appl.,* 35, 255, 1984.
81. **Schroeder, L. A.,** Protein limitation of a tree leaf feeding Lepidopteran, *Entomol. Exp. Appl.,* 41, 115, 1986.
82. **Scriber, J. M.,** Cyanogenic glycosides in *Lotus corniculatus.* Their effect upon growth, energy budget, and nitrogen utilization of the southern armyworm, *Spodoptera eridania, Oecologia,* 34, 143, 1978.
83. **Scriber, J. M.,** Effects of leaf-water supplementation upon post-ingestive nutritional indices of forb-, shrub-, vine-, and tree-feeding Lepidoptera, *Entomol. Exp. Appl.,* 25, 240, 1979.
84. **Scriber, J. M.,** Sequential diets, metabolic costs, and growth of *Spodoptera eridania* (Lepidoptera: Noctuidae) feeding upon dill, lima bean and cabbage, *Oecologia,* 51, 175, 1981.
85. **Scriber, J. M.,** The behavior and nutritional physiology of southern armyworm larvae as a function of plant species consumed in earlier instars, *Entomol. Exp. Appl.,* 31, 359, 1982.
86. **Scriber, J. M.,** Host-plant suitability, in *Chemical Ecology of Insects,* Bell, W. J. and Cardé, R. T., Eds., Chapman & Hall, London, 1984, 159.
87. **Scriber, J. M. and Slansky, F., Jr.,** The nutritional ecology of immature insects, *Ann. Rev. Entomol.,* 26, 183, 1981.
88. **Sell, C. R., Weiss, M. A., Moffitt, H. R., and Burditt, A. K., Jr.,** An automated technique for monitoring carbon dioxide respired by insects, *Physiol. Entomol.,* 10, 317, 1985.
89. **Siegert, K. and Ziegler, R.,** Sauerstoffverbrauch und respiratorische Quotienten bei *Manduca sexta* (Lepidoptera: Sphingidae), *Verh. Dtsch. Zool. Ges.,* 1982, 332.
90. **Simpson, S. J.,** Changes in the efficiency of utilisation of food throughout the fifth-instar nymphs of *Locusta migratoria, Entomol. Exp. Appl.,* 31, 265, 1982.
91. **Simpson, S. J. and Abisgold, J. D.,** Compensation by locusts for changes in dietary nutrients: behavioural mechanisms, *Physiol. Entomol.,* 10, 443, 1985.
92. **Simpson, S. J., Simmonds, M. S. J., and Blaney, W. M.,** A comparison of dietary selection behaviour in larval *Locusta migratoria* and *Spodoptera littoralis, Physiol. Entomol.,* 13, 225, 1988.
93. **Simpson, S. J. and Simpson, C. L.,** The mechanisms of nutritional compensation by phytophagous insects, in *Insect-Plant Interactions,* Vol. 1, Bernays, E. A., Ed., CRC Press, Boca Raton, FL, 1989, III.
94. **Slansky, F.,** Insect nutrition, an adaptationist's perspective, *Fla. Entomol.,* 65, 45, 1982.
95. **Slansky, F., Jr.,** Food utilization by insects. Interpretation of observed differences between dry weight and energy efficiencies, *Entomol. Exp. Appl.,* 39, 47, 1985.
96. **Slansky, F., Jr. and Feeny, P.,** Stabilization of the rate of nitrogen

accumulation by larvae of the cabbage butterfly on wild and cultivated food plants, *Ecol. Monogr.*, 47, 209, 1977.

97. **Slansky, F., Jr. and Rodriguez, J. G.,** Eds., *Nutritional Ecology of Insects, Mites, Spiders and Related Invertebrates*, John Wiley & Sons, New York, 1987.
98. **Slansky, F., Jr. and Scriber, J. M.,** Food consumption and utilization, in *Comprehensive Insect Physiology, Biochemistry and Pharmacology,* Vol. 4, Kerkut, G. A. and Gilbert, L. I., Eds., Pergamon Press, New York, 1985, 87.
99. **Soo Hoo, C. F. and Fraenkel, G.,** The consumption, digestion and utilization of food plants by a polyphagous insect, *Prodenia eridania* (Cramer), *J. Insect Physiol.*, 12, 711, 1966.
100. **Southwood, T. R. E.,** *Ecological Methods,* 2nd ed., Chapman & Hall, London, 1978, 456.
101. **Stroshine, R. and Martins, J.,** Varietal differences in high temperature drying of maize in the Midwestern United States, in *Drying '86* Vol. 2, Mujumdar, A. S., Ed., Hemisphere Publishing, Washington, D.C., 1986, 470.
102. **Taylor, W. E. and Bardner, R.,** Energy relationships between larvae of *Phaedon cochleariae* (Coleoptera: Chrysomelidae) or *Plutella maculipennis* (Lepidoptera: Plutellidae) and radish or turnip plants, *Entomol. Exp. Appl.,* 13, 403, 1970.
103. **Uritani, I. and Asahi, T.,** Respiration and related metabolic activity in wounded and infected tissues, in *The Biochemistry of Plants*, Vol. 2, Stumpf, P. K. and Conn, E. E., Eds., Academic Press, London, 1980, 463.
104. **Van Es, A. H.,** Maintenance, in *Handbuch der Tierernährung,* Vol. 2, Lenkeit, W. and Breirem, K., Eds., Parey, Hamburg, 1972, 1.
105. **Van Loon, J. J. A.,** A flow-through respirometer for leaf chewing insects, *Entomol. Exp. Appl.,* 49, 265, 1988.
106. **Van Loon, J. J. A.,** Sensory and Nutritional Effects of Amino Acids and Phenolic Plant Compounds on the Caterpillars of Two *Pieris* Species, Thesis, Agricultural University Wageningen, 1988.
107. **Van Loon, J. J. A.,** unpublished data, 1988.
108. **Van Soest, P. J.,** *Nutritional Ecology of the Ruminant,* O&B Books, Corvallis, OR, 1983, 39.
109. **Van't Hof, H. M. and Martin, M. M.,** The effect of diet water content on energy expenditure by third-instar *Manduca sexta* larvae (Lepidoptera, Sphingidae), *J. Insect Physiol.,* 35, 433, 1989.
110. **Waldbauer, G. P.,** The consumption and utilization of food by insects, *Adv. Insect Physiol.,* 5, 229, 1968.
111. **Waller, D. A. and Jones, C. G.,** Measuring herbivory, *Ecol. Entomol.,* 14, 479, 1989.
112. **Weinstein, L. H. and Porter, C. A.,** Changes in free amino acid and amide levels of leaf pieces, detached leaves, detached plants and intact plants of tobacco at different times, *Contrib. Boyce Thompson Inst.,* 21, 387, 1962.
113. **Welter, S. C.,** Arthropod impact on plant gas exchange, in *Insect-Plant Interactions,* Vol. 1, Bernays, E. A., Ed., CRC Press, Boca Raton, 1989, 136.
114. **Wiegert, R. G. and Petersen, C. E.,** Energy transfer in insects, *Ann. Rev. Entomol.,* 28, 455, 1983.
115. **Wightman, J. A.,** Respirometry techniques for terrestrial invertebrates and their application to energetics studies, *N. Z. J. Zool.,* 4, 453, 1977, 107.
116. **Wightman, J. A.,** Why insect energy budgets do not balance, *Oecologia,* 550, 166, 1981.

117. **Woodring, J. P., Roe, R. M., and Clifford, C. W.,** Relation of feeding, growth, and metabolism to age in the larval, female house cricket, *J. Insect Physiol.,* 23, 207, 1977.
118. **Woodring, J. P., Clifford, C. W., and Beckman, B. R.,** Food utilization and metabolic efficiency in larval and adult house crickets, *J. Insect Physiol.,* 25, 903, 1979.
119. **Ziegler, R.,** Metabolic energy expenditure and its hormonal regulation, in *Environmental Physiology and Biochemistry of Insects,* Hoffman, K. H., Ed., Springer-Verlag, Berlin, 1984, 95.

4 The Influence of Plant Chemistry on Aphid Feeding Behavior

Clytia B. Montllor
Department of Entomology
University of California, Berkeley
Albany, California

TABLE OF CONTENTS

I. INTRODUCTION

An understanding of aphid-host plant relationships requires some fundamentally different techniques from those used for chewing insects because of the selective way in which aphids feed. For both chewing and sap-feeding insects close observations of behavior are needed to find out which host factors are relevant to the interaction. Detailed observations of feeding behavior are more difficult with sap-feeders, e.g., aphids, however, because these insects are essentially internal feeders. Probing, i.e., inserting, the stylets into plant tissue, does not necessarily imply ingestion, although it may include it; both phenomena are critical components of an aphid's relationship to the plant on which it finds itself. Moreover, research in this area is hindered by the difficulty of determining the chemical composition not only of aphids' food (predominantly phloem), but of the tissues encountered on the way to the vascular tissue. Because of their specialized mode of feeding, then, it is important to understand and to experimentally verify precisely which tissues are encountered in probing, as well as how the chemical components encountered during aphids' probes affect their behavior and physiology. To some extent, the relatively large review literature on aphids and their host plants obscures the fact that there are enormous gaps in our knowledge of some fundamental facts precisely in these areas. Table 1 summarizes the many levels at which chemical information may be available to probing aphids.

Earlier reviews have discussed aphid probing[96,97,142,177,208,216] and the importance of plant chemicals to aphids;[46,47,72,142,222,223] an ecological perspective is given by Risebrow and Dixon.[182] This chapter is an attempt to integrate and update what is known about aphid feeding behavior (and, where appropriate, performance) as it is affected by plant chemistry.

II. THE FOOD OF APHIDS

It is well known that the primary food of aphids is the phloem sap. The predominace of ingestion from phloem during prolonged feeding is attested to by the sugary nature of honeydew, by histological examinations of probed tissues, and by electronic monitoring (in conjunction with histology; see Section IV). It has been suggested that some aphids may have primary feeding sites other than phloem. For example, histological studies of barley probed for 48 h by *Schizaphis graminum* biotype B[188] and of beans probed 24 to 48 h by *Myzus ornatus*[113] showed a majority of stylet sheaths ending in mesophyll. Both studies concluded that mesophyll tissue was the preferred food source for these aphids on these apparently suitable hosts. However, the critical measurement, i.e., the time spent ingesting from mesophyll compared to phloem, could not be compared, and in one of these studies,[113] only four mesophyll cells in over 400 tissue sections appeared to have been penetrated.

Several authors have included a measurement of "mesophyll ingestion" or "nonphloem ingestion" (NPI) in their feeding behavior studies. Measurements of so-called NPI, determined electronically (see Section IV.C),[126] have been

Table 1
LEVELS AT WHICH PLANT CHEMISTRY MAY AFFECT PROBING APHIDS

Plant surface
Detection of surface chemicals
Antennal and/or tarsal receptors?
Cibarial sensilla
Stylet pathway
Inter- or intracellular ("intramural?") penetration
Nutrient or 2° chemical concentrations and gradients
Plant intercellular (middle lamella) chemistry
Aphid salivary interactions
Plant cell damage/response
Food source
Nutrient and 2° chemical content of sap
Aphid salivary interactions
Plant cell damage/response

made for several aphids, including *S. graminum*, probing wheat,[160,185], barley,[171] and sorghum,[31] and for the hop aphid, *Phorodon humuli*, on hop.[42] Mean durations of such NPI "bouts" ranged from 21 to 48 min for *S. graminum* on wheat,[185] 15 to 51 min on barley,[171] and 27 to 37 min for *P. humuli* on hop;[42] in none of these cases was mean duration of NPI bouts correlated with host suitability. Total durations of NPI tended to be greater on unsuitable hosts for *S. graminum* on sorghum[31] and for some wheat cultivar comparisons,[160] though the correlation was not consistent for all wheat cultivars.[185] For *S. graminum* on barley[171] and *P. humuli* on hop[42] there was no such correlation between host resistance and duration of NPI events.

Although it is not clear what is being measured by NPI, and in particular whether it primarily represents ingestion, the length of time which aphids spend in this activity (up to 3 h in a 12-h period[160]) makes it a parameter worth further study. It is possible that several behaviors are represented by this "ingestion" waveform, but it is particularly important to know whether ingestion from tissues other than phloem occurs, and if so, the relevant tissues should be identified.

The evidence for prolonged feeding from mesophyll tissues remains somewhat tenuous. Plant physiological studies indicate that localized damage may alter the permeability of mesophyll cell membranes,[232] perhaps allowing solute leakage from such cells into the apoplast; however, the relationship of such processes to aphid feeding is completely unknown.

Even if it is assumed that the primary food of aphids is the phloem sap, it is difficult to obtain this fluid and to determine its definitive chemical composition. Phloem sap will exude from shallow incisions in many woody plants and a few herbaceous species, but the purity of such sap is questionable.[167] One promising technique, first introduced by Kennedy and Mittler[85] involves the use of "exuding stylets". Cutting the stylets of an ingesting aphid allows sap to exude from the stylet stumps. This method has been refined by using lasers or high frequency radio microcautery.[14,44,57,133] These techniques have been useful in the determination of some phloem components and of phloem osmolarity, as well as in the study of translocation and phloem physiology;[57] however, these techniques have not been used extensively to definitively determine what compounds might make certain plants more suitable or acceptable as hosts. Exceptions include recent studies by Weibull,[234-236] who monitored amino acid levels in phloem sap collected from aphid stylets in oats and barley of different ages. These levels were positively correlated with the relative growth rate of *Rhopalosiphum padi* on plants of the same ages. He also compared the concentration and composition of free amino acids in phloem sap from resistant and susceptible (to *R. padi*) accessions of oats and barley, and found that resistance could be related to composition (e.g., glutamic acid was usually higher in resistant plants), but not to overall levels of amino acids. Such correlations, however, do not exclude the possibility that other factors may account for resistance of the accessions tested.

Another technique for obtaining phloem sap was introduced by King and Zeevart.[93] This method used ethylenediaminetetraacetic acid (EDTA), a chelating agent, to prevent callose formation at the cut surfaces of leaves or petioles. Phloem sap would then exude into the water in which the cut ends were placed. This technique has been used with apparent success to study translocation in grasses, for example, by Tully and Hanson.[219] The EDTA technique has not been used extensively in aphid studies. Febvay et al.[54] recently measured the amino acid and sugar content of phloem sap with respect to aphid resistance (see Section VI.A), although quantification was not possible because the rate and degree of sap leakage into the EDTA solution was not known. In general, because of technical difficulties, studies of food quality for aphids have relied on whole leaf analyses. These are crude approximations, and may not accurately reflect the composition, to say nothing of concentration, of amino acids in the phloem.[167] The usefulness of such approximations needs to be considered more carefully in future investigations.

Ingestion from xylem elements has also been claimed for aphids.[127,203] Electronic monitoring of feeding aphids showed a correlation between the recording of a low frequency waveform and the location of the stylet tips, examined with transmission electron microscopy (TEM), in xylem elements. A corroborating positive correlation was found between the occurrence of this electronically recorded pattern, called G, and the degree of starvation or desiccation of a given individual.[203] For *Nasonovia ribis-nigri* and *Myzus persicae* feeding on lettuce, this pattern occurred in 25% of the aphids monitored, and where it occurred, its mean total duration was 55 min out of a total assay of 4 h.[154] The significance of such putative feeding from xylem is unknown, as it has been so little studied. Montllor and Tjallingii[154] found no correlation between occurrence/duration of pattern G (assumed to be ingestion from xylem) and host suitability of lettuce to *M. persicae* or *N. ribis-nigri*. For these two aphids, ingestion from xylem appears to be a facultative behavior, related perhaps to their physiological state.

III. RELEVANCE OF LOCATION AND DISTRIBUTION OF PLANT COMPOUNDS

Several considerations are especially important to an understanding of the information that aphids receive upon probing and feeding, and of how they respond to that information. As is always the case in studying the effects of plant secondary compounds on insects, it is important that the way in which the compounds are tested should be relevant, as far as can be known, to the whole plant situation. The concentration of the compound(s) is of obvious importance, although in many cases the concentration in particular plant tissues is not known. Many compounds are stored in specialized organelles or in particular tissues and may not be encountered during probing. On the other hand, when such localized compounds are encountered by aphids, they will necessarily occur at much greater concentrations than in the plant tissue as a whole. In the case of translocated substances, concentrations are also likely to be different in phloem than in the rest of the plant, and gradients may occur in the process of phloem loading of secondary compounds, just as they probably do for sugars and amino acids.[39]

It is impossible to generalize about the location of secondary compounds in plant organs and tissues. Secondary compounds are often said to be stored primarily in vacuoles and in "external" tissues, e.g., epidermal cells.[123,181,244] Although several workers have suggested that aphids avoid such compounds in epidermis or mesophyll by means of intercellular probing, this is highly unlikely, given the complex behavioral reactions that aphids have toward the more peripheral plant tissues (see Section V.B). Theoretically, only small amounts of such compounds need to be taken up into the stylets to be perceived by the cibarial sensilla (Section V.A), and therefore for aphids to behaviorally discriminate among plants based on the presence of such compounds. Analogous situations using artificial diets have shown that aphids may be deterred or stimulated to probe into, and/or ingest from, acceptable diets depending on the susbstances that are included in solutions laid on top of them. For example, *Amphorophora agathonica* was deterred from probing into a sucrose solution when a solution of phlorizin was layered on top,[148] and *M. persicae, S. graminum,* and *Acyrthosiphon pisum* ingested differentially from diets overlayed with various mono- and polysaccharides that were postulated to be important in host discrimination[30].

Some chemicals known to be stored in particular plant tissues, such as quinolizidine alkaloids in epidermal cells of legumes,[244] are also known to be translocated throughout the plant, and therefore to occur in phloem sieve elements.[123,245,247] McKey[123] discusses the possibility that other compounds (some of which may require metabolic transformation to make them more water soluble), including monoterpenes, cyanogenic glycosides, lectins, and non-protein amino acids, may be translocated through the plant. Raven[181] gives additional examples of phenolics such as salicylic acid and cardenolides occurring in the phloem of various plant species. Several plant secondary substances (or their metabolites) have been recovered from honeydew or cornicle secretions of aphids or other phloem-feeding Homoptera, giving incontrovertible evidence that they have been ingested. These include indolizidine, pyrrolizidine, quinolizidine and other alkaloids;[51,94,147,246] pinitol;[27] glucosinolates (see reference to Weber et al. in Reference 147); iridoid glycosides;[163] and cardenolides.[147]

Many plant compounds are stored and transported in different forms. Glycosylation, for example, is a common solution to problems of solubility, which is a prerequisite for translocation. Glycosylation may also influence the relative toxicity of compounds to both the plant and the insect.[123] Flavanone glycosides were inactive as feeding deterrents to *S. graminum* and *M. persicae* at concentrations up to 1% in diet while the corresponding aglycones were deterrents at a range of concentrations (e.g., ED_{50} of approximately 0.1%).[49] Similar findings have been reported by Todd et al.[217]

The occurrence of secondary compounds in xylem has also been reported;[181] however, it is difficult to judge the importance of xylem constituents to aphids since the ingestion of xylem contents has been very little studied (see Section II). It has been assumed that xylem contents are irrelevant to the probing and/or feeding process. For example, the fact that nicotine is translocated in the xylem, and not in phloem, has been suggested as an explanation for the ability of *M. persicae* to feed with impunity on tobacco.[66] In this case, given 72 h, aphids apparently did not ingest any of the radioactive nicotine from tobacco plants that had their cut ends immersed in a solution of labeled nicotine; however, it is probably not safe to assume that xylem contents are generally irrelevant to probing aphids, and it would be interesting to study the phenomenon of xylem ingestion by aphids on a plant species known to transport secondary chemicals in the xylem.

Localized differences in concentrations of nutrients and secondary compounds within plants may affect aphid behavior, in as yet undefined ways. The distribution of nutrients and secondary compounds within plants is important in the commonly observed phenomenon of selective herbivory by chewing insects.[61,87,162] The parts of a given plant that aphids colonize may differ among species. For example, three aphid species feeding on pecan become highly selective for feeding sites after the early instars. Each species prefers particular leaflet positions, vein sizes, and/or areas on the leaf.[209,210] Lowe[113,114] described the feeding sites of another three species on broadbean, which characteristically chose different sized veins of this host on which to feed.

Though morphlogical features of the aphid and the plant likely restrict certain species to particular plant parts,[229] it is possible that plant chemistry also plays a role. Aphids and other homopterans often grow and reproduce differently on different plant parts, or leaves of different ages; this is usually accounted for by differences in nutrient levels.[1,77,109,230] Newbery[159] measured both coccid population size and the concentration of soluble nitrogen in the leaves of its host. He found that, while senescing leaves supported larger coccid populations, the level of soluble nitrogen in these leaves was lower than in younger leaves. Although there was no positive correlation with soluble nitrogen, coccid population size was positively correlated with the amount of nitrogen exported from senescing leaves. It is also possible, however, that coccids avoid the younger green leaves of their *Euphorbia* hosts due to the higher concentration of resinous compounds in the waxes of these leaves.[159] Phenolic gradients within developing leaf blades of cottonwood, *Populus angustifolia*, have been correlated with microhabitat suitability to *Pemphigus betae*, a gall-forming aphid[258]. The phenomenon of differential choice and/or suitability of plant parts is an interesting one that deserves further critical study.

IV. MEASUREMENTS OF APHID BEHAVIOR IN RELATION TO PLANT CHEMISTRY

In order to begin feeding on a suitable host plant, an aphid must make a series of hierarchical decisions; a combination of aphid behaviors and plant factors (many of them chemical; see Table 1) determines host suitability. The first step may involve orientation to the plant from some distance. Though host plant finding per se is a topic beyond the scope of this chapter, it is worth noting that a variety of plant characteristics are thought to be relevant to searching aphids. Visual orientation has long been established,[146] but surprisingly little work has been done on olfactory responses leading to orientation to the host plant. Pettersson found that female sexuals of *R. padi* could discriminate the odor of their winter host in an olfactometer,[172] and viviparous females of *B. brassicae* responded to the odor of buds and flowers of their summer host.[173] Chapman et al.[33] provided the first good evidence that an aphid, *Cavariella aegipodii*, is attracted to host odors, though the experiment did not determine whether odor-conditioned visual orientation or anemotaxis was involved. Visser and Taanman[228] have unequivocally demonstrated an odor-conditioned anemotaxis by *Cryptomyzus korschelti*. These latter aphids responded to odors of their host plant, but not to certain nonhost odors, by walking upwind.

In the next step, aphids are said to "settle". (Plant surface factors and their effects on settling and feeding are discussed in Section V.A.) Behaviorally, settling may involve simply initiating so-called test probes, or walking to the preferred feeding site, e.g., the abaxial surface of the leaf. Characteristics of the first probe and/or walking behavior are therefore criteria that have been used to evaluate the initial reaction of the aphid to a plant. Settling occurs when initial reactions to physical and chemical plant characteristics are favorable, and implies some degree of acceptance. As an aphid settles and continues probing, it eventually finds its food source, usually the phloem, and begins ingesting from it, though small amounts of other plant fluids are almost certainly ingested prior to reaching the phloem.[127,128,212] The measurement of such behaviors has relied most heavily on observations of aphids on artificial diets, histological techniques, and the use of electronic monitors of feeding.

A. ARTIFICIAL DIETS

The duration of probes and length of the stylet sheath, or depth of penetration, in artificial diets, and the relative proportion of aphids settling (in choice and no-choice situations) on diets containing test materials, have been used extensively[142] to determine whether plant compounds act as deterrents or phagostimulants to aphids. Diet uptake has also been measured by weighing diets before and after feeding, or by using radioactive tracers.[72,108]

The meaning of some of these criteria may be open to interpretation. For example, a relatively deep probe into an artificial diet may indicate the presence of a probing stimulant, an ingestion deterrent, or perhaps the lack of an ingestion stimulant. An increase in the number of probes into a medium has been interpreted as indicating that the test substance was "stimulating",[95] but in other cases is correlated with "dissatisfaction" on a host.[218] This is certainly true on whole plants; for example, the number of stylet insertions and withdrawals by *S. graminum* into an unsuitable host was twice as great as on a suitable host,[31] and Holbrook[75] found that a large number of probes in whole plants was one of two reliable indicators of unsuitability of *Solanum* hosts for *M. persicae*. It is probable that behavioral criteria indicating acceptability on artificial diets are not always similar to those on plants. All things being equal, however, behavior on artificial diets can provide a reasonable initial comparative approach between factors thought to be important to aphids probing in the whole plant.

B. HISTOLOGY

Aphids and other sap-feeding Homoptera deposit a salivary sheath as the stylets penetrate plant tissue.[135] These easily stained sheaths leave a record of where the stylets have been. Pollard[177] discusses many studies using histological techniques to determine, for example, the preferred feeding sites of aphids by determining the termination of stylet sheaths, or of the stylet tips themselves. More recent histological work using electron microscopy has included studies on the stylet pathway (see Section V.B) and on the damage caused by aphid probing.[4,19] Aphid behavioral or physiological reactions to plant damage, which have not been much studied,[138] could be important to our understanding of aphid/plant compatibility. Nielson and Don[161] suggested that phytoalexins produced as an initial reaction to probing by *Therioaphis maculata* might account for the subsequent differential behavior of these aphids on alfalfa clones; however, there is no firm experimental evidence to support such a hypothesis in this or other aphid/plant systems. Phenomena that are well-known plant reactions to pathogen attack, such as phytoalexin production or hypersensitive responses, could profitably be studied in the context of aphid "parasitism".[29] The elucidation of plant factors that affect and are affected by aphid probing need to be further studied using histological examination of the stylet pathway along with histochemistry of probed and surrounding tissues.

C. ELECTRONIC MONITORING

Since the early 1980s, there has been a good deal of interest in the electronic monitoring of aphid probing behavior. This technique was pioneered by McLean and Kinsey.[125-127] Improvements and modifications were subsequently made,[23,92,124,129] including DC variants[189,212,214,216] of the original AC systems. Both AC and DC systems are currently used. In both, an aphid is made part of an electrical circuit by attaching a gold wire to the dorsum, and applying voltage to the medium that the aphid will probe (e.g., the plant or artificial diet). The aphid completes the primary circuit when it inserts its stylets into the medium. Changes in aphid resistance, and varying potentials in the aphid and plant[214] result in voltage fluctuations which are amplified and recorded on tape or paper. Although results of studies using both AC and DC systems are discussed in this chapter, a much-needed detailed comparison of the two systems cannot be included here.

Electronic monitoring of aphid feeding with emphasis on the AC system has been reviewed,[208] and Tjallingii[216] has also recently discussed it, especially with reference to the DC system which he has used extensively. The technique has been used to monitor specific behaviors of aphids (for example, number of probes, time taken to reach the phloem, duration of phloem ingestion) in response to whole plants that differ in some known, but mostly unknown, ways. It has been used most often to investigate host plant resistance or suitability to aphids[2,31,67,75,84,151,195] and/or to investigate feeding differences between aphid biotypes with respect to host suitability.[150,160,161,171,185] The technique has also been used to study virus/vector/host plant relationships.[73,74,105,152,153,198,220]

A few studies have assessed electronically monitored behavioral responses to particular plant compounds in artificial diets. Such studies yield the same kind of information as might be obtained by direct observations of aphid behavior, such as duration of "test" probes and number of probes into diet, but also include the more difficult to observe temporal measurements such as the total time spent ingesting the diet. For example, Campbell et al.[30] found that aphids settled differentially on diets that were overlaid by various plant mono- and polysaccharides. Based on settling preference, these compounds were classified as stimulatory or deterrent. Subsequent electronic monitoring of aphids showed that those probing a deterrent overlay made more but shorter (duration) probes, and ingested for far shorter times, presumably from the underlying diet, compared to those probing through the stimulatory overlay. Other studies have described an increase in the number of probes and a decrease in the duration of ingestion from artificial diets when gramine[262] or hydroxamic acids[8] were added to the diet.

Though the use of whole plants introduces variables beyond the compound(s) being tested, or those thought to be relevant, other parameters, such as time to reach the phloem, number of contacts with phloem by aphid stylets, and parameters relating to phloem acceptance can be compared. In the study on aphid probing into diets including gramine mentioned above, the authors also compared the probing behavior of *S. graminum* and *R. padi* on whole barley seedlings of varieties with and without gramine.[262] The resultant differences in probing parameters, including increased time to reach the phloem and complete lack of phloem ingestion from the plants containing gramine, were attributed to the presence of this indole alkaloid, although the other possible differences between gramine-free and gramine-containing varieties of barley were not addressed.

The ways in which feeding behavior of aphids differs on plants that are known to differ in suitability as hosts can give important clues as to the factors involved in host suitability. Behavior does not necessarily differ on good and poor hosts, though one might expect to find some correlates between behavioral characteristics and performance parameters. Van Emden[222] has suggested that host selection behavior and subsequent performance are likely to be favored by the same plant characteristics. In virtually every case where electronically monitored behavior has been compared on hosts that differ in suitability, differences in behavioral parameters have been found. One exception is a study on *Chaetosiphon fragaefolii* on resistant and susceptible strawberries by Shanks and Chase,[195] who concluded that the "probing behavior of aphids cannot be predicted on the basis of a plant's susceptibility or resistance". It is possible that the parameters measured and the small sample sizes (five aphids per treatment in some cases) did not allow any differences to be detected.

V. PLANT FACTORS INFLUENCING APHID PROBING

A. PLANT SURFACES

Of particular interest in the context of this chapter are chemosensory responses to surface chemicals (volatile or nonvolatile), which might affect probing and ingestion, rather than the earlier step of host finding. Prior to settling, aphids wave the antennae, touching the plant surface, and often drag the tarsi along the leaf surface. Anderson and Bromley[5] suggest that the antennal receptors of aphids are suitable for the detection of plant volatiles, and electroantennograms show aphids do respond to common plant volatiles.[257] The tarsi may have chemoreceptors[5] but this has yet to be determined. Tarsal and antennal contact chemoreceptors might play a part in detecting even nonvolatile components of the plant surface, but almost nothing is known of this possibility.[13]

Though Tarn and Adams[208] have suggested that receptors at the tip of the proboscis (labium) of *M. persicae* are chemoreceptors, the prevailing view is that the aphid labium is equipped only with mechanoreceptors, based on the two species examined, *Aphis nerii*[13] and *B. brassicae*.[213,240] This is in contrast to other Homoptera, including many members of the suborder Auchenorrhyncha (e.g., leafhoppers);[13] in the suborder Sternorrhyncha, which includes aphids, evidence of chemosensory hairs on the apex of the labium has been found only recently in six species of Aleyrodidae.[231] In spite of the apparent lack of labial contact chemoreceptors, it is evident from observations of aphids on their host plants that decisions as to host suitability can be made after very short first probes. Probing and uptake of very small quantities of fluid, or of solids solubilized by the saliva, probably provide chemical cues via a row of chemoreceptors both anterior and posterior to the precibarial valve.[129,241] Backus[13] suggested that the anterior sensilla could provide a first, qualitative test of the fluid taken up, while the posterior sensilla could further discriminate once the fluid is allowed past the precibarial valve. At least partial "extravasation" (alternately "egestion"[68] or "regurgitation"[69]) of materials tasted in this way can be accomplished by contracting the muscles of the cibarial pump while leaving the precibarial valve open,[129] allowing for tasting without commitment to ingestion. Solubilized chemicals on the surfaces of plants may be tasted in this way.

Compounds extracted from leaf surfaces, and compounds added to leaf or artificial surfaces affect aphid behavior; such studies show the potential importance of surface chemicals in settling and feeding. Compounds on plant surfaces include those found in epicuticular waxes, and leaf exudates, especially from glandular trichomes. For example, *p*-hydroxybenzaldehyde, which is a major component of the epicuticular wax of sorghum,[253] is a deterrent to *S. graminum* (ED_{50} = 0.13%) when incorporated into artificial diets.[52] Exudates of glandular trichomes from *Solanum* deter feeding by *Macrosiphum euphorbiae*[62,63] and *M. persicae*.[62] Though the criteria used are variable, and sometimes not defined, both positive and negative aphid reactions to surface extracts and to chemicals applied to surfaces have been reported. These are summarized in Table 2.

B. MESOPHYLL FACTORS

Chemical and physical attributes of mesophyll tissue are likely to affect the progress of the stylets to the plant vascular tissue. Such mesophyll factors, or "pathway" factors, are discussed below with respect to the chemicals encountered by the stylets, and to the consequences for reaching the phloem, the major food source for aphids.

1. The Stylet Pathway

Of special importance in the understanding of what plant factors are relevant to probing by aphids is the stylet pathway. This has been recently reiterated by a number of authors.[46,202,156] Along this pathway, a wide variety of plant compounds may be encountered and assessed, if fluid uptake occurs.[127,128] Though some aphids may take only a matter of minutes to reach the phloem,[177] others, such as *S. graminum*, may take an average of up to 3 h before reaching the phloem, even on a suitable host plant.[150,151,153] Because of the long periods of probing that may occur prior to contact with and ingestion from the phloem, and because generally aphids can discriminate at least between hosts and nonhost plants before reaching the phloem, the notion that only compounds found in the phloem are of relevance to probing aphids is untenable. Whether a more fine-tuned discrimination, e.g., between plant parts or alternate hosts, requires tasting of the phloem has not been studied. The degree to which plant compounds, be they found in cytoplasm, vacuoles, vascular tissues, bound to cell walls, etc., are actually encountered is one of the most pressing questions in this field. As Klingauf (Reference 96, see p. 217) has pointed out, "it is amazing that there is so little information about the promoting or inhibiting stimuli in stylet penetration by the aphids." Griffiths et al.[65] also suggested more than a decade ago that the study of host selection in aphids would benefit from knowledge of the chemistry of tissues between the leaf surface and the phloem. Such knowledge is still rather scarce; however, even if the chemical nature of specific leaf tissues were known, and if the nature of the symplast and the apoplast were completely defined chemically, we would be left with the problem of determining what compounds influence the course of penetration of the stylets, because the exact course itself is in most cases also undefined.

Pollard[177] has reviewed the evidence relating to the route taken by the stylets of many aphid species on their host plants. Stylet penetration was classified as predominantly inter- or intracellular, and different aphid species were pre-

Table 2
PLANT SURFACE EXTRACTS AND/OR COMPOUNDS APPLIED TO SURFACES THAT AFFECT APHID BEHAVIOR POSITIVELY (+) OR NEGATIVELY (–)

Aphid	Compound/Source	Criterion	Ref.
Aphis fabae	Phenolics/host surface	Test probes into diet (+)	79
Aphis pomi and Rhopalosiphum insertum	Phlorizin/host	Probes on silica gel (+)	95
Acyrthosiphon	Waxes/host surface	Settling on non-host (+)	99
Hyadaphis erysimi	Sinigrin/host	Settling on non-host (+)	158
Brevicoryne brassicae	Sinigrin/host	Probing artificial substrate (+)	239
Myzus persicae	Host extract	Settling on non-host (+)	207
	Fatty acids	Settling on leaves (–)	197
	Fatty acids	Settling on artificial substrate (–)	64
	Neem and aphid extracts	Settling on substrate (–)	65
	Trichome exudate	Probing on host (–)	103
Acyrthosiphon spartii	Sparteine/host	Settling on non-host (+)	201
Macrosiphum euphorbiae	Trichome exudate/ resistant host	Settling, probing on host cultivars (–)	63
	Surface extract/ resistant host	Settling on artificial substrate (–)	63
Chaetosiphon fragaefolii	Host extracts (various solvents)	Probing artificial substrate (+)	196

sumed to use one or the other mode. Most of the histological studies reviewed by Pollard were based on light microscope examinations of stylets or stylet sheaths in plant tissues. Studies based on examinations with the light microscope may lack the resolution to clearly define the stylet pathway.[19,89,90,202] From recent electron microscope (EM) studies a clearer, but more complicated, picture of the stylet pathway has begun to emerge.

According to an early EM study of aphid penetration of plant tissue by Evert et al.,[53] *Rhopalosiphum maidis* probed primarily intercellularly through the epidermis and mesophyll of barley, except close to the vascular bundles, where some mesophyll cells were punctured. In some cases, stylets appeared to enter cell walls, but without piercing the plasmalemma; however, salivary sheath material occasionally ruptured the plasmalemma and was found inside cells. Using the electron microscope, Lopez-Abella and Bradley[112] found that on tobacco, probes by *M. persicae* which had been considered completely intercellular were actually partly intracellular — some cell walls were apparently breached by salivary sheath material. In these species, as well as in *Logistigma caryae* on linden[53] and *Brevicoryne brassicae* on cabbage,[89,202] penetration by stylets between the cell wall and the plasmalemma, referred to by Spiller et al.[202] as "intramural-extracellular", appears to be a fairly common phenomenon, observed only with the greater resolution of the electron microscope. Such a phenomenon somewhat blurs the distinction between inter- and intracellular penetration, especially when salivary sheath material appears able to occasionally enter the protoplast while the stylets stay primarily outside it. In any case, the term "intracellular" should probably be reserved for cases in which the stylets themselves actually penetrate the plasmalemma, enter the protoplast, and traverse the cell(s).

Electron microscopy of the stylet pathway may help to explain penetration phenomena recorded during the electronic monitoring of probing aphids[91] (see Section IV.C). Tjallingii[215] has described short drops in electrical potential lasting a few seconds during monitoring of nine aphid species on their host plants. These so-called "potential drops" are extremely common and occur in species that have been described as probing predominantly intra- or intercellularly, i.e., in which the stylets presumably pass predominantly either through or between cells. For *B. brassicae* on Brussels sprouts short potential drops occur every 25 to 35 s over periods of 5 to 30 min.[215] For *M. persicae* and *N. ribis-nigri* probing susceptible lettuce host plants, short potential drops occur on average once per minute, during certain prolonged phases of probing.[154] Tjallingii[215] hypothesized that these drops in potential represent membrane potentials, measured when the stylets, acting as microelectrodes, are inserted into the protoplast. Kimmins[89] has suggested that "intramural" stylet penetration may lead to brief stylet punctures of the plasmalemma, which may account for the short potential drops; subsequent rapid repair of the membrane might occur, leaving no histological trace of the very short intracellular intrusions. Based on the data for *M. persicae*, *B. brassicae,* and *N. ribis-nigri*, which is probably typical of aphids probing favorable hosts, such a scenario would require that almost all of the stylet penetration through mesophyll be of the intramural type, since short potential drops are a characteristic feature of this phase of probing. So far, the histological evidence from other species is not able to confirm a preponderance of "intramural" penetration; however, too few EM studies have been carried out to generalize.

The significance of the short potential drops so commonly recorded using the DC electronic monitoring system is not yet completely clear. Montllor and Tjallingii[263] measured the magnitude of short potential drops made by *M. persicae* and *N. ribis-nigri* on three lettuce cultivars and found that they differed significantly, even within a cultivar, depending on the aphid/cultivar pair. For example, for *N. ribis-nigri* and *M. persicae* on the same cultivar, mean magnitudes were –37 and –100 mV, respectively. Conversely, mean magnitudes were –37 and –148 mV for *N. ribis-nigri* probing two different suitable cultivars.

Potential drops of an intermediate duration are also occasionally made during the course of probing. Potential drops lasting 1 to 2.5 min each were suggested to indicate ingestion from mesophyll parenchyma cells, based on histological evidence and the nature of the electrically recorded signal.[91] Very little is actually known about so-called "mesophyll ingestion" during probing (see Section II). Presumably it does occur, at least by way of sampling, and the contents of such cells are likely to be important in the process of host acceptance and/or phloem finding.

2. Reaching the Phloem

Reaching the phloem is of critical importance for most aphids on their host plants (see Section II). The time it takes for an aphid's stylets to reach the phloem has been measured electronically using the AC system as the time elapsed before the occurrence of a waveform (x-wave[126,127]) correlated with phloem contact. (A criterion for phloem contact has not yet unequivocally been defined using the DC system.[154]) This time varies from 5 min for *A. pisum* on broadbean[127] to several hours for *S. graminum* on sorghum.[150,151] Other estimates of the time taken to reach the phloem, based on histological studies, range from 3 min (*M. persicae* on sugar beet) to 3 h *(Megoura viciae* on beans).[177] Individual variation between aphids on the same plant species can be very wide.[177]

Differences in time to reach the phloem, measured electronically, have been reported between biotypes probing plant varieties that differed in their susceptibility to the biotypes. For example, biotype E of *S. graminum*, which grew well on a particular sorghum variety, was able to reach the phloem in a mean of 193 min, compared to 362 min for biotype C, which grew poorly on the same variety.[150] These data have been interpreted as indicating relative difficulty in penetrating plant tissue en route to the phloem for the biotype C aphids.[45] The unpublished work of Dreyer, Jones, and Campbell[264] describes a similar phenomenon and explanation for *A. pisum* probing different alfalfa varieties. The time between the initial probe and the probe that actually leads to the phloem may include a number of separate probes not leading to the phloem, as well as time spent not probing, especially on resistant plants. Although *S. graminum* biotype E aphids reached the phloem sooner than biotype C aphids (193 vs. 362 min) after being placed on the plant, this difference did not appear to be related to the ability to reach the phloem with the stylets (about 40 min for each biotype), but rather to a greater number of interruptions in probing on the part of biotype C.

Further evidence that the time taken to reach the phloem from the start of an assay is not necessarily correlated with aphid performance was provided by the observation of the feeding behavior of *S. graminum* on virus-infected and on healthy oats: although aphids grew larger on infected oats, and fed for twice as long from the phloem of the virus-infected plants during 7 h assays, there was no difference in the time taken (approximately 3 h) to reach the phloem of healthy as compared to diseased plants.[153] The time taken to reach the phloem by *T. maculata* on resistant alfalfa[161] or by *P. humili* on resistant hop[42] was not different from that on susceptible varieties.

The number of stylet contacts with phloem made during a given access period has been measured electronically (number of x-waves) and histologically (number of stylet sheaths ending in phloem). Histological studies of sorghum probed by *R. maidis*[88] and a planthopper[58] showed fewer salivary sheaths terminating in the phloem of less suitable varieties or growth stages. McMurtry and Stanford[131] found that when they stained stylet sheaths left in alfalfa tissue after feeding by *T. maculata*, 36% (46 of 128) ended in the vascular bundle on resistant alfalfa compared to 65% (91 of 150) on susceptible alfalfa. These studies might indicate that host suitability was related to the insects' ability to locate the phloem. Lowe and Russell,[115] on the other hand, found no correlation between the proportion of stylet sheaths of *M. persicae* or *Aphis fabae* which reached the phloem of resistant or susceptible sugar beet plants, indicating that susceptibility was not related to the ability to reach the phloem.

Kennedy et al.[84] found that although the proportion of stylet sheath branches of *Aphis gossypii* ending in phloem was smaller on resistant muskmelon plants than on susceptible (34 vs. 49%), the total number of branches in phloem was actually greater on the resistant plants (85 vs. 69). This was because more branches were made in resistant plant tissue. *A. gossypii*, then, appears to make a greater number of stylet contacts with the phloem of resistant hosts. Several studies on *S. graminum* probing sorghum, wheat, and oats have determined that these aphids also make more contacts with the phloem on unsuitable plants,[151,153,185] measured as number of x-waves during the assay period; such contacts were not necessarily followed by ingestion from phloem (see Section V.C). *P. humili* also made more phloem contacts (x-waves) on resistant than on susceptible hop.[42]

It is not yet possible to generalize as to the relative importance of phloem vs. pathway factors in plant resistance to aphids. It is clear that in many cases aphids repeatedly interrupt their ingestion probes to the phloem on unsuitable hosts, so the acceptability of the phloem per se is evidently of importance in some cases. Possible pathway factors, i.e., factors encountered on the way to the phloem (or other food source), have been poorly defined so far, and have not been explicitly investigated, with the exception of of work by Campbell and Dreyer on the importance of plant matrix polysaccharides to probing aphids (Section VI.C). Although host suitability to aphids does not appear in most cases to be related specifically to the ability to reach the phloem or to time taken to reach it, other pathway factors may also influence feeding behavior. The effects of plant chemicals on these and other important parameters related to the stylet pathway remain largely unstudied.

C. PHLOEM FACTORS

Once an aphid's stylets have reached the phloem, it is characteristics of phloem tissue and sap that are important in determining subsequent feeding behaviors. A significant electronically measured probing parameter which often differs between aphids on suitable vs. unsuitable hosts is the amount of time spent ingesting from the phloem: on suitable hosts, aphids consistently spend a significantly greater proportion of time ingesting from the phloem than on less suitable hosts.[31,42,84,150-153,160,161] The plant factors that influence this parameter are virtually unknown. The measurement of phloem ingestion is complicated by the fact that the rate of sap ingestion, which may be influenced by the plant,[12] cannot be measured by this method. In addition, reduced ingestion from phloem during an assay of at most 24 h may come about in different ways. The duration of phloem ingestion is related to the time taken to initiate probing, if the assay does not start at the beginning of the first probe, as well as to the time taken to reach and/or "accept" the phloem once probing is initiated. The time taken to initiate probing is not often measured in electronic monitoring studies; however, the time taken to reach the phloem, and the time taken to accept it (committed phloem ingestion, CPI; see below) are now frequently measured. Electronic monitoring may be especially useful in helping to identify and separate plant factors that influence probing before the aphid reaches the phloem (i.e., pathway factors, referred to above), and those that are specifically related to ingestion of phloem sap.

CPI was defined by Montllor et al.[150] as ingestion from phloem lasting at least 15 min, based on the occurrence of waveforms indicating sieve element penetration and ingestion.[127] The period of 15 min was chosen based on observations of electronically monitored feeding behavior of hundreds of individuals of *S. graminum* on sorghum: aphids that had ingested for 15 min appeared to continue ingesting for very long periods of time from the phloem. In fact, no objective criteria were used to define CPI, and the 15 min definition of CPI it is not necessarily appropriate for all aphids on all hosts. Dorschner and Baird[42] defined "prolonged" phloem ingestion as lasting at least 60 min for *P. humili*, but again, no criterion was given for choosing that period of time.

Use of the CPI parameter presumes that bouts of phloem ingestion are not randomly distributed, but tend to be either very short (not "committed"), or much longer. A more objective criterion might be arrived at by measuring individual "bouts" of ingestion from phloem and plotting the distribution of bout durations as a log survivor function, as Slater[200] did to describe the feeding pattern of birds (see Reference 199 for an application of Slater's method to an analysis of feeding in a grasshopper). Figure 1 shows such a plot for *S. graminum* aphids probing resistant and susceptible sorghum. On susceptible plants, it is clear that there were several short (<15 min) bouts of ingestion (18 of 35), but that bouts lasting at least 15 min tended to go on for long periods, i.e., > 6 h. Fifteen minutes appears to be an appropriate criterion for CPI for these aphids. On resistant sorghum, the great majority of bouts lasted <15 min (88/103), but ingestion that lasted at least 15 to 60 min tended to go on for 2 to 6 h. The CPI criterion is somewhat more ambiguous here, but the concave curve indicates that the longer the ingestion bout, the more likely an aphid will continue to ingest (for up to 6 h). Distributions of ingestion bouts were markedly different on suitable (susceptible) and unsuitable (resistant) hosts (Figure 1), as implied previously for *S. graminum*[150-153] (see below). The fact that the CPI criterion itself (e.g., 15 vs. 60 min) may be different for aphids probing suitable vs. unsuitable hosts is interesting, but has not yet been investigated.

The time taken to establish CPI has emerged as a significant factor in the feeding behavior of *S. graminum* on sorghum,[150,151] barley,[171] wheat,[160,185] and

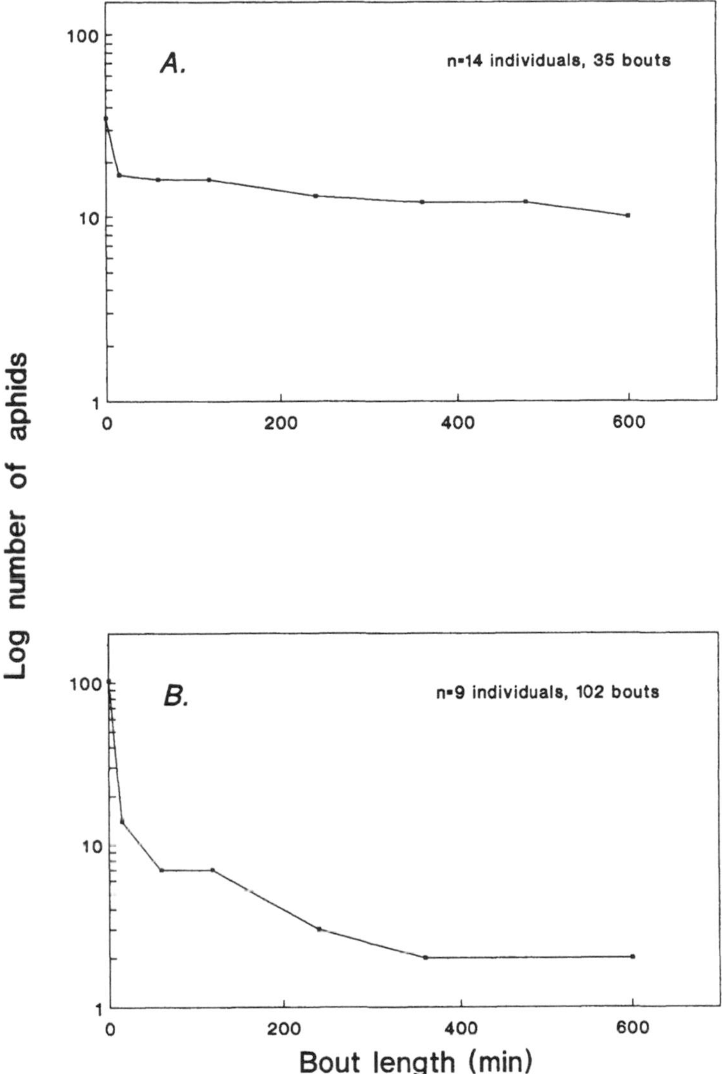

FIGURE 1. Log survivor functions for electronically monitored ingestion bouts of 9 to 14 aphids (*Schizaphis graminum*) on (A) susceptible, and (B) resistant sorghum plants. Each point gives the number of bouts longer than the time given on the X-axis.[200] Changes in slope might be used to more objectively assess a criterion for CPI[150] (see text).

oats.[153] For these aphids, the time elapsed before CPI began was up to several hours longer on unsuitable compared to more suitable hosts. The hop aphid also took twice as long on resistant plants to establish "prolonged" phloem ingestion (lasting > 60 min).[42]

In order to further define and understand factors affecting probing on different hosts, data relating to phloem ingestion or acceptance may be looked at in other ways. For example, it is important to consider the portion of the assay occurring after the phloem has been reached and/or CPI has been established, in order to separate phloem-related factors from those occurring en route to the feeding site (pathway factors). For *S. graminum* the proportion of remaining assay time spent ingesting from phloem, after the first bout of committed phloem ingestion, was significantly reduced from 83% on susceptible sorghum plants to 60 to 66% on resistant plants. This was primarily due to interruptions of ingestion bouts.[151] Throughout the assay period of 24 h, 77 to 85% of all ingestion bouts lasted less than 15 min on resistant plants. Reduced phloem ingestion from the resistant host was not, therefore, only attributable to the longer time aphids took to reach and ingest from phloem on these plants.

The number of contacts with the phloem made by *S. graminum* is consistently greater on relatively poorer hosts of sorghum,[151] wheat,[185] and oats.[153] Ryan et al.[185] calculated that given the greater number of total probes and the greater number of probes leading to phloem contact for aphids on resistant wheat, the probability of finding the phloem on a resistant plant was similar to that on a susceptible plant; however, aphids had a significantly lower probability of "accepting" the phloem (i.e., of ingesting longer than 15 min) on resistant plants. Similarly, 24% of all stylet contacts with phloem of healthy oats were followed by relatively long (>15 min) ingestion periods, compared to 53% on the much more suitable virus-infected oats.[153] Contacts with phloem on the poorer host were, in other words, half as likely to lead to phloem acceptance, using CPI as a measure of acceptance.

In the studies cited above, aphids probing a poor host could be characterized fairly consistently as having all or most of the following characteristics: (1) taking longer to establish CPI, (2) making more separate probes, (3) making more contacts with phloem, and (4) ingesting for shorter periods from phloem, both in terms of individual feeding bouts and in terms of total phloem ingestion. Montllor et al.,[150] however, found that feeding behavior could be significantly altered by a period of conditioning of 1 to 3 d on a resistant host. Although host status evidently did not change (i.e., plants remain resistant in spite of this change in behavior), probing characteristics of conditioned aphids on the resistant host, compared to those of naive aphids, were similar to those associated with a highly suitable host. In this case, the aforementioned correlations between aphid performance parameters and feeding behavioral characteristics did not hold. The nature and extent of such a conditioning phenomenon is not known, but may turn out to be an important aspect of the relationship between aphid probing and plant characteristics.

The relationship of phloem finding, acceptance, and ingestion to plant chemistry must be investigated separately in each aphid/plant combination. In no case have the various aspects of probing behavior on plants, including electronically monitored probing, yet been definitively explained by plant nutrients, secondary substances, or structural components.

VI. APHID REACTIONS TO PLANT CHEMICALS

Many studies of aphid feeding behavior and performance have focused on the importance of nutrients and/or secondary plant compounds. Morphological and structural features of plants, e.g., the distance from epidermis to vascular bundle, or the presence of highly lignified tissues have been given some attention, but appear not to be of great importance in most cases.[30] Gradients in pH[39,156] were long ago suggested to be important in phloem finding,[55], but little recent work has followed up this suggestion. Delrot[39] gives general pH values of 8.0 to 8.5 in the sieve tube, 5.0 in the apoplast, and 7.3 in a parenchyma cell. Schmidt[192] measured the pH of various sorghum cells and found means of 6.5 for mesophyll, 5.6 for xylem contents, and 8.0 for phloem cells. There were no differences in these values for plants that were resistant or susceptible to *S. graminum*, which prefers an artificial diet at a pH of around 8 or higher.[52,192]

Kennedy and Booth[83] postulated the dual discrimination theory of host selection by aphids 40 years ago, proposing that aphids discriminate between hosts on the basis of both nutritional and plant secondary substances. There is now abundant evidence in the literature that both kinds of substances affect the behavior and performance of aphids, just as is true for other insects.[18] Almost 20 years ago, Van Emden[222] discussed the paucity of evidence concerning the relative importance of major nutrients and of secondary plant compounds in determining host suitability to aphids. Since then, many studies correlating aphid performance and/or behavior with plant nutrients or secondary compounds have been published and many discrepancies in such simple correlations exist. Host suitability is not likely to be governed by a single factor.

A. NUTRIENTS

Most of the literature relating aphids and plant nutrient levels does not deal specifically with feeding behavior. Therefore, many of the studies discussed here are more related to aphid performance (growth and reproduction), but are included in recognition of the emphasis in the literature on the importance of nutrients to aphid feeding. In fact, many of these studies present problems that might be useful to consider in future work, including more behavioral studies, and are therefore important to consider here.

Some direct measurements of the effects of sugars and amino acids on aphid feeding behavior have been made. The classical nutritional work of Mittler and Dadd[141-145] has clearly shown that *M. persicae* detects and responds to various amino acids in artificial diets. Srivastava and Auclair[204] showed that *A. pisum* ingested two to five times more from diets containing amino acids, compared to those that lacked them. Sucrose has also long been known to be phagostimulatory for aphids when incorporated into artificial diets.[142] It is possible that sucrose gradients in intercellular spaces of plants may also affect probing. *R. maidis* probing the abaxial surface of sorghum leaves, to which a sucrose solution had been applied on the opposite surface, made more branched probes than aphids probing untreated leaves.[88] In addition, only on sucrose-treated leaves did any probes lead to the adaxial epidermis.

Various aphid species have been reared for ten or more generations on artificial diets with 4 to 6% (w/v) free amino acids.[142] In shorter term experiments, *M. persicae* ingested most and grew optimally on diets containing 3 to 4% amino acids when tested on a range of concentrations from 0 to 4.8%.[141] Plant levels of free amino acids are generally thought to fall at or below these levels.[167] Recent studies using the exuding stylet technique (see Section II) to collect and analyze phloem sap for free amino acids have reported higher levels of free amino acids in phloem. Fukumorita and Chino[60] reported up to 8% (w/v) free amino acids in rice sap, and recent studies by Weibull and co-workers[234-237] give much higher figures for other grasses — up to 27% in barley. These latter levels seem unreasonably high and it is possible that evaporation from droplets at the ends of severed stylets was a significant problem (droplets were apparently collected every 10 to 20 min).[237]

These recent studies notwithstanding, because of the generally low levels of nitrogen and the high carbon:nitrogen ratio in phloem, sap-feeders are presumed to be nitrogen-limited. Brodbeck and Strong[22] compared published values for concentrations of amino acids in phloem and leaves of a range of plant species with levels found to be nutritionally adequate (in artificial diets) for five phloem feeders and five leaf feeders. Although leaf feeders and phloem feeders (including three aphid species) had similar requirements for individual amino acids, levels in phloem were consistently lower than those in leaves. Although these data do not purport to relate the needs of particular insects with nutritional supplies in their own hosts, the general pattern supports the notion that phloem feeders may face more severe nutritional limitations than foliage feeders. This assumption has triggered a fairly extensive body of work correlating nitrogen levels in plants to homopteran outbreaks,[22,242] population fluctuations,[165] and to various measurements of insect fitness (for aphids, see References 12, 119, and 227). Although other essential elements such as sodium, calcium, and boron are also in short supply in the phloem relative to the whole plant,[181], they have not received much attention. Vitamins in whole plants have also largely been neglected.[222] Other nutrients have not been much studied with respect to host suitability, but thus far neither levels of sterols[32] nor of cyclitols (e.g., myoinositol)[27] have been correlated with plant susceptibility to aphids.

1. Relevant Measurements of Nitrogen

The amounts and forms of dietary nitrogen required by herbivores in general are not well known.[22,122] Plant nitrogen content is typically measured in one or more of the following ways:

- Total nitrogen. This includes all nitrogen molecules in the plant, soluble and insoluble, from free amino acids to the nitrogen in nucleic acids, and that in secondary compounds such as alkaloids or cyanogenic compounds. Of total plant nitrogen, 85 to 90% is bound in proteins and free amino acids.[11,122]
- Amino nitrogen. Includes polypeptides and proteins, nonprotein amino acids, free amino acids.
- Soluble nitrogen. This includes free amino acids, amides, nitrates, ammonium, small peptides. Soluble nitrogen often accounts for a small percentage of the total nitrogen of whole leaves; for example, in Sitka spruce this proportion is 3 to 17%, and depends, furthermore, on the season.[165]
- Free amino acids.

Measurements of plant nitrogen that have been correlated with aphid (or other homopteran) performance run the gamut from total nitrogen in whole plants to free amino acids in phloem sap. Aphids utilize free amino acids, and may be able to utilize small peptides, both in artificial diets[104] and in the plant.[176] Therefore, determination of free amino acids, and perhaps soluble nitrogen, in the phloem sap appears to be the most relevant measurement of the nitrogen available to phloem-sucking insects.[12,40] The relevance of other measurements of plant nitrogen will depend on the relationship between these and the actual utilizable nitrogen in the phloem. The nitrogenous constituents of phloem include mostly amides, free amino acids, and perhaps some peptides.[167] Although the amino fraction of whole leaves often differs significantly from that of phloem exudates,[167] few studies have attempted to show what the relationship between amino acid levels in phloem and whole leaves is in particular species or under particular physiological conditions. For a more general discussion of appropriate measures of nitrogen in relation to insect nutrition, see Brodbeck and Strong.[22]

The literature relating aphid performance to dietary nitrogen is voluminous, spanning more than 3 decades, and is not comprehensively reviewed here, but because of the importance of this topic, and some of the problems involved in its study, some representative studies are discussed.

It has long been clear that absolute levels of free amino acids may not be the most important factor to aphid behavior or performance (for examples, see Reference 142) and recently there has been a reemphasis in the literature on the importance of nitrogen quality, as reflected in amino acid composition, as opposed to quantity, to insect nutrition more generally.[22,82]

Prestidge and McNeill[180] give many examples of the performance of sap-feeders correlated with the levels of particular amino acids. Some included measurement of behavior. Parry,[166] for example, studied populations of *Elatobium abietinum* on Sitka spruce. In summer, aphids preferred the needles of previous years over the current year's growth in choice tests, and they began ingesting from these preferred needles sooner, as measured by uptake of ^{32}P from potted plants. When survival of aphids was compared on shoots into which various amino acids had been infiltrated, Parry found that an increase in amino acid levels made no difference; however, aphids feeding on shoots to which isoleucine, methionine, and/or histidine had been added survived longer than controls. At least some of the difference in performance of aphids on plants that differ in amino acid composition may have to do with the fact that some amino acids are more phagostimulatory than others,[142] but to a large extent actual mechanisms are unknown.

Recent studies have corroborated the importance of amino acid balance to feeding aphids. Weibull[234] simultaneously monitored phenological variation in the amino acid content of phloem sap, collected via aphid stylets (see Section II), and population growth of *R. padi* on plants at three growth stages. On both oats and barley, the relative growth rate of aphids was positively correlated with the total free amino acid content of plants; however, the amino acid profile, but not absolute levels, were correlated with resistance of barley and oats accessions to *R. padi*. Glutamate, for example, was high in most of the resistant plants. (It should be noted that in this study amino acid levels were found to be exceptionally high in all accessions: 14 to 27% with one exception at 7% w/v). Febvay et al.[54] analyzed four clones of *Medicago sativa* for sugars and amino acids in the sap, using the EDTA method (see Section II). The sugar/amino acid ratios in the clones most susceptible and most resistant to *A. pisum* were 0.6 and 3.6, respectively; the reproductive rate of these aphids on the susceptible clone was six times that on the resistant clone. The authors attempted to simulate the macronutrient status of these four clones in artificial diets and found that the diet that reflected the sugar and amino acid composition of the susceptible clone could not support reproduction of adult aphids. In fact, all aphids died on these diets. Aphid reproduction was best on a diet that mimicked the sugar/amino acid ratio of the resistant plant and the amino acid profile of the susceptible plant. The authors concluded that amino acid balance was an important factor in host suitability, but that the macronutrients found in the sap were not the principal or only factor of resistance in these clones.

2. Experimental Manipulations

The role of nitrogen nutrition in determining host suitability to aphids also has been studied via the use of nitrogen fertilizers in field and laboratory conditions. In addition, physiological factors in plants that are associated with a change in nitrogen status have been studied with respect to their effects on aphids. Such factors include various plant stresses. These experimental studies are often hard to interpret. Nitrate fertilizers, for example, are used quite differently by different plants. Nitrates may be reduced to organic compounds or stored in the inorganic form in the roots; or nitrates or organic nitrogen may be transported to leaves.[168] Therefore, fertilizers may enhance the nitrate levels in some plants, while boosting other, more utilizable, nitrogen sources in others. Klingauf[97] also discusses some problems with the interpretation of fertilizer treatments, e.g., the importance of potassium treatment to nitrogen status of the fertilized plant, and to aphid behavior. Conversely, added nitrogen may lead to profound changes in the concentration and distribution of other minerals (P, K, Ca, Mg) in plants;[101] effects of such changes on feeding aphids are unknown.

Plant stress, while often leading to higher levels of free amino acids in some plants, must have other physiological effects that are almost never measured with respect to aphid performance. Some specific studies using fertilizers and stressed plants are discussed below.

a. Fertilizers

In some cases a linear increase in leaf soluble nitrogen, total nitrogen, and free amino acids has been associated with increasing levels of nitrogen fertilizer.[134,175] In other cases, however, seasonal fluctuations and variations in irrigation or age of the leaf or plant preclude a constant relationship between leaf nitrogen and fertilizer levels,[134,179] and therefore, assumptions as to the levels of nitrogen in foliage of fertilized plants are even more difficult to make. Fertilizers may also change the composition of free amino acids in the phloem; for example, there was a significant increase in aspartate and glutamate and their amides, but not in other free amino acids, in the sap of oats and barley given supplemental nitrates.[234] Since in several cases it is composition and not absolute levels of free amino acids that are thought to be important to feeding aphids (see above), such changes would be important to monitor.

Many reports on the effects of fertilizer treatments of plants on homopteran fitness are ambiguous or contradictory. For example, Pfeiffer and Burts[174] found that egg and nymph densities of psyllids were positively correlated with the levels of nitrogen applied to trees; however, the rate of development was equally high on plants with no fertilizer and those treated with high levels of urea, but was low on trees given intermediate levels. In a later study, Pfeiffer and Burts[175] found that psyllids contained more protein per milligram and weighed more when raised on pears with a lower level of nitrgen fertilizer. They suggested that the psyllids were adjusting to the lower level of nitrogen in their food by feeding at a faster rate from low-nitrogen trees. Whether aphids and other phloem-feeding Homoptera compensate behaviorally and/or physiologically to low-nutrient diets appears to be unpredictable,[234] creating another variable that hinders the clear interpretation of these kinds of experiments.

Van Emden and Bashford[226] produced "resistant" and "susceptible" plants of the same variety of Brussels sprouts by treating them with low- or high-nitrogen fertilizers, respectively. Regressions of individual amino acids on aphid growth for *M. persicae* and *B. brassicae* on brussels sprouts have identified particular amino acids that are positively or negatively correlated with growth of each aphid, in spite of a lack of correlation between total soluble nitrogen and growth of the aphids.

b. Plant Stress

Additional empirical (but still equivocal) evidence for the importance of soluble nitrogen, especially free amino acids, to sap-feeding insects comes from the work on the effects of host plant stress on these insects (see, for instance, Reference 243). ("Stress" is used here in its broadest sense, and includes physiological, environmental, and chemical factors that impinge on plants.) Brodbeck and Strong[22] suggested that phloem feeders might be more affected by plant stress because the composition of free and not of protein amino acids would be most likely to change as a result of the stress. The improved performance of homopterans on herbicide- or bioregulator-treated (Reference 25 and references therein), water-stressed,[139,242] senescent or excised (Reference 151 and references therein), and diseased[153,220] plants; plants previously exposed to aphids;[43,56,86] and plants subjected to air pollution[20,41,164] (see also Reference 36 for lack of a pollutant effect on aphids) has been attributed in many cases to the increases in "available nitrogen" that result from these stresses. In some cases, however, the amino acid status of the plant may change as a result of the stress, for example, in sycamores previously infested with *Drepanosiphum platanoidis,* without a concomitant effect on aphid performance.[238] In several studies of this sort, insect performance is measured without simultaneously measuring plant nitrogen, for example, studies by Wratten[255] and Montllor and Gildow.[153] Other studies have made measurements that are probably irrelevant to aphids, e.g., the protein levels in herbicide-treated corn.[164]

Some alternative explanations of stress-related changes in aphid behavior and performance have been put forward. Dreyer et al.[48] suggested that the feeding behavior and performance of *S. graminum* was affected on sorghum treated with the bioregulator CCC due to an alteration in the structure and composition of the plant middle lamellar polysaccharides ("pectins"; see Section IV.C). The same phenomenon for *M. persicae* on CCC-treated Brussels sprouts had been previously explained as due to changes in amino acid composition.[221]

High levels of proline are associated with water stress in plants[206] and are possibly phagostimulatory,[104] but the growth and reproduction of aphids on a barley mutant with six times the normal proline content was not enhanced, nor were the plants made more susceptible.[21] Zúñiga and Corcuera[260] suggest that accumulation of proline does not seem to account for the improved survivorship of *S. graminum* and *R. padi* on stressed compared to healthy barley. Instead, they posit that the accumulation of glycine-betaine in stressed tissues reduces the normally harmful effects of gramine, levels of which are unaffected by water stress. The mechanism is unknown, but the authors believe that it may be related to the ability of glycine-betaine to reverse the inhibitory effects of gramine on mitochondria, as previously shown in a bacterial system.

Studies of aphid performance on such stressed plants do not explain whether the physiological changes induced thereby in the plants are perceived by aphids as they probe and feed; however, in some cases the feeding behavior of aphids has been examined on such plants. Montllor et al.[151] studied the feeding behavior of *S. graminum* on excised leaves of resistant varieties of sorgum in which free amino acid levels were determined to be significantly higher than in intact leaves (due to premature senescence).[16] Many aspects of the feeding behavior of *S. graminum*, which was electronically monitored for 24 h, were significantly changed on excised leaves. These changes included more rapid establishment of phloem ingestion, and fewer, longer probes in the phloem sieve elements.

A similar study of the feeding behavior of *S. graminum* and *R. padi* on oats showed dramatic changes of the same kind in the probing of the former aphid when plants were infected with a virus.[153] *R. padi*, a more significant vector of the virus in question, showed little behavioral response to virus infection in oats[153] or barley.[220] It is known that this particular virus, barley yellow dwarf virus, causes an increase in free amino acids in infected plants,[3,120] but it is not certain whether the "improved" feeding behavior of *S. graminum* on infected oats or on excised sorghum leaves could be attributed to such a cause. Alternative explanations have not been extensively investigated, but there is some evidence that other factors that may affect feeding behavior could be involved. Lowe and Strong[116] found that the reproductive rate of *M. persicae* on virus-infected *Zinnia* is lower than that on healthy plants, an exception to the rule of improved aphid performance on virus-infected tissues. The authors cited a possible change in amino acid composition due to disease as responsible for this detrimental effect, but they also found that *in vitro* pectinase digestion of healthy leaf tissue released more than twice as many cells as a similar treatment of infected leaves. This implied that there might be an increase in the amount (or change in composition?) of intercellular pectins in infected leaves. Such an increase might impair the probing of *M. persicae*, leading in turn, to less feeding and the observed lower reproduction on infected plants. Whether changes in the intercellular matrix of infected oats might account for the changes in probing behavior of *S. graminum* described by Montllor and Gildow[153] has not been investigated. For a discussion of the role of plant intercellular polysaccharides in aphid probing behavior, see Section VI.C.

B. PLANT SECONDARY COMPOUNDS

Many plant compounds have been shown to be deterrent, toxic, or phagostimulatory to aphids. By adding compounds to artificial diets, or noting correlations between concentrations of given substances in plants and aphid performance, several kinds of plant secondary compounds have been imputed to be significant in aphid-plant interactions. Often, the exact nature of the significance is not made explicit. Many studies do not, for example, clearly distinguish between deterrence and toxicity of the substances being tested. Thus, if aphids die on artificial diets to which a secondary compound has been added, toxicity is often invoked. Argandoña et al.[8] found that the mortality after 24 h of aphids on diets containing 6 and 8 m*M* of a hydroxamic acid, DIMBOA (2,4-dihydroxy-7-methoxy-2H-1,4- benzoaxin-3(4H)-one), was the same as for aphids starved over the same period. In addition, a toxic effect was postulated, based on exposure of aphids to diets containing different concentrations of DIMBOA for 12 h before transferring them to control diets. Subsequently, mortality was highest for aphids that had subsisted on diets with intermediate levels of DIMBOA of 3 to 4 m*M*, indicating that where uptake was not completely hindered (as it was at the highest concentrations), toxicity resulted. In this case, both deterrence and toxicity were explicitly and separately examined. In the plant resistance literature, it is the analagous concepts of nonpreference and antibiosis that are often not properly distinguished. Studies of feeding behavior and aphid performance can be designed to clarify the factors that make plants acceptable to and/or suitable to aphids.

1. Stimulants and Deterrents in Artificial Diets

Mittler[142] has recently reviewed the use of artificial diet techniques in the study of virtually every aspect of aphid biology. One of the most extensive uses of artificial diets has been as a medium in which to test compounds for effects on feeding and/or settling. Interest in the effects of plant secondary compounds on host discrimination, feeding, and performance of aphids has led to a number of such studies.

Schoonhoven and Derksen-Koppers[193] tested 24 plant secondary compounds, incorporated into artificial diets, for their effects on the feeding and survival of *M. persicae*. Several, such as atropine, phlorizin, quinine, and tannic acid completely deterred feeding in choice tests, and all aphids died after 2 to 4 d on these diets, probably due to starvation. All but 2 of the 24 compounds were at least somewhat deterrent, and none was phagostimulatory. The authors concluded that food selection for this polyphagous aphid relies on the presence of nutrients and the absence of deterrents. Junde and Lidao[80] also

Table 3
STIMULANTS AND DETERRENTS FOR APHIDS PROBING ARTIFICIAL DIETS (EXCLUDING SUCROSE, PROTEIN AMINO ACIDS)

Compound	Aphid (response)	Ref.
Phenolics and related cpds		
Various	*Schizaphis graminum* (–)	217
	Myzus persicae (–)	78
	Acyrthosiphon pisum (–)	49, 52
	Aphis fabae (+)	79
Phlorizin	*M. persicae* (–)	148
	Aphis pomi (+)	148
	Amphorophora agathonica (–)	148
	Rhopalosiphum insertum (+)	95
	A. pisum (–)	95
Catechin	*Macrosiphum rosae* (–)	137, 169
"Condensed tannins"	*M. rosae* (+)	169
Coumarins	*A. pisum* (–/0)	50
Alkaloids	*A. pisum* (–)	51
	S. graminum (–)	37
	Rhopalosiphum maidis (–)	259
	Rhopalosiphum padi (–)	259
Hydroxamic acids	*S. graminum* (–)	8, 38
	R. maidis	110
Sinigrin	*Hyadaphis erysini* (+)	158
	Brevicoryne brassicae (+)	155, 233
Mono-, polysaccarides	*S. graminum* (4+/8–)	30
	A. pisum (3+/5–)	30
	M. persicae (2+/4–)	30
Nonprotein amino acids	*M. persicae* (+/0/–)	205
Glucose esters of fatty acids of trichomes	*Macrosiphum euphorbiae* (–)	62
Diterpene acids	*S. graminum* (–)	184

found that six plant secondary compounds, including sinigrin, tomatine, and phlorizin, deterred feeding by *M. persicae* on artificial diets. More generally, published studies show that the majority of compounds tested with a variety of aphids have had negative effects on aphid feeding (Table 3); however, this may reflect an interest on the part of researchers in compounds that might confer resistance to host plants.

Aphids may respond differently to compounds depending on the medium in which they are presented. For example, *M. persicae* ingested more from 20% sucrose solutions when sinigrin was added but did not distinguish between complex diets with or without sinigrin.[102] A two- to threefold increase in the concentration of sucrose or amino acids in artificial diets significantly decreased the deterrence of the phenolic catechin to feeding by *Macrosiphum rosae.*[169] The milieu, and therefore the location, of the compound in the plant can influence an aphid's perception of it (see Section III), and tests with artificial diets may be inadequate in mimicking the whole-plant situation. Other compounds, such as phlorizin, may be unstable when mixed into a complex diet, yielding anomalous results. Phloretin, the aglycone that results from mixing phlorizin into a holidic diet, elicits no response from *Aphis pomi*, while phlorizin is a probing stimulant.[95] Thus, the form in which the compound is presented to the probing aphid is evidently important. It is now well known that phenolic compounds, which have frequently been tested as feeding deterrents to aphids,[193] may complex with constituents of artificial diets and/or become oxidized.[17,169] The possibility of interactions between test compounds and aphid diet components has not been investigated to any great extent.

With these caveats, a list of plant secondary compounds found to elicit behavioral responses in aphids probing artificial diets is given in Table 3. Herbach[72] discusses many of these; Dreyer and Campbell[47] discuss some of these cases as well, but conclude that most of the compounds tested are irrelevant to the interaction between aphids and plants because they may not be in the tissues on which the aphids feed (see Sections II and III).

2. Secondary Compounds in Plants

Many studies, especially those concerned with plant resistance to aphids, have noted correlations between aphid behavior or performance and the presence or concentrations of particular secondary compounds in plants. This section contains a discussion of some of the more extensively studied examples.

a. Hydroxamic Acids

Long et al.[110] found a negative correlation between levels of the hydroxamic acid, DIMBOA, and the size of populations of *R. maidis* on different lines of corn. Beck et al.[15] obtained similar results. It has also been suggested that DIMBOA can account for the resistance of some wheat cultivars to the greenbug, *S. graminum.*[8-10] DIMBOA, which is deterrent to *S. graminum*[8] and to *R. maidis*[110] in artificial diets, was found in higher concentrations in lateral veins dissected from corn leaves, than in whole leaves, and since no DIMBOA was detected in xylem exudate or in guttation drops of corn, Argandoña and Corcuera[6,7] concluded that it must occur in the phloem. On the other hand, Molyneux et al.[147] refer to their own analyses showing a lack of DIMBOA or its metabolites in the honeydew of *S. graminum* feeding on wheat. They concluded that DIMBOA is not translocated in the phloem and therefore could not contribute to aphid resistance. Moreover, since DIMBOA levels differ not only with respect to plant tissues within a leaf (e.g., lateral veins compared to whole leaf), but also between leaves of different ages or plants of different ages,[6] it would be important to correlate preferences for feeding sites (e.g., young vs. old leaves) and for different aged plants with these varying levels of DIMBOA. At present, the role of DIMBOA in host resistance to aphids is unclear.

b. Alkaloids

Levels of the indole alkaloid gramine in barley cultivars were found to be inversely proportional to susceptibility of plants to *R. padi* and *S. graminum*. Gramine levels were also lower in older plants, which were reportedly more susceptible to *S. graminum*.[261] The location of gramine in plants was not determined, but Molyneux et al.[147] found no gramine or metabolites in the honeydew of *S. graminum* feeding on barley. It is not clear whether the aphids were feeding on a barley cultivar known to contain gramine; some lack it altogether.[259] Other factors that may co-vary with gramine levels in plants of different ages were also not taken into consideration. Therefore, as with DIMBOA, the influence of gramine on aphid/host compatibility is inconclusive.

Another class of alkaloids, the quinolizidine alkaloids found throughout the Leguminosae, have been studied extensively with respect to aphid/plant relations. These compounds are important in the acceptability of lupines (*Lupinus* spp.) and brooms (*Cytisus* or *Genista* spp.) to several aphid species.[201,245,246] Lupines containing quinolizidine alkaloids are resistant to the pea aphid, *A. pisum*, whereas alkaloid-free varieties are susceptible. Moreover, if alkaloids were infiltrated into the leaves of susceptible varieties, most aphids confined to such treated leaves died within 1 week as a result of either starvation or toxicity (Wegorek and Krzymanska, cited in Reference 47). On the other hand, the early work of Smith[201] established that one of these alkaloids, sparteine, was a feeding stimulant for *Acyrthosiphon spartii*, a specialist on broom: aphids fed selectively from parts of the plant, e.g., developing seed pods, where alkaloid levels were highest.

Quinolizidine alkaloids accumulate in plant epidermal tissues[244] but also are translocated in the phloem of lupines[247] and of broom.[245] *Aphis cytisorum* prefer broom and lupine plants with intermediate levels of alkaloids, while *Macrosiphum albifrons*, strictly lupine specialists, prefer alkaloid-rich ("bitter") plants of *L. albus* over sweet varieties.[246] *A. cytisorum* and *M. albifrons* feeding on alkaloid- containing hosts contain alkaloids at levels of 0.5 to 1.8 mg/g fresh weight.[245,246] Alkaloid levels in the phloem of *Lupinus* spp. have been estimated to be at 1 to 5 mg/ml, depending on the plant part.[245,247] Therefore, aphids appear to be able to sequester these compounds at high levels with respect to the concentrations found in their food. *M. albifrons* excretes host alkaloids, but in different relative concentrations from those found in the plant.[246]

c. Phenolics

Although some simple phenolics are toxic or deterrent to aphids when incorporated into artificial diets (see Table 2), the role of such compounds in the plant in mediating aphid behavior and fitness is ambiguous in most cases. In general, aphids appear to lack or have relatively low activities of gut enzymes which are considered to be useful in detoxifying phenolics (e.g., *trans*-epoxide hydrolase).[157] The effects of phenolics on aphids and other phloem-feeding insects feeding on artificial diets may be compared, with some caution, to effects on insect populations on plants in the field. Several sorghum phenolics, including *p*-hydroxybenzoic acid, *p*-hydroxybenzaldehyde, and the cyanogen, dhurrin, were deterrent to *S. graminum* when incorporated into artificial diets.[52]

The phloem-feeding planthopper *Peregrinus maidis* produced less honeydew when fed on artificial diets containing mixtures of phenolic acids than on control diets, made fewer probes that ended in the phloem, and more probes to tissues other than phloem, on sorghum varieties with relatively higher levels of phenolics.[58] While sorghum plants with high levels of phenolic acids had smaller populations of *P. maidis* in the field, no such effect on the corn leaf aphid, *R. maidis*, was apparent.[254] Beck et al.[15] also found no correlation between total phenol or *o*-dihyroxybenzene levels in various corn lines and natural infestations of *R. maidis*. No similar field observations are available for *S. graminum*. According to Fisk,[58] *P. maidis* probes intracellularly, and Pollard[177] cites Brandes as showing that *R. maidis* also probes primarily intracellularly, at least on corn. Evert et al.[53] found, however, using electron microscopy, that *R. maidis* probed mostly intercellularly on barley. If the latter observations are accurate, the path of stylet penetration of these two insects might account for the difference in correlation between foliar phenolics and insect numbers in the field, but this is purely conjectural.

No generalizations can be made about the location of phenolic compounds in plants: they occur in many tissues (see Section III). In sorghum, dhurrin is known to occur primarily in the vacuoles of epidermal cells,[100,187] and *p*-hydroxybenzaldehyde is a major constituent of the epicuticular wax.[253] Woodhead and Cooper-Driver[252] reported four phenolic acids in the hydrolyzed cell wall fraction of sorghum. Campbell and Dreyer[28] hydrolyzed pectic fractions from sorghum but found none of the sorghum phenolics known to affect aphid behavior when incorporated into diets. These fractions presumably did not include cell walls. Whether the phenolic compounds that typically occur bound to cell walls (see below) affect aphid probing is unknown.

The cell walls of many plants have a variety of phenolic constituents.[70,71,76] Treatment of some cell walls with polysaccharases releases these as free phenolic acids and aldehydes. It remains to be seen whether aphid polysaccharases[45] might similarly release such compounds during probing, and whether aphids might then react to them. Some plant phenolics, such as *p*-hydroxybenzoic acid, are potent enzyme inhibitors[29] and might further act to reduce the activity of aphid polysaccharases, which are thought to be important in aphid probing (see Section VI.C). Miles[138] has suggested that the salivary sheath may absorb and thereby inactivate some plant phenolics by becoming "tanned". Aphid watery saliva contains polyphenoloxidases (PPO)[136], which are capable of oxidizing phenolics to more reactive quinones; however, the specific interaction between these enzymes and plant phenols in the living system is unknown, as is their role in mediating aphid/plant compatibility. A detoxicative function of salivary enzymes has been recently reported in the rose aphid.[140,169,170] *M. rosae* produces a salivary catechol oxidase that polymerizes catechol (deterrent to the aphid at concentrations found in rose buds); the oxidative condensation product is actually somewhat phagostimulatory for this aphid in artificial diet.[169,170] A salivary peroxidase was also found in *M. rosae*, *A. gossypii*, and *M. persicae*, which oxidized the terpenoid gossypol, and the alkaloid hordenine in artificial diets.[140]

Plant oxidases have also been postulated to play a role in aphid feeding.

Pollard[177] suggested that aphid PPO, when added to the plant's own PPO, may result in the oxidation of plant phenols to nontoxic end products, but gave no evidence. Leszczynski et al.[107] found higher peroxidase and PPO activity in wheat plants, especially of susceptible varieties, which had been infested with aphids. They proposed that these plant enzymes might partially detoxify the plant phenolics, levels of which had previously been positively correlated with plant resistance.[106]

In other aphid/plant systems, correlations between aphid behavior or performance and levels of simple phenolics have been found. Cole[35] extracted free phenolic acids from lettuce roots of six cultivars that differed in their resistance to the lettuce root aphid *Pemphigus bursarius*. In particular, the concentration of isochlorogenic acid was significantly higher in resistant than in susceptible cultivars and closely correlated with aphid performance; however, this compound was not assayed for toxicity or deterrence to the aphid. SenGupta and Miles[194] found a negative correlation between the ratio of phenolics to amino nitrogen in several apple varieties and the susceptibility to the woolly apple aphid *Eriosoma lanigerum*. Phenolic levels did not vary as greatly as nitrogen levels and, looked at separately, were not correlated with susceptibility of the apple varieties. The authors suggest that at low levels of amino nitrogen, however, aphids would consume more and thereby take in more of the potentially toxic phenolic compounds, assuming they occur in the phloem. Leszczynski et al.[107] found that the population size of two grain aphids was negatively correlated with total phenolic, and especially free phenolic, levels in four wheat cultivars. When eight radioactive phenolics were added individually to the water of experimental plants, five decreased uptake of label by *R. padi* at the lower concentration (62.5 mg/l), and all reduced uptake at the higher concentration (125 mg/l), relative to control plants. Actual phenolic levels in treated plants were not determined, and so could not be compared with natural levels in the wheat cultivars.

Another system in which plant phenols seem to play an important role is that of *Pemphigus* gall aphids on cottonwood, their primary host plant. These aphids are known to be highly selective in their choice of leaves for oviposition; females inducing galls at the preferred sites, the bases of the largest leaves in recently opened buds, have the greatest potential fitness.[248-250] In addition, these preferred leaves have the lowest phenol levels within the bud.[258] In general, total phenol concentrations in this host plant are negatively correlated with aphid numbers, but whether these compounds actually affect survivorship of the aphids was not determined.

Increased phenolic production in response to the feeding of chewing insects has been frequently documented, although their importance as plant defenses has been challenged;[59] however, very little work has been done on the short-term induction of these compounds in response to aphid feeding. Woodhead[251] found that infection with fungi and feeding by certain insects (stemborer and shootfly) led to increased levels of phenolics in several varieties of sorghum in the field; feeding by *R. maidis* and *P. maidis* (aphid and planthopper, respectively) had no apparent effect on phenolic levels in laboratory-grown plants. Whether aphids will induce and/or react to the same plant "defenses" as do chewing insects is very much an open question. Artificial wounding of birch trees was reported to have no effect on an aphid, but decreased the palatability of foliage to two lepidopteran species, although increased levels of phenolics were not specifically implicated.[256] Ciepiela[34] reported that phenylalanine and tyrosine, both aromatic amino acids and precursors to plant phenolics, increased in resistant wheat as a result of infestation with *Sitobion avenae*. Furthermore, activities of the enzymes involved in the biosynthesis of phenolics from these amino acids were higher in resistant than in susceptible plants, and also increased as a result of aphid infestation in resistant plants. In earlier studies, Leszczynski[106] found a similar phenomenon for wheat infested by *R. padi*, but Montllor[149] found no difference in activity of one of these enzymes, phenylalanine ammonia lyase (PAL), in different varieties of sorghum, or as a result of infestation with *S. graminum*.

Resistant spruce trees apparently produce phenolics in response to the feeding of a gall-forming aphid, *Chermes abietis*, creating a hypersensitive response (rapid cell death) that prevents gall formation.[183] Miles[137] also suggested that the rose aphid, *M. rosae*, might be sensitive to the capacity for sepals to accumulate catechin, possibly in response to aphid feeding. Subsequent work showed that catechin was more likely decreased as a result of aphid feeding, or as a result of polymerization by an aphid salivary oxidase.[169,170] *Acyrthosiphon kondoi* and *A. pisum* caused increased levels of the flavonoid coumesterol to accumulate (up to 90 ppm) in aphid-damaged leaves and stems of alfalfa.[81] This effect, moreover, was more pronounced in susceptible than in resistant varieties.[111] Dreyer et al.[50] reported, however, that coumestrol was not deterrent to pea aphids when added to artificial diets at up to 1000 ppm; longer term effects were not tested.

C. OTHER FACTORS: THE PECTIN-PECTINASE HYPOTHESIS

A recent series of papers by Campbell, Dreyer, and co-workers[24,28-30,45-47] has put forth a "pectin-pectinase hypothesis" of aphid-host plant compatability. Working mainly with *S. graminum* on sorghum, they proposed that characteristics of and interactions between the intercellular matrix polysaccharides (generically, pectins) of plants and the polysaccharases of aphids in some cases determine the relative compatability of aphid biotypes and plant genotypes. Such pectin-degrading enzymes have long been assumed to occur in aphid saliva, and Ma et al.[118] have recently detected pectinesterase and polygalacturonase activity in the salivary secretions (as opposed to whole-body extracts) of probing *S. graminum*. No activity was found in similar secretions from *R. padi* or *R. maidis*.

In vitro biochemical evidence indicates the potential importance of pectin-pectinase interactions in determining host susceptibility. Extracts of two biotypes of *S. graminum* depolymerized a variety of matrix polysaccharides, mostly at similar rates,[28] but the rate of depolymerization of matrix polysaccharides from two sorghum varieties was related to the relative susceptibility of these varieties to the two biotypes.[45] Campbell and Dreyer[28] also suggested that the composition of the plant intercellular matrix might influence aphid probing, irrespective of rate of depolymerization by aphid enzymes. They found an association between the uronic acid and the neutral sugar content of matrix polysaccharides from five sorghum varieties and the performance of *S. graminum* biotypes on these varieties. Preliminary analyses of Ryan et al.[186] also suggest a difference in the relative amounts of galactose and arabinose in pectic fractions from near-isogenic lines of wheat that differ in resistance to this same aphid. The implication is that these compounds are characteristic of the host, and are important in host discrimination, just as has been proposed for nutrients and plant secondary compounds.[83]

S. graminum probing susceptible sorghum treated with the plant growth regulator CCC behaved as though they were on resistant plants: they spent less time probing and ingesting from phloem, and more time before reaching the phloem.[48] Clearly, the probing parameters measured are all interrelated, and a primary behavioral factor contributing to this induced "resistance" cannot be identified. The authors suggested, however, that the increased levels of methoxy-pectins found in treated susceptible plants inhibited access by aphids to the phloem of these plants.[48] A more recent unpublished study[264] of *A. pisum* probing alfalfa cultivars showed a similarly extended time to reach the phloem and reduced rate of population increase, for aphids on CCC-treated susceptible plants; however, there was no increase in extractable pectin or in methoxy content in treated alfalfa plants. Instead, the neutral sugar composition of treated susceptible plants became more similar to that of resistant plants. This points, again, to the importance of the recognition of the intercellular matrix rather than its role in preventing access to the phloem.

That aphids respond behaviorally to plant matrix polysaccharides has been shown by Campbell et al.[30] by overlaying artificial diets with 18 mono- and polysaccharides occurring naturally in the intercellular matrix of plants. They found that *M. persicae* responded discriminatively, either positively or negatively, to 6 of these, while *A. pisum* and *S. graminum* responded to 8 and 12, respectively. In all cases, more of the compounds deterred rather than stimulated the aphids (see Table 3). The authors suggested that the polyphagous *M. persicae* (which was affected by one of seven polysaccharides) may be less discriminating among the polysaccharides than the more host-restricted *S. graminum* and *A. pisum* (each affected by six of seven polysaccharides), indicating a role for host-specific matrix polysaccharides in host selection.

Further, that aphids actually ingest matrix polysaccharides, or fragments thereof, is indicated by the presence of oligosaccharides, probably breakdown products of pectic polysaccharides, in the honeydew of *S. graminum*.[24] Although it was suggested that these fragments are ingested from the phloem, which stained positively for "pectin", it seems just as likely, and more relevant to the pectin hypothesis, that many are ingested as the aphid probes more peripheral tissues. In any case, further details of the composition of the plant matrix and of aphid behavioral reactions to it clearly will be worthwhile. One of the more interesting suggestions that has emerged from this work is that these aphid polysaccharases are synthesized by the aphids' bacterial intracellular symbiotes,[26] implying a significant role for such symbiotes in the probing behavior of aphids.

VII. CONCLUSIONS

Correlations between plant chemicals and aphid performance, i.e., growth and reproduction, have been extensively studied. Fewer studies have focused on the effects of plant chemistry on probing and feeding behavior per se, because of difficulties inherent in the measurement of such activities, and of plant chemicals, in the internal tissues of plants. Such studies are of critical importance in understanding the chemical nature of aphid-plant interactions.

Investigations of the effects of plant compounds on aphids have emphasized both nutrients, especially nitrogen, and plant secondary chemicals, such as phenolics and alkaloids. It is clear that both categories are important, and that their relative importance differs in different cases. Because chemical factors co-vary in plants (e.g., nitrogen and secondary chemicals), and because of the need to verify that aphids encounter the chemicals in question, and at what concentrations, simple correlations are not appropriate; however, the role of secondary chemicals to probing aphids should not be dismissed as a generally unimportant phenomenon, in spite of recent criticism.

A role has also been established for polysaccharides of the plant intercellular matrix, but effects of other components of this matrix, e.g., wall-bound phenolics, have not yet been studied. This part of the plant may be especially important to aphids that probe primarily intercellularly. A great deal of work on matrix chemistry has been done with respect to relative resistance to plant pathogens, such as fungi. Such work may serve as a model to researchers interested in chemical and molecular components of the aphid-plant system.

Categorization of probing modes as "intercellular" or "extracellular" is no longer as straightforward as it was once believed, based on recent transmission electron microscope (TEM) studies of the stylet pathway. Evidence of the piercing of cells as the stylets proceed "intercellularly" make it important for researchers to reasses notions of which tissues contribute to the chemical information available to aphids. The chemical nature of all probed tissues, including cuticle, epidermis, mesophyll, and xylem, as well as of phloem sap, needs to be understood.

Much more detailed information on the stylet pathway would be very helpful. This will require a combination of techniques, including TEM, histochemistry, and electronic monitoring of probing. Improvements in the techniques for collection and analysis of phloem sap, and the possibility of chemical analysis of single cells will add materially to our understanding.

Ecological considerations have not been discussed here. The possible role of bacterial and other microsymbionts in mediating aphid-plant interactions has been mentioned, but little work has been done so far. The predominance of host plant specialization among aphids, the phenomenon of host alternation as a way of life for many aphids, the rapid formation of aphid biotypes in many species, and the role of host chemistry in protection from predators, are all important topics, which at some level are connected to feeding behavior. Although a great deal of work has been published in the area of aphid-plant interactions, we now appear to be poised for some important breakthroughs in our understanding of the chemical nature of this interaction, as it is mediated behaviorally.

ACKNOWLEDGMENTS

Thanks to B. C. Campbell, T. E. Mittler, and M. G. Kinsey for many discussions and for reading a previous draft of this manuscript, and to R. F. Chapman, who made insightful suggestions on a portion of the manuscript. Thanks also to W. F. Tjallingii, who has broadened my view of aphid feeding behavior. Finally, a special thanks to the editor of this series, E. A. Bernays.

REFERENCES

1. **Acreman, T. M. and Dixon, A. F. G.,** The role of awns in the resistance of cereals to the grain aphid, *Sitobion avenae, Ann. Appl. Biol.*, 109, 375, 1986.
2. **Adams, J. B. and Wade, C. V.,** Aphid behavior and host-plant preference demonstrated by electronic patterns of probing and feeding, *Am. Pot. J.*, 53, 261, 1976.
3. **Ajayi, O.,** The effect of barley yellow dwarf virus on the amino acid composition of spring wheat, *Ann. Appl. Biol.*, 108, 145, 1986.
4. **Al-Mousawi, A. H., Richardson, P. E., and Burton, R. L.,** Ultrastructural studies of greenbug (Hemiptera, Aphididae) feeding damage to susceptible and resistant wheat cultivars, *Ann. Entomol. Soc. Am.*, 76, 964, 1983.
5. **Anderson, M. and Bromley, A. K.,** Sensory system, in *Aphids, Their Biology, Natural Enemies and Control,* Vol. 2A, Minks, A. K., and Harrewijn, P., Eds., Elsevier, Amsterdam, 1987, 153.
6. **Argandoña, V. H. and Corcuera, L. J.,** Distribution of hydroxamic acids in *Zea mays* tissue, *Phytochemistry,* 24, 177, 1985a.
7. **Argandoña, V. H. and Corcuera, L. J.,** Distribution of hydroxamic acids in maize tissues, *Plant Physiol.*, Suppl., 77, 1985.
8. **Argandoña, V. H., Corcuera, L. J., Neimeyer, H. M., and Campbell, B. C.,** Toxicity and feeding deterrency of hydroxamic acids from Graminae in synthetic diets against the greenbug, *Schizaphis graminum, Entomol. Exp. Appl.*, 34, 134, 1983.
9. **Argandoña, V. H., Luza, J. G., Niemeyer, H. M., and Corcuera, L. J.,** Role of hydroxamic acids in the resistance of cereals to aphids, *Phytochemistry,* 19, 1665, 1980.
10. **Argandoña, V. H., Niemeyer, H. M., and Corcuera, L. J.,** Effect of content and distribution of hydroxamic acids in wheat on infestation by the aphid *Schizaphis graminum, Phytochemistry,* 20, 673, 1981.
11. **Årgren, G. I.,** Limits to plant

production, *J. Theor. Biol.*, 113, 89, 1985.

12. **Auclair, J. L.,** Aphid feeding and nutrition, *Ann. Rev. Entomol.*, 8, 439, 1963.
13. **Backus, E. A.,** Sensory systems and behaviours which mediate hemipteran plant-feeding, a taxonomic overview, *J. Insect Physiol.*, 34, 151, 1988.
14. **Barlow, C. A. and McCully, M. E.,** The ruby laser as an instrument for cutting the stylets of feeding aphids, *Can. J. Zool.*, 50, 1497, 1972.
15. **Beck, D. L., Dunn, G. M., Routley, D. G., and Bowman, J. S.,** Biochemical basis of resistance in corn to the corn leaf aphid, *Crop Sci.*, 23, 995, 1983.
16. **Beevers, L.,** Senescence, in *Plant Biochemistry*, Bonner, J. and Varner, J. E., Eds., Academic Press, New York, 1976, 771.
17. **Berenbaum, M.,** Postingestive effects of phytochemicals on insects, on paracelsus and plant products, in *Insect-Plant Interactions*, Miller, J. R. and Miller, T. A., Eds., Springer-Verlag, New York, 1986, 121.
18. **Bernays, E. A. and Chapman, R. F.,** Plant chemistry and acridoid feeding behavior, in *Biochemical Aspects of Plant and Animal Coevolution*, Harborne, J. B., Ed., Academic Press, London, 1978, 141.
19. **Brzezina, A. S., Spiller, N. J., and Llewellyn, M.,** Mesophyll cell damage of wheat plants caused by probing of the aphid, *Metopolophium dirhodum, Entomol. Exp. Appl.*, 42, 195, 1986.
20. **Bolsinger, M. and Fluckiger, W.,** Enhanced aphid infestation at motorways, the role of ambient air pollution, *Entomol. Exp. Appl.*, 45, 237, 1987.
21. **Bright, S. W. J., Lea, P. J., Kueh, J. S. H., Woodcock, C., Holloman, D. W., and Scott, G. C.,** Proline content does not influence pest and disease susceptibility of barley, *Nature,* 295, 592, 1982.
22. **Brodbeck, B. and Strong, D.,** Amino acid nutrition of herbivorous insects and stress to host plants, in *Insect Outbreaks*, Barbosa, P. and Schultz, J. C., Eds., Academic Press, San Diego, CA, 1987, 347.
23. **Brown, C. M. and Holbrook, F. R.,** An improved electronic system for monitoring feeding of aphids, *Am. Pot. J.,* 53, 457, 1976.
24. **Campbell, B. C.,** Host-plant oligosaccharins in the honeydew of *Schizaphis graminum*(Rondani) (Insecta, Aphididae), *Experientia,* 42, 451, 1986.
25. **Campbell, B. C.,** The effects of plant growth regulators and herbicides on host-plant quality to insects, in *Plant Stress-Insect Interactions*, Heinrichs, E. A., Ed., John Wiley & Sons, New York, 1988, 205.
26. **Campbell, B. C.,** On the role of microbial symbiotes in herbivorous insects, in *Insect-Plant Interactions*, Vol. 1, Bernays, E. A., Ed., CRC Press, Boca Raton, FL, 1989, 1.
27. **Campbell, B. C. and Binder, R. G.,** Alfalfa cyclitols in the honeydew of an aphid, *Phytochemistry,* 23, 1786, 1984.
28. **Campbell, B. C. and Dreyer, D. L.,** Host-plant resistance of sorghum, differential hydrolysis of sorghum pectic substances by polysaccharases of greenbug biotypes (*Schizaphis graminum*) (Homoptera, Aphididae), *Arch. Insect Biochem. Physiol.,* 2, 203, 1985.
29. **Campbell, B. C. and Dreyer, D. L.,** The role of plant matrix polysaccharides in aphid-plant interactions, in *Aphid-Plant Genotype Interactions*, Campbell, R. K., and

Eikenbary, R. D. Eds., Elsevier, Amsterdam, 1990, 149.

30. **Campbell, B. C., Jones, K. C., and Dreyer, D. L.,** Discriminative behavioral responses by aphids to various plant matrix polysaccharides, *Entomol. Exp. Appl.*, 41, 17, 1986.
31. **Campbell, B. C., McLean, D. L., Kinsey, M. G., Jones, K. C., and Dreyer, D. L.,** Probing behavior of the greenbug (*Schizaphis graminum,* biotype C) on resistant and susceptible varieties of sorghum, *Entomol. Exp. Appl.,* 31, 140, 1982.
32. **Campbell, B. C. and Nes, W. D.,** A reappraisal of sterol biosynthesis and metabolism in aphids, *J. Insect Physiol.,* 29, 149, 1983.
33. **Chapman, R. F., Bernays, E. A., and Simpson, S. J.,** Attraction and repulsion of the aphid, *Cavariella aegopodii,* by plant odors, *J. Chem. Ecol.*, 7, 881, 1987.
34. **Ciepiela, A.,** Biochemical basis of winter wheat resistance to the grain aphid, *Sitobion avenae, Entomol. Exp. Appl.,* 51, 269, 1989.
35. **Cole, R. A.,** Phenolic acids associated with the resistance of lettuce cultivars to the lettuce root aphid, *Ann. Appl. Biol.,* 105, 129, 1984.
36. **Coleman, J. S. and Jones, C. G.,** Acute ozone stress on eastern cottonwood (*Populus deltoides* Bartr.) and the pest potential of the aphid, *Chaitophorus populicola* Thomas (Homoptera, Aphididae), *Environ. Entomol.,* 17, 207, 1988.
37. **Corcuera, L. J.,** Effects of indole alkaloids from Graminae on aphids, *Phytochemistry,* 23, 539, 1984.
38. **Corcuera, L. J., Argandoña, V. H., Peña, G. F., Perez, F. J., and Niemeyer, H. M.,** Effect of a benzoxazinone from wheat on aphids, in *Proc. 5th Int. Symp. Insect-Plant Relationships*, Pudoc, Wageningen, 1982, 33.
39. **Delrot, S.,** Phloem loading, apoplastic or symplastic? *Plant Physiol. Biochem.*, 25, 667, 1987.
40. **Dixon, A. F. G.,** Quality and availability of food for a Sycamore aphid population, in *Animal Populations in Relation to their Food Resources*, Watson, A., Ed., British Ecological Society, London, 1970, 271.
41. **Dohmen, G. P., McNeill, S., and Bell, J. N. B.,** Air pollution increases *Aphis fabae* pest potential, *Nature,* 307, 52, 1984.
42. **Dorschner, K. W. and Baird, C. R.,** Electronically monitored feeding behavior of *Phorodon humili* (Homoptera, Aphididae) on resistant and susceptible hop genotypes, *J. Insect Behav.,* 2, 437, 1989.
43. **Dorschner, K. W., Ryan, J. D., Johnson, R. C., and Eikenbary, R. D.,** Modification of host nitrogen levels by the greenbug (Homoptera, Aphididae), its role in resistance of winter wheat to aphids, *Environ. Entomol.*, 16, 1007, 1987.
44. **Downing, N. and Unwin, D. M.,** A new method for cutting the mouthparts of feeding aphids and for collecting plant sap, *Physiol. Entomol.*, 2, 275, 1977.
45. **Dreyer, D. L. and Campbell, B. C.,** Association of the degree of methylation of intercellular pectin with plant resistance to aphids and with induction of aphid biotypes, *Experientia,* 40, 224, 1984.
46. **Dreyer, D. L. and Campbell, B. C.,** Chemical basis of host-plant resistance to sap-feeding insects, *Rev. Latinoam. Quím.*, 17, 204, 1986.
47. **Dreyer, D. L. and Campbell, B. C.,** Chemical basis of host-plant resistance to aphids, *Plant Cell Environ.*, 10, 353, 1987.
48. **Dreyer, D. L., Campbell, B. C.,**

and Jones, K. C., Effect of bioregulator-treated sorghum on greenbug fecundity and feeding behavior, implications for host-plant resistance, *Phytochemistry,* 23, 1593, 1984.

49. **Dreyer, D. L. and Jones, K. C.,** Feeding deterrency of flavonoids and related phenolics towards *Schizaphis graminum* and *Myzus persicae*, aphid feeding deterrents in wheat, *Phytochemistry,* 20, 2489, 1981.
50. **Dreyer, D. L., Jones, K. C., Jurd, L., and Campbell, B. C.,** Feeding deterrency of some 4-hydroxycoumarins and related compounds, relationship to host-plant resistance of alfalfa towards pea aphid (*Acyrthosiphon pisum*), *J. Chem. Ecol.*, 13, 925, 1987.
51. **Dreyer, D. L., Jones, K. C., and Molyneux, R. J.,** Feeding deterrency of some pyrrolizidine, indolizidine, and quinolizidine alkaloids towards pea aphid (*Acyrthosiphon pisum*) and evidence for phloem transport of indolizidine alkaloid swainsonine, *J. Chem. Ecol.,* 11, 1045, 1985.
52. **Dreyer, D. L., Reese, J. C., and Jones, K. C.,** Aphid feeding deterrents in sorghum, bioassay, isolation and characterization, *J. Chem. Ecol.*, 7, 273, 1981.
53. **Evert, R. F., Eschrich, W., Eichornt, S. E., and Limbach, S. T.,** Observations on penetration of barley leaves by the aphid *Rhopalosiphum maidis*(Fitch), *Protoplasma,* 77, 95, 1973.
54. **Febvay, G., Bonnin, J., Rahbe, Y., Bournoville, R., Delrot, S., and Bonnemain, J. L.,** Resistance of different lucerne cultivars to the pea aphid *Acyrthosiphon pisum*, influence of phloem composition on aphid fecundity, *Entomol. Exp. Appl.*, 48, 127, 1988.
55. **Fife, J. M. and Frampton, V. L.,** The pH gradient extending from the phloem into the parenchyma of the sugar beet and its relation to the feeding behavior of *Eutettix tenellus*, *J. Agric. Res.,* 53, 581, 1936.
56. **Fisher, M.,** The effect of previously infested spruce needles on the growth of the green spruce aphid, *Elatobium abietinum*, and the effect of the aphid on the amino acid balance of the host plant, *Ann. Appl. Biol.*, 111, 33, 1987.
57. **Fisher, D. B. and Frame, J. M.,** A guide to the use of the exuding-stylet technique in phloem physiology, *Planta,* 161, 385, 1987.
58. **Fisk, J.,** Effects of HCN, phenolic acids and related compounds in *Sorghum bicolor* on the feeding behavior of the planthopper *Peregrinus maidis*, *Entomol. Exp. Appl.*, 27, 211, 1980.
59. **Fowler, S. V. and Lawton, J. H.,** Rapidly induced defenses and talking trees, the devil's advocate position. *Am. Nat.*, 126, 181, 1985.
60. **Fukumorita, T. and Chino, M.,** Sugar, amino acid and inorganic contents in rice phloem sap, *Plant Cell Physiol.*, 23, 273, 1982.
61. **Gall, L. F.,** Leaflet position influences caterpillar feeding and development, *Oikos,* 49, 172, 1987.
62. **Goffreda, J. C., Mutschler, M. A., Avé, D. A., Tingey, W. M., and Steffens, J. C.,** Aphid deterrence by glucose esters in glandular trichome exudate of the wild tomato, *Lycopersicon pennellii*, *J. Chem. Ecol.*, 15, 2135, 1989.
63. **Goffreda, J. C., Mutschler, M. A., and Tingey, W. M.,** Feeding behavior of potato aphid affected by glandular trichomes of wild tomato, *Entomol. Exp. Appl.,* 48, 101, 1988.
64. **Greenway, A. R., Griffiths, D. C.,**

and Lloyd, S. L., Response of *Myzus persicae* to components of aphid extracts and to carboxylic acids, *Entomol. Exp. Appl.*, 24, 369, 1978.

65. **Griffiths, D. C., Greenway, A. R., and Lloyd, S. L.**, The influence of repellent materials and aphid extracts on settling behaviour and larviposition of *Myzus persicae*(Sulzer) (Hemiptera, Aphididae), *Bull. Entomol. Res.*, 68, 613, 1978.
66. **Guthrie, F. E., Campbell, W. V., and Baron, R. L.**, Feeding sites of the green peach aphid with respect to its adaptation to tobacco, *Ann. Entomol. Soc. Am.*, 55, 42, 1962.
67. **Haniotakis, G. E. and Lange, W. H.**, Beet yellow virus resistance in sugar beets, mechanisms of resistance, *J. Econ. Entomol.*, 67, 25, 1974.
68. **Harris, K. F.**, An ingestion-egestion hypothesis of noncirculative virus transmission, in *Aphids as Virus Vectors*, Harris, K. F. and Maramarosch, K., Eds., Academic Press, New York, 1977, 165.
69. **Harris, K. F. and Bath, J. E.**, Regurgitation by *Myzus persicae* during feeding, its likely function in transmission of nonpersistent plant viruses, *Ann. Entomol. Soc. Am.,* 66, 793, 1973.
70. **Harris, P. J. and Hartley, R. D.**, Phenolic constituents of the cell wall of monocotyledons, *Biochem. Syst. Ecol.*, 8, 153, 1980.
71. **Hartley, R. D. and Keene, A. S.**, Aromatic aldehyde constituents of graminaceous cell walls, *Phytochemistry,* 23, 1305, 1984.
72. **Herbach, E.**, Role of semiochemicals in aphid-plant relationships. II. Allelochemicals, *Agronomie(Paris),* 5, 375, 1985.
73. **Hodges, L. R. and McLean, D. L.**, Correlation of transmission of bean yellow mosaic virus with salivation activity of *Acyrthosiphon pisum*(Homoptera, Aphididae), *Ann. Entomol. Soc. Am.*, 62, 1398, 1969.
74. **Holbrook, F. R.**, Transmission of potato leafroll virus by the green peach aphid, *Ann. Entomol. Soc. Am.*, 72, 830, 1978.
75. **Holbrook, F. R.**, An index of acceptability to green peach aphids for *Solanum* germplasm and for a suspected non-host plant, *Am. Pot. J.*, 57, 1, 1980.
76. **Ishii, S.**, Cell wall cementing materials of grass leaves, *Plant Physiol.*, 76, 959, 1984.
77. **Jepson, P. C.**, A controlled environment study of the effect of leaf physiological age on the movement of apterous *Myzus persicae* on sugar-beet plants, *Ann. Appl. Biol.*, 103, 173, 1983.
78. **Jones, K. C. and Klocke, J. A.**, Aphid feeding deterrency of ellagitannins, their phenolic hydrolysis products and related phenolic derivatives, *Entomol. Exp. Appl.*, 44, 229, 1987.
79. **Jördens-Rötger, D.**, The role of phenolic substances for host-selection behavior of the black bean aphid, *Aphis fabae*, *Entomol. Exp. Appl.*, 26, 49, 1979.
80. **Junde, Q. and Lidao, K.**, The influence of secondary plant substances on the growth and development of *Myzus persicae* of Beijing, *Entomol. Exp. Appl.*, 35, 17, 1984.
81. **Kain, W. M. and Biggs, D. R.**, Effect of pea aphid and bluegreen lucerne aphid (*Acyrthosiphon* spp.) on coumesterol levels in herbage of lucerne (*Medicago sativa*), *N. Z. J. Agric. Res.*, 23, 563, 1980.
82. **Karowe, D. N. and Martin, M. M.**, The effects of quantity and quality of diet nitrogen on the growth, efficiency of food utilization, nitrogen

budget, and metabolic rate of fifth-instar *Spodoptera eridania* larvae (Lepidoptera, Noctuidae), *J. Insect Physiol.*, 35, 699, 1989.

83. **Kennedy, J. S. and Booth, C. O.,** Host alternation in *Aphisfabae* Scop. I. Feeding preferences and fecundity in relation to the age and kind of leaves, *Ann. Appl. Biol.*, 38, 25, 1951.
84. **Kennedy, G. G., McLean, D. L., and Kinsey, M. G.,** Probing behavior of *Aphis gossypii* on resistant and susceptible muskmelon, *J. Econ. Entomol.*, 71, 13, 1978.
85. **Kennedy, J. S. and Mittler, T. E.,** A method of obtaining phloem sap via the mouthparts of aphids, *Nature,* 171, 528, 1953.
86. **Kidd, N. A. C., Lewis, G. B., and Howell, C. A.,** An association between two species of pine aphid, *Schizolachnus pineti* and *Eulachnus agilis*, *Ecol. Entomol.,* 10, 427, 1985.
87. **Kimmerer, T. A. and Potter, D. A.,** Nutritional quality of specific leaf tissues and selective feeding by a specialist leafminer, *Oecologia,* 71, 548, 1987.
88. **Kimmins, F.,** The probing behavior of *Rhopalosiphum maidis*, in *5th Int. Symp. Insect-Plant Relationships*, Pudoc, Wageningen, 1982, 411.
89. **Kimmins, F.,** Ultrastructure of the stylet pathway of *Brevicoryne brassicae* in host plant tissue, *Brassica oleracea*, *Entomol. Exp. Appl.*, 41, 283, 1986.
90. **Kimmins, F. M.,** Transmission electron microscopy (TEM), in *Aphids, Their Biology, Natural Enemies and Control*, Vol. 2B, Minks, A. K. and Harrewijn, P., Eds., Elsevier, Amsterdam, 1988, 47.
91. **Kimmins, F. M. and Tjallingii, W. F.,** Ultrastructure of sieve element penetration by aphid stylets during electrical recording, *Entomol. Exp. Appl.,* 39, 135, 1985.
92. **Kimsey, R. B. and McLean, D. L.,** Versatile electronic measurement system for studying probing and feeding behavior of piercing and sucking insects, *Ann. Entomol. Soc. Am.*, 80, 118, 1987.
93. **King, R. W. and Zeevart, J. A. D.,** Enhancement of phloem exudation from cut petioles by chelating agents, *Plant Physiol.,* 53, 96, 1974.
94. **Kiss, A.,** Host-selection behavior of nymphs of *Vanduzeea arquata* and *Enchenopa binotata* discrimination of whole twigs, leaf extracts and sap exudates, *Entomol. Exp. Appl.,* 36, 169, 1984.
95. **Klingauf, F.,** Die Wirkung des Glucosids Phlorizia auf des Wirtswahlverhalten von *Rh. insertum* (Walk.) und *A. pomi* DeGeer (Homoptera, Aphididae), *Z. Angew. Entomol.*, 68, 41, 1971.
96. **Klingauf, F. A.,** Host plant finding and acceptance, in *Aphids, Their Biology, Natural Enemies and Control*, Vol. 2A, Minks, A. K. and Harrewijn, P., Eds., Elsevier, Amsterdam, 1987, 209.
97. **Klingauf, F A.,** Feeding, adaptation and excretion, in *Aphids, Their Biology, Natural Enemies and Control*, Vol. 2A, Minks, A. K. and Harrewijn, P., Eds., Elsevier, Amsterdam, 1987, 225.
98. **Klingauf, F., Nöcker-Wenzel, K., and Klein, W.,** Influence of some wax components of *Vicia faba* L. on the food plant selection behaviour of *Acyrthosiphon pisum, Z. Planzenkr. Planzensch.*, 78, 641, 1971.
99. **Klingauf, F., Nöcker-Wenzel, K., and Röttger, U.,** The role of cuticle waxes in insect infestation behaviour, *Z. Pflanzenkr. Pflanzensch.,* 85, 228, 1978.
100. **Kojima, M., Poulton, J. E., Thayer, S. S., and Conn, E. E.,**

Tissue distributions of dhurrin and of enzymes involved in its metabolism in leaves of *Sorghum bicolor*, *Plant Physiol.*, 63, 1022, 1979.

101. **Kullman, A., Ogunlela, V. V., and Geisler, G.,** Concentrations and distribution of some mineral elements in oilseed rape (*Brassica napus* L.) plants in relation to nitrogen supply, *J. Agron. Crop Sci.,* 163, 225, 1989.
102. **Kunkel, H.,** Membrane feeding systems in aphid research, in *Aphids as Virus Vectors*, Harris, K. F. and Maramorosch, K., Eds., Academic Press, New York, 1977, 311.
103. **Lapointe, S. L. and Tingey, W. M.,** Feeding response of the green peach aphid (Homoptera, Aphididae) to potato glandular trichomes, *J. Econ. Entomol.*, 77, 386, 1984.
104. **Leckstein, P. M. and Llewellyn, M.,** The role of amino acids in diet intake and selection and the utilization of dipeptides by *Aphis fabae*, *J. Insect Physiol.*, 20, 877, 1974.
105. **Leonard, S. H. and Holbrook, F. R.,** Minimum acquisition and transmission times for potato leaf roll virus by the green peach aphid, *Ann. Entomol. Soc. Am.*, 71, 493, 1978.
106. **Leszcznski, B.,** Changes in phenols content and metabolism in leaves of susceptible and resistant winter wheat cultivars infested by *Rhopalosiphum padi* (L.), (Hom., Aphididae), *Z. Angew. Entomol.,* 100, 343, 1985.
107. **Leszczynski, B., Warchol, J., and Niraz, S.,** The influence of phenolic compounds on the preference of winter wheat cultivars by cereal aphids, *Insect Sci. Appl.,* 6, 157, 1985.
108. **Lewis, A. C. and van Emden, H. F.,** Assays for insect feeding, in *Insect-Plant Interactions*, Miller, J. R. and Miller, T. A., Eds., Springer-Verlag, New York, 1986, 95.
109. **Llewellyn, M. and Qureshi, A. L.,** The energetics of *Megoura vicae* reared on different parts of the broad bean plant (*Vicia faba*), *Entomol. Exp. Appl.*, 26, 127, 1979.
110. **Long, B. J., Dunn, G. M., Bowman, J. S., and Routley, D. G.,** Relationship of hydroxamic acid content in corn and resistance to the corn leaf aphid, *Crop Sci.*, 17, 55, 1977.
111. **Loper, G. M.,** Effect of aphid infestation on the coumesterol content of alfalfa varieties differing in aphid resistance, *Crop Sci.,* 8, 104, 1968.
112. **Lopez-Abella, D. and Bradley, R. H. E.,** Aphids may not acquire and transmit stylet-borne viruses while probing intercellularly, *Virology,* 39, 338, 1969.
113. **Lowe, H. J. B.,** Interspecific differences in the biology of aphids (Homoptera, Aphididae) on leaves of *Vicia faba*. I. Feeding behaviour, *Entomol. Exp. Appl.*, 10, 347, 1967.
114. **Lowe, H. J. B.,** Interspecific differences in the biology of aphids (Homotera, Aphididae) on leaves of *Vicia faba*. II. Growth and excretion, *Entomol. Exp. Appl.*, 10, 413, 1967.
115. **Lowe, H. J. B. and Russell, G.,** Probing by aphids in the leaf tissues of resistant and susceptible sugar beet, *Entomol. Exp. Appl.*, 17, 468, 1974.
116. **Lowe, S. and Strong, F. E.,** The unsuitability of some viruliferous plants as hosts for the green peach aphid, *Myzus persicae*, *J. Econ. Entomol.*, 56, 307, 1963.
117. **Lyth, M.,** Hypersensitivity in apple to feeding by *Dysaphis plantaginea,* effects on aphid biology, Ann. *Appl. Biol.*, 107, 155, 1985.
118. **Ma, R., Reese, J. C., Black, W. C., IV, and Bramel-Cox, P.,** Detection

of pectinesterase and polygalacturonase from salivary secretions of living greenbugs, *Schizaphis graminum* (Homoptera, Aphididae), *J. Insect Physiol.*, in press.

119. **MacKauer, M. and Way, M. J.,** *Myzus persicae* Sulz., an aphid of world importance, in *Studies in Biological Control,* Delucchi, V. L., Ed., Cambridge University Press, London, 1976, 51.
120. **Markkula, M. and Laurema, S.,** Changes in the concentration of free amino acids in plants induced by virus diseases and the reproduction of aphids, *Ann. Agric. Fenn.*, 3, 265, 1964.
121. **Maxwell, R. C. and Harwood, R. F.,** Increased reproduction of pea aphids on broad beans treated with 2,4-D, *Ann. Entomol. Soc. Am.,* 53, 199, 1960.
122. **McClure, M. S.,** Competition between herbivores and increased resource heterogeneity, in *Variable Plants and Herbivores in Natural and Managed Systems,* Denno, R. F. and McClure, M. S., Eds., Academic Press, New York, 1983, 125.
123. **McKey, D.,** The distribution of secondary compounds within plants, in *Herbivores, their Interactions with Secondary Plant Metabolites,* Rosenthal, G. A. and Janzen, D. H., Eds., Academic Press, New York, 1979, 56.
124. **McLean, D. L.,** An electrical measurement system for studying aphid probing behavior, in *Aphids as Virus Vectors,* Harris, K. F. and Maramarosch, K., Eds., Academic Press, New York, 1977, 277.
125. **McLean, D. L. and Kinsey, M. G.,** A technique for electronically recording aphid feeding and salivation, *Nature,* 202, 1358, 1964.
126. **McLean, D. L. and Kinsey, M. G.,** Identification of electronically recorded curve patterns associated with aphid salivation and ingestion, *Nature,* 205, 1130, 1965.
127. **McLean, D. L. and Kinsey, M. G.,** Probing behavior of the pea aphid, *Acyrthosiphon pisum.* I. Definitive correlation of electronically recorded waveforms with aphid probing activities, *Ann. Entomol. Soc. Am.*, 60, 400, 1967.
128. **McLean, D. L. and Kinsey, M. G.,** Probing behavior of the pea aphid, *Acyrthosiphon pisum.* II. Comparisons of salivation and ingestion in host and non-host plant leaves, *Ann. Entomol. Soc. Am.*, 61, 730, 1968.
129. **McLean, D. L. and Kinsey, M. G.,** The precibarial valve and its role in the feeding behavior of the pea aphid *Acyrthosiphon pisum, Bull. Entomol. Soc. Am.*, 30, 26, 1984.
130. **McLean, D. L. and Weigt, W. A.,** An electronic measuring system to record aphid salivation and ingestion, *Ann. Entomol. Soc. Am.*, 61, 180, 1968.
131. **McMurtry, J. A. and Stanford, E. H.,** Observations of feeding habits of the spotted alfalfa aphid on resistant and susceptible alfalfa plants, *J. Econ. Entomol.,* 53, 714, 1960.
132. **McNeill, S. and Southwood, T. R. E.,** The role of nitrogen in the development of insect/plant relationships, in *Biochemical Aspects of Plant and Animal Coevolution,* Harborne, J. B., Ed., Academic Press, London, 1978, 77.
133. **Mentink, P. J. M., Kimmins, F. M., Harrewijn, P., Dieleman, F. L., Tjallingii, W. F., van Rheenen, B., and Eenink, A. H.,** Electrical penetration graphs combined with stylet cutting in the study of host plant resistance to aphids, *Entomol. Exp. Appl.*, 36, 210, 1984.
134. **Metcalf, J. R.,** Studies on the effect of the nutrient status of sugar-cane on the fecundity of *Saccharosydne*

saccharivora (Westw.) (Homoptera, Delphacidae), *Bull. Entomol. Res.*, 60, 309, 1970.

135. **Miles, P. W.,** Insect secretions in plants, *Ann. Rev. Phytopathol.*, 6, 137, 1968.
136. **Miles, P. W.,** Interactions between plant phenols and salivary phenolases in the relationship between plants and Hemiptera, *Entomol. Exp. Appl.*, 12, 736, 1969.
137. **Miles, P. W.,** Dynamic aspects of the chemical relation between the rose aphid and rose buds, *Entomol. Exp. Appl.*, 37, 129, 1985.
138. **Miles, P. W.,** Feeding process of Aphidoidea in relation to effects on their food plants, in *Aphids, Their Biology, Natural Enemies and Control*, Vol. 2A, Minks, A. K. and Harrewijn, P., Eds., Elsevier, Amsterdam, 1987, 321.
139. **Miles, P. W., Aspinall, D., and Rosenberg, L.,** Performance of the cabbage aphid, *Brevicoryne brassicae* (L.), on water-stressed rape plants, in relation to changes in their chemical composition, *Aust. J. Zool.*, 30, 337, 1982.
140. **Miles, P. W. and Peng, Z.,** Studies on the salivary physiology of plant bugs, detoxification of phytochemicals by the salivary peroxidase of aphids, *J. Insect Physiol.*, 35, 865, 1989.
141. **Mittler, T. E.,** Effect of amino acid and sugar concentrations on the food uptake of the aphid *Myzus persicae*, *Entomol. Exp. Appl.*, 10, 39, 1967.
142. **Mittler, T. E.,** Application of artificial feeding techniques for aphids, in *Aphids, Their Biology, Natural Enemies and Control*, Vol. 2B, Minks, A. K. and Harrewijn, P., Eds., Elsevier, Amsterdam, 1988, 145.
143. **Mittler, T. E. and Dadd, R. H.,** Studies on artificial feeding and the aphid *Myzus persicae* (Sulzer). I. Relative uptake of water and sucrose solutions, *J. Insect Physiol.*, 9, 623, 1963.
144. **Mittler, T. E. and Dadd, R. H.,** Gustatory discrimination between liquids by the aphid *Myzus persicae* (Sulzer), *Entomol. Exp. Appl.*, 7, 315, 1964.
145. **Mittler, T. E. and Dadd, R. H.,** Differences in the probing responses of *Myzus persicae* (Sulzer) elicited by different feeding solutions behind a parafilm membrane, *Entomol. Exp. Appl.*, 8, 107, 1965.
146. **Moericke, V.,** Über die Lebensgewohnheiten der geflügelten Blattläuse (Aphidina) unter besonderer Berucksichtigung des Verhaltens beim Landen, *Z. Angew. Entomol.*, 37, 29, 1955.
147. **Molyneux, R. J., Campbell, B. C., and Dreyer, D. L.,** Honeydew analysis for detecting phloem transport of plant natural products, implications for host-plant resistance to sap-sucking insects, *J. Chem. Ecol.*, 16, 1899, 1990.
148. **Montgomery, M. E. and Arn, H.,** Feeding response of *Aphis pomi*, *Myzus persicae* and *Amphorophora agathonica* to phlorizin, *J. Insect Physiol.*, 20, 413, 1974.
149. **Montllor, C. B.,** Host-Plant Suitability for Aphids, Studies on Feeding Behavior and Plant Response, Ph.D. dissertation, University of California, Berkeley, 1985.
150. **Montllor, C. B., Campbell, B. C., and Mittler,T. E.,** Natural and induced differences in the probing behavior of the greenbug, *Schizaphis graminum*, in relation to resistance in sorghum, *Entomol. Exp. Appl.*, 34, 99, 1983.
151. **Montllor, C. B., Campbell, B. C., and Mittler, T. E.,** Responses of *Schizaphis graminium* (Homoptera, Aphididae) to leaf excision in resistant and susceptible

sorghum, *Ann. Appl. Biol.*, 16, 189, 1990.

152. **Montllor, C. B. and Gildow, F. E.,** Barley yellow dwarf virus in oats: effects on feeding behavior of two grain aphids, *Plant Physiol.*, Suppl., 75, 1358, 1985.
153. **Montllor, C. B. and Gildow, F. E.,** Feeding responses of two grain aphids to barley yellow dwarf virus-infected oats, *Entomol. Exp. Appl.*, 42, 63, 1986.
154. **Montllor, C. B. and Tjallingii, W. F.,** Stylet penetration by two aphid species on susceptible and resistant lettuce, *Entomol. Exp. Appl.*, 52, 103, 1989.
155. **Moon, M. S.,** Phagostimulation of a monophagous aphid, *Oikos,* 18, 96, 1966.
156. **Mullin, C. A.,** Adaptive divergence of chewing and sucking arthropods to plant allelochemicals, in *Molecular Aspects of Insect-Plant Associations*, Brattsten, L. B. and Ahmad, S., Eds., Plenum, New York, 1986, 175.
157. **Mullin, C. A. and Croft, B. A.,** *Trans*-epoxide hydrolase, a key indicator enzyme for herbivory in arthropods, *Experientia,* 40, 176, 1984.
158. **Nault, L. R. and Styer, W. E.,** Effects of sinigrin on host selection of aphids, *Entomol. Exp. Appl.*, 15, 423, 1972.
159. **Newbery, D. McC.,** Interactions between the coccid, *Icerya seychellarum* (Westw.) and its host tree species on Aldabra atoll. I. *Euphorbia pyrifolia* Lam., *Oecologia (Berlin),* 46, 171, 1980.
160. **Niassy, A., Ryan, J. D., and Peters, D. C.,** Variations in feeding behavior, fecundity and damage of biotypes B and E of *Schizaphis graminum* (Homoptera, Aphididae) on three wheat genotypes, *Environ. Entomol.*, 16, 1163, 1987.
161. **Nielson, M. W. and Don, H.,** Probing behavior of biotypes of the spotted alfalfa aphid on resistant and susceptible alfalfa clones, *Entomol. Exp. Appl.*, 17, 447, 1974.
162. **Niemelä, P., Tuomi, J., and Siren, S.,** Selective herbivory on mosaic leaves of variegated *Acer pseudoplatanus, Experientia,* 40, 1433, 1984.
163. **Nishida, R. and Fukami, H.,** Host-plant iridoid-based chemical defense of an aphid, *Acyrthosiphon nipponicus*, against ladybird beetles, *J. Chem. Ecol.,* 15, 1837, 1989.
164. **Oka, I. N. and Pimentel, D.,** Herbicide (2,4-D) increased insect and pathogen pests on corn, *Science,* 193, 239, 1976.
165. **Parry, W. H.,** The effects of nitrogen levels in Sitka spruce needles on *Elatobium abietinum* populations in north-eastern Scotland, *Oecologia,* 15, 305, 1974.
166. **Parry, W. H.,** The effect of needle age on the acceptability of Sitka spruce needles to the aphid *Elatobium abietinum* (Walker), *Oecologia,* 23, 297, 1976.
167. **Pate, J. S.,** Transport and partitioning of nitrogenous solutes, *Ann. Rev. Plant Physiol.*, 31, 313, 1980.
168. **Pate, J. S.,** Patterns of nitrogen metabolism in higher plants and their ecological significance, in *Nitrogen as an Ecological Factor*, Lee, J. A., McNeill, S., and Rorison, I. H., Eds., Blackwell Scientific, Oxford, 1983, 225.
169. **Peng, Z. and Miles, P. W.,** Acceptability of catechin and its oxidative condensation products to the rose aphid, *Macrosiphum rosae, Entomol. Exp. Appl.*, 47, 255, 1988.
170. **Peng, Z. and Miles, P. W.,** Studies on the salivary physiology of plant bugs, function of the catechol oxidase of the rose aphid, *J. Insect*

Physiol., 34, 1027, 1988.

171. **Peters, D. C., Kerns, D. Puterka, G. J., and McNew, R.,** Feeding behavior, development, and damage by biotypes B, C, and E of *Schizaphis graminum* (Homoptera, Aphididae) on 'Winter Malt' and 'Post' barley, *Environ. Entomol.*, 17, 503, 1988.
172. **Pettersson, J.,** Studies on *Rhopalosiphum padi* (L.). I. Laboratory studies on olfactometric responses to the winter host *Prunuspadus* L, *Lantbr. Hogsk. Ann.,* 36, 381, 1970.
173. **Pettersson, J.,** Olfactory reactions of *Brevicoryne brassicae* (L.) (Homoptera, Aphididae), *Swed. J. Agric. Res.*, 3, 95, 1973.
174. **Pfeiffer, D. G. and Burts, E. C.,** Effects of tree fertilization on numbers and development of pear psylla and on fruit damage, *Environ. Entomol.*, 12, 895, 1983.
175. **Pfeiffer, D. G. and Burts, E. C.,** Effect of tree fertilization on protein and free amino acid content and feeding rate of pear psylla (Homoptera, Psyllidae), *Environ. Entomol.*, 13, 1487, 1984.
176. **Poehling, H. M. and Dörfer, K.,** Uptake and utilization of proteins, peptides and amino acids by *Aphis fabae,* Abstr. 17th Int. Congr. Entomology, Hamburg, 1984, 143.
177. **Pollard, D. G.,** Plant penetration by feeding aphids (Hemiptera, Aphidoidea), a review, *Bull. Entomol. Res.*, 62, 631, 1973.
178. **Prestidge, R. A.,** Instar duration, adult consumption, oviposition and nitrogen utilization efficiencies of leafhoppers feeding on different quality food (Auchenorrhyncha, Homoptera), *Ecol. Entomol.*, 7, 91, 1982.
179. **Prestidge, R. A.,** The influence of nitrogen fertilizer on the grassland Auchenorrhyncha (Homoptera), *J. Appl. Ecol.*, 19, 735, 1982.
180. **Prestidge, R. A. and McNeill, S.,** The role of nitrogen in the ecology of grassland Auchenorrhyncha, in *Nitrogen as an Ecological Factor,* Vol. 22, Lee, J. A., McNeill, S., and Rorison, I. H., Eds., Blackwell Scientific, Oxford, 1983, 257.
181. **Raven, J. A.,** Phytophages of xylem and phloem, a comparison of animal and plant sap-feeders, *Adv. Ecol. Res.*, 13, 136, 1983.
182. **Risebrow, A. and Dixon, A. F. G.,** Nutritional ecology of phloem-feeding insects, in *Nutritional Ecology of Insects, Mites, Spiders, and Related Invertebrates,* Slansky, F., Jr. and Rodriguez, J. G., Eds., John Wiley & Sons, New York, 1987, 421.
183. **Rohrfritsch, O.,** A "defense" mechanism of *Picea excelsa* L. against the gall former *Chermes abietis* L. (Homoptera, Adelgidae), *Z. Angew. Entomol.*, 92, 18, 1981.
184. **Rose, A. F., Jones, K. C., Haddon, W. F., and Dreyer, D. L.,** Grindelane diterpenoid acids from *Grindelia humilis,* feeding deterrency of diterpene acids towards aphids, *Phytochemistry,* 20, 2249, 1981.
185. **Ryan, J. D., Dorschner, K. W., Girma, M., Johnson, R. C., and Eikenbary, R. D.,** Feeding behavior, fecundity and honeydew production of two biotypes of greenbug (Homoptera, Aphididae) on resistant and susceptible wheat, *Environ. Entomol.*, 16, 757, 1987.
186. **Ryan, J. D., Mort, A. J., and Johnson, R. C.,** Middle lamellar pectins and greenbug resistance in wheat. *Plant Physiol.*, Suppl., 80, 18, 1986.
187. **Saunders, J. A. and Conn, E. E.,** Presence of the cyanogenic glucoside dhurrin in isolated vacuoles from *Sorghum, Plant Physiol.*, 61, 154, 1978.

188. **Saxena, P. N. and Chada, H. L.,** The greenbug, *Schizaphis graminum.* I. Mouth parts and feeding habits, *Ann. Entomol. Soc. Am.*, 64, 897, 1971.
189. **Schaefers, G. A.,** The use of direct current for electronically recording aphid feeding and salivation, *Ann. Entomol. Soc. Am.*, 59, 1022, 1966.
190. **Schalk, J. M., Kinder, S. D., and Manglitz, G. R.,** Temperature and the preference of the spotted alfalfa aphid for resistant and susceptible alfalfa plants, *J. Econ. Entomol.*, 62, 1000, 1969.
191. **Scheller, H. V. and Shukle, R. H.,** Feeding behavior and transmission of barley yellow dwarf virus by *Sitobion avenae* on oats, *Entomol. Exp. Appl.*, 40, 189, 1986.
192. **Schmidt, D. J.,** The Physiology of Greenbug Feeding Behavior and the Effect of Salivary Toxins on Sorghum Plants, Ph. D. dissertation, Kansas State University, Manhattan, 1987.
193. **Schoonhoven, L. M. and Derksen-Koppers,** Effects of some allelochemics on food uptake and survival of a polyphagous aphid, *Myzus persicae*, *Entomol. Exp. Appl.*, 19, 52, 1976.
194. **SenGupta, G. C. and Miles, P. W.,** Studies on the susceptibility of varieties of apple to the feeding of two strains of woolly aphis (Homoptera) in relation to the chemical content of the tissues of the host, *Aust. J. Agric. Res.*, 26, 157, 1975.
195. **Shanks, C. H., Jr. and Chase, D.,** Electrical measurement of feeding by the strawberry aphid on susceptible and resistant strawberries and non-host plants, *Ann. Entomol. Soc. Am.*, 69, 784, 1976.
196. **Shanks, C. H. and Finnigan, B.,** Probing behavior of the strawberry aphid, *Ann. Entomol. Soc. Am.*, 63, 734, 1970.
197. **Sherwood, M. H., Greenway, A. R., and Griffiths, D. C.,** Response of *Myzus persicae*(Sulzer) (Hemiptera, Aphididae) to plants treated with fatty acids, *Bull. Entomol. Res.*, 71, 133, 1981.
198. **Shukle, R. H., Lampe, D. J., Lister, R. M., and Foster, J. E.,** Aphid feeding behavior, relationship to barley yellow dwarf virus resistance in *Agropyron* species, *Phytopathology,* 77, 725, 1987.
199. **Simpson, S. J.,** Patterns in feeding, a behavioural analysis using *Locusta migritoria* nymphs, *Physiol. Entomol.*, 7, 325, 1982.
200. **Slater, P. J. B.,** The temporal patterns of feeding in the zebra finch, *Anim. Behav.*, 122, 506, 1974.
201. **Smith, B. D.,** Effect of the plant alkaloid sparteine on the distribution of the aphid *Acyrthosiphon spartii* (Koch), *Nature,* 212, 213, 1966.
202. **Spiller, N. J., Kimmins, F. M., and Llewellyn, M.,** Fine structure of aphid stylet pathways and its use in host plant resistance studies, *Entomol. Exp. Appl.*, 38, 293, 1985.
203. **Spiller, N. J., Tjallingii, W. F., and Llewellyn, M. J.,** Xylem ingestion by aphids, in *Proc. 6th Int. Symp Insect-Plant Relationships*, Dr. W. Junk, Dordrecht, 1987, 411.
204. **Srivastava, P. M. and Auclair, J. L.,** Effect of amino acid concentration on diet uptake and performance by the pea aphid, *Acyrthosiphon pisum* (Homoptera, Aphididae), *Can. Entomol.*, 106, 149, 1974.
205. **Srivastava, P. N., Lambein, F., and Auclair, J. L.,** Nonprotein amino acid-aphid interaction, Phagostimulatory effects and survival of the pea aphid, *Acyrthosiphon pisum*, *Entomol. Exp. Appl.*, 48, 109, 1988.
206. **Stewart, G. R. and Larher, F.,** Accumulation of amino acids and re-

lated compounds in relation to environmental stress, in *The Biochemistry of Plants*, Vol. 5, Academic Press, New York, 1980, 609.

207. **Tamaki, G., Butt, B. A., and Landis, B. J.,** Arrest and aggregation of male *Myzus persicae* (Hemiptera, Aphididae), *Ann. Entomol. Soc. Am.*, 63, 955, 1970.
208. **Tarn, T. R. and Adams, J. B.,** Resistance for the green peach aphid in some potato species, *Am. Pot. J.*, 50, 383, 1973.
209. **Tedders, W. L.,** Important Biological and Morphological Characteristics of the Foliar-Feeding Aphids of Pecan, USDA Tech. Bull. 1579, USDA, Washington, D.C., 1978.
210. **Tedders, W. L. and Thompson, J. M.,** Histological Investigation of Stylet Penetration and Feeding Damage to Pecan Foliage by Three Aphids (Hemiptera (Homoptera), Aphididae), Misc. Publ. No. 12, Entomological Society of America, 1981, 69.
211. **Tjallingii, W. F.,** A preliminary study of host selection and acceptance behaviour in the cabbage aphid, *Symp. Biol. Hung.*, 16, 283, 1976.
212. **Tjallingii, W. F.,** Electronic recording of penetration behavior by aphids, *Entomol. Exp. Appl.*, 24, 521, 1978.
213. **Tjallingii, W. F.,** Mechanoreceptors of the aphid labium, *Entomol. Exp. Appl.*, 24, 531, 1978.
214. **Tjallingii, W. F.,** Electrical nature of recorded signals during stylet penetration by aphids, *Entomol. Exp. Appl.*, 38, 177, 1985.
215. **Tjallingii, W. F.,** Membrane potentials as an indication for plant cell penetration by aphid stylets, *Entomol. Exp. Appl.*, 38, 187, 1985.
216. **Tjallingii, W. F.,** Electrical recording of stylet penetration activities, in *Aphids, Their Biology, Natural Enemies and Control*, Vol. 2B, Minks, A. K. and Harrewijn, P., Eds., Elsevier, Amsterdam, 1988, 95.
217. **Todd, G. W., Getahun, A., and Cress, D. C.,** Resistance in barley to the greenbug, *Schizaphisgraminum*. I. Toxicity of phenolic and flavonoid compounds and related substances, *Ann. Entomol. Soc. Am.*, 64, 718, 1971.
218. **Traynier, R. M. M. and Hines, E. R.,** Probes by aphids indicated by stain induced fluorescence in leaves, *Entomol. Exp. Appl.*, 45, 198, 1988.
219. **Tully, R. E. and Hanson, A. D.,** Amino acids translocated from turgid water-stressed barley leaves. I. Phloem exudation studies, *Plant Physiol.,* 64, 460, 1979.
220. **Ullman, D. E., Qualset, C. O., and McLean, D. L.,** Feeding responses of *Rhopalosiphum padi* (Homoptera, Aphididae) to barley yellow dwarf virus resistant and susceptible varieties, *Environ. Entomol.*, 17, 988, 1988.
221. **Van Emden, H. F.,** Plant resistance to aphids induced by chemicals, *J. Sci. Food Agric.*, 20, 385, 1969.
222. **Van Emden, H. F.,** Aphids as phytochemists, in *Phytochemical Ecology*, Harborne, J. B., Ed., Academic Press, New York, 1972, 25.
223. **Van Emden, H. F.,** Insects and secondary plant substances, an alternative viewpoint with special reference to aphids, in *Biochemical Aspects of Plant and Animal Coevolution*, Harborne, J. B., Ed., Academic Press, New York, 1978, 309.
224. **Van Emden, H. F. and Bashford, M. A.,** A comparison of the reproduction of *Brevicoryne brassicae* and *Myzus persicae* in relation to soluble nitrogen concentration and leaf age (leaf position) in the Brus-

sels sprout plant, *Entomol. Exp. Appl.*, 12, 351, 1969.

225. **Van Emden, H. F. and Bashford, M. A.,** The performance of *Brevicoryne brassicae* and *Myzus persicae* in relation to plant age and leaf amino acids, *Entomol. Exp. Appl.*, 14, 349, 1971.
226. **Van Emden, H. F. and Bashford, M. A.,** The effect of leaf excision on the performance of *Brevicoryne brassicae* and *Myzus persicae* in relation to the nutrient treatment of the plants, *Physiol. Entomol.*, 1, 67, 1976.
227. **Van Emden, H. F., Eastop, V. F., Hughes, R. D., and Way, M. J.,** The ecology of *Myzus persicae*, *Ann. Rev. Entomol.*, 14, 197, 1969.
228. **Visser, J. H. and Taanman, J. W.,** Odour-conditioned anemotaxis of apterous aphids (*Cryptomyzus korschelti*) in response to host plants, *Physiol. Entomol.*, 12, 473, 1987.
229. **Völkl, W.,** Resource partitioning in a guild of aphid species associated with creeping thistle *Cirsium arvense*, *Entomol. Exp. Appl.*, 51, 41, 1989.
230. **Walker, G. P.,** Probing behavior of *Aphis helianthi* (Homoptera, Aphididae) and its preference for *Pittosporum tobira* leaves of different ages, *Pan-Pac. Entomol.*, 63, 258, 1987.
231. **Walker, G. P. and Gordh, G.,** The occurrence of apical labial sensilla in the Aleyrodidae and evidence for a contact chemosensory function, *Entomol. Exp. Appl.*, 51, 215, 1989.
232. **Walker-Simmons, M., Holländer-Czytko, H., Andersen, J. K., and Ryan, C. A.,** Wound signals in plants, a systemic plant wound signal alters plasma membrane integrity, *Proc. Natl. Acad. Sci.*, 81, 373, 1984.
233. **Wearing, C. H.,** Responses of aphids to pressure applied to liquid diet behind parafilm membrane, Longevity and larviposition of *Myzuspersicae*(Sulz.) and *Brevicorynebrassicae* (L.) (Homoptera, Aphididae) feeding on sucrose and sinigrin solutions, *N. Z. J. Sci.*, 11, 105, 1968.
234. **Weibull, J.,** Seasonal changes in free amino acids of oat and barley phloem sap in relation to plant growth stage and growth of *Rhopalosiphum padi*, *Ann. Appl. Biol.*, 111, 729, 1987.
235. **Weibull, J.,** Free amino acids of phloem sap from oats and barley resistant to *Rhopalosiphum padi* (L.). *Phytochemistry,* 27, 2069, 1988.
236. **Weibull, J.,** Resistance in the wild crop relatives *Avena macrostachya* and *Hordeum bogdani* to the aphid *Rhopalosiphum padi*, *Entomol. Exp. Appl.*, 48, 225, 1988.
237. **Weibull, J., Brishammer, S., and Pettersson, J.,** Amino acid analysis of phloem sap from oats and barley, a combination of aphid stylet excision and high performance liquid chromotography, *Entomol. Exp. Appl.*, 42, 27, 1986.
238. **Wellings, P. W. and Dixon, A. F. G.,** Sycamore aphid numbers and population density. III. The role of aphid-induced changes in plant quality, *J. Anim. Ecol.*, 56, 161, 1987.
239. **Wensler, R. J. D.,** Mode of host selection by an aphid, *Nature,* 195, 830, 1962.
240. **Wensler, R. J. D.,** The fine structure of distal receptors on the labium of the aphid, *Brevicoryne brassicae* L., *Cell Tissue Res.*, 181, 409, 1977.
241. **Wensler, R. J. D. and Filshie, B. K.,** Gustatory sense organs in the food canal of aphids, *J. Morphol.*, 129, 473, 1969.

242. **White, T. C. R.,** An index to measure weather-induced stress of trees associated with outbreaks of psyllids in Australia, *Ecology,* 50, 905, 1969.
243. **White, T. C. R.,** The abundance of invertebrate herbivores in relation to the availability of nitrogen in stressed food plants, *Oecologia (Berlin),* 63, 90, 1984.
244. **Wink, M.,** Storage of quinolizidine alkaloids in epidermal tissues, *Z. Naturforsch.,* 41c, 375, 1986.
245. **Wink, M., Hartmann, T., Witte, L., and Rheinheimer, J.,** Interrelationship between quinolizidine alkaloid-producing legumes and infesting insects, exploitation of the alkaloid-containing phloem sap of *Cytisus scoparius* by the broom aphid *Aphis cytisorum, Z. Naturforsch.*, 37c, 1081, 1982.
246. **Wink, M. and Römer, P.,** Acquired toxicity — the advantages of specializing on alkaloid-rich lupins to *Macrosiphonalbifrons* (Aphididae), *Naturwissenschaften,* 73, 210, 1986.
247. **Wink, M. and Witte, L.,** Turnover and transport of quinolizidine alkaloids. Diurnal fluctuations of lupanine in the phloem sap, leaves and fruits of *Lupinusalbus* L., *Planta,* 161, 519, 1984.
248. **Witham, T. G.,** Habitat selection by *Pemphigus* aphids in response to resource limitation and competition, *Ecology,* 59, 1164, 1978.
249. **Witham, T. G.,** Territorial behavior of *Pemphigus* gall aphids, *Nature,* 279, 324, 1979.
250. **Witham, T. G.,** The theory of habitat selection, examined and extended using *Pemphigus* aphids, *Am. Nat.*, 115, 449, 1980.
251. **Woodhead, S.,** Environmental and biotic factors affecting phenolic content of different cultivars of *Sorghum bicolor, J. Chem. Ecol.*, 7, 1035, 1981.
252. **Woodhead, S. and Cooper-Driver, G.,** Phenolic acids and resistance to insect attack in *Sorghum bicolor, Biochem. Syst. Ecol.*, 7, 309, 1979.
253. **Woodhead, S., Galleffi, C., and Bettolo, G. B. M.,** p-Hydroxybenzaldehyde as a major constituent of the epicuticular wax of seedling *Sorghum bicolor, Phytochemistry,* 21, 455, 1982.
254. **Woodhead, S., Padgham, D. E., and Bernays, E. A.,** Insect feeding on different sorghum cultivars in relation to cyanide and phenolic acid content, *Ann. Appl. Biol.*, 95, 151, 1980.
255. **Wratten, S. D.,** Aggregation in the birch aphid *Euceraphis punctipennis* (Zett.) in relation to food quality, *J. Anim. Ecol.*, 43, 191, 1974.
256. **Wratten, S. D., Edwards, P. J., and Dunn, I.,** Wound-induced changes in the palatability of *Betula pubescens* and *B. pendula, Oecologia,* 61, 372, 1984.
257. **Yan, F. S. and Visser, J. H.,** Electroantennogram responses of the cereal aphid *Sitobion avenae* to plant volatile compounds, in *Proc. 5th Int. Symp. Insect-Plant Relationships*, Pudoc, Wageningen, 1982, 387.
258. **Zucker, M. V.,** How aphids choose leaves, the roles of phenolics in host selection by a galling aphid, *Ecology,* 63, 972, 1982.
259. **Zúñiga, C. E. and Corcuera, L. J.,** Effect of gramine on the resistance of barley seedlings to the aphid *Rhopalosiphum padi, Entomol. Exp. Appl.*, 40, 259, 1986.
260. **Zúñiga, G. E. and Corcuera, L. J.,** Glycine-betaine in wilted barley reduces the effects of gramine on aphids, *Phytochemistry,* 26, 3197, 1987.

261. **Zúñiga, C. E., Salgado, M. S., and Corcuera, L. J.,** Role of an indole alkaloid on the resistance of barley seedlings to aphids, *Phytochemistry,* 24, 945, 1985.
262. **Zúñiga, G. E., Varanda, E. M., and Corcuera, L. J.,** Effect of gramine on the feeding behavior of the aphids *Schizaphis graminum*and *Rhopalosiphum padi, Entomol. Exp. Appl.*, 47, 161, 1988.
263. **Montllor, C. and Tjallingii, W. F.,** unpublished data.
264. **Dreyer, D. L., Jones, K. C., and Campbell, B. C.,** unpublished.

5 How Insect Herbivores Find Suitable Host Plants: The Interplay between Random and Nonrandom Movement

William F. Morris
Center for Population Biology
University of California
Davis, California

Peter M. Kareiva
Department of Zoology
University of Washington
Seattle, Washington

TABLE OF CONTENTS

I. INTRODUCTION

Laboratory studies often give the impression that phytophagous insects are extremely effective at orienting to stimuli that signal food plant quality; indeed, a vast literature surrounds host plant finding in insects that emphasizes taxis and orientation.[1] On the other hand, when biologists follow the movements of individual insects under natural conditions, they usually obtain a very different impression of the sophistication with which insects search; in particular, field observations often reveal phytophagous insects wandering past food plants,[23] departing high-quality plants,[51] or simply bumbling about in a senseless manner.[15,16] Thus, one is faced with two contrasting views of herbivore searching: (1) the "laboratory view", which emphasizes orderly search and highly directed motion, and (2) the "field view", which emphasizes randomness and seemingly erratic behavior. This chapter utilizes a mathematical model to unify these two points of view, and to provide one framework in which to quantify the role of both random and directed movement in herbivore search. The contribution of random movement to searching effectiveness has been underemphasized, partly because laboratory studies may detect striking orientation behaviors that are greatly diminished under natural conditions with many competing stimuli. Thus, one of the main messages of the chapter is that the random element of herbivore search can be extremely important; in fact the model demonstrates how the addition of random movement to directed search can actually enhance the effectiveness of the searching process. Finally, the model is used to derive suggestions for ways in which field biologists might dissect the components of behavior that allow insects to concentrate on the best food plants.

The first step is to delineate three major ways in which insects alter their movement in response to plant suitability, thereby introducing nonrandomness into the insects' patterns of movement. The authors then present a mathematical model that encompasses these three behaviors and that explicitly incorporates spatial variation in plant suitability. The model is used to explore the joint impact of random and nonrandom movement on the ability of herbivores to find food. In order to link abstract reasoning to the practical concerns of empiricists, are reviewed experimental studies of movement, emphasizing those factors that model implicates as important ingredients of insect searching behavior. Finally, guidelines are presented by which future studies can use models to make quantitative predictions about the outcome of host plant search by herbivorous insects.

II. HOW MOVEMENT PATTERNS CHANGE IN RESPONSE TO SPATIAL VARIATION IN FOOD PLANT QUALITY

The pattern with which an insect (or any organism) moves can be decomposed into a few simple behavioral characteristics: (1) the proportion of time spent moving; (2) the rate of movement when actually in motion; and (3) the direction of movement. An insect searches randomly for food plants if the frequency, rate, and orientation of movement are unrelated to the suitability of plants within the perceptual range of the insect. In contrast, searching is nonrandom if any or all of these three components of movement are sensitive to plant suitability. This provides a convenient framework with which to classify sources of nonrandomness in insect movement.

The simplest form of nonrandom search involves cases in which both the rate and the direction of movement are entirely random, but in which herbivores either settle down or remain mobile, depending on the quality of the plants they encounter. This mechanism, which is labeled as "nonrandom settlement", requires only that an insect responds to information (e.g., olfactory or "taste" cues[37,41,92]) provided by the plant on which it is currently resting; moreover, the insect's response includes only two alternatives: to either "halt" or "not halt" movement. Since even the simplest microrganisms are capable of moving or not moving depending on local conditions, it is reasonable to assume that "nonrandom settlement" is part of the behavioral repertoire of most herbivorous insects.

A second, slightly more complex nonrandom behavior involves continuous movement, but at a rate that declines in the immediate vicinity of superior food plants. Such an adjustment in movement rate could be accomplished by altering the frequency of between-plant hops or flights, or the velocity of motion, or how often an individual reverses direction (so that it covers the same ground repeatedly). A classic example of adjusting local rates of net movement is "area restricted search", which refers to an increase in turning frequency and/or a decrease in velocity in the neighborhood of favorable resource stimuli. Area restriced search has been reported for numerous phytophagous insects[28,75] as well as for a wide diversity of other organisms (e.g. chemotactic bacteria,[56] insect predators,[19,20,34,39,54] flies searching for sugar droplets,[40,72] and foraging bumblebees[46]). In one sense, area restricted search is simply a refined version of nonrandom settlement, with the major difference being that instead of an "all or nothing" response (movement or settlement), the individual exhibits a graded response. Importantly, in both "nonrandom settlement" and "area restricted search", it is the local conditions (i.e., what the insect experiences at its current position) that influence behavior.

Finally, herbivores may use cues about the relative suitability of neighboring plants (e.g., odors[58,71,98] or visual cues[57,73,77,82,83]) to alter their direction of movement. This mechanism, referred to as "orientation", requires a relatively well-developed sensory ability, since the insect must be able to detect spatial gradients in plant suitability or in indicators of plant suitability. In other words, the behavior of the insect is affected not only by local conditions, but also by conditions at neighboring points in space.

Although theoretical biologists have developed several models that translate the above suites of behaviors into equations describing patterns of movement,[27,56,76] these models have typically focused on only a single form of nonrandom behavior at a time. By restricting models to single-factor sources of nonrandomness, it is often possible to generate analytical solutions, or to at least obtain numerical generalizations; however, one cannot rely on such simplified models if one is interested in the relative contributions of different mechanisms of nonrandom behavior to overall searching efficiency. Consequently, a less elegant framework is adopted that allows the generation of quantitative comparisons of particular behavioral strategies, but does not produce general conclusions.

III. A MATHEMATICAL DESCRIPTION OF HERBIVORE SEARCH

Constructing a model of herbivore foraging movement necessitates a quantitative measure of plant suitability. Following the lead of Edelstein-Keshet[29] and Edelstein-Keshet and Rausher,[30] plant suitability is represented with a dimensionless variable Q that is constrained to vary between 0 to 1, and that is meant to measure the relative fitness of herbivores feeding on a particular plant. Plants with a Q value of 0 are completely unsuitable (i.e., herbivores feeding on that species are unable to grow and reproduce). In contrast, herbivores feeding on plants with $Q = 1$ grow and reproduce at their maximum rate. Hosts of intermediate quality have Q values that lie between 0 and 1, with their exact Q given as "observed intrinsic rate of multiplication" divided by the "maximum rate of multiplication" (which by definition is obtained on the highest quality plants). A heterogeneous population of plants characterized by an array of Q values could represent different species, different varieties of the same species, different genotypes of the same species, or simply plants that differ because they grow on different soils or have been stressed to differing degrees. The key idea is that in order to draw quantitative conclusions about searching efficiency, one must first define the possible plant choices along a well-defined and biologically meaningful scale.

The next important distinction to be made in the model is to divide all insects into one of two categories: those in the process of moving, whose density at position x and time t is represented by the variable $M(x,t)$, and sessile individuals, with density $S(x,t)$. Having defined the three key variables, Q, M, and S, it is now possible to write a general model for insect foraging in heterogeneous environments:

$$\frac{\partial M}{\partial t} = F(M,S,Q) + \frac{\partial}{\partial x}[D(Q)\frac{\partial M}{\partial x} + Y(Q)M\frac{\partial Q}{\partial x}] \quad (1)$$

$$\frac{\partial S}{\partial t} = -F(M,S,Q) \quad (2)$$

The derivative, $\partial M/\partial t$, which describes changes in the density of mobile insects at a particular point in space, has been written so that the three forms of nonrandom search identified in the previous section can be associated with three distinct terms on the right side of Equation 1. Each of the three terms describe how a population (or the probability density for the location of an individual forager) will change when confronted by spatial variation in Q; each term can be formally derived by extrapolating patterns of individual behavior to the population level.[76] Specifically, the function F (M,S,Q) represents transition of mobile insects into the sessile class, depending on the suitability Q(x) of the plant at position x; this term arises when insects settle on plants in accord with the quality of those plants. The remaining two terms are both components of the "spatial flux" ($\partial/\partial x$), or flow of individuals along the x-dimension. The first component of the flux, D(Q) $\partial M/\partial x$, represents motion whose direction is random but whose rate, D(Q), varies with plant quality; this term arises when insects adjust their rate of movement either in response to the quality of plants averaged over the local area or in response to the quality of the plant on which they are directly sitting. The last contribution to the flux of individuals, Y M $\partial Q/\partial x$, represents tactic movement toward more suitable plants. This taxis term can arise if individual insects orient so that they tend to move toward plants of higher quality. Less obviously, a taxis term is also obtained when the instantaneous rate of random motion is affected by local plant suitability, even though there is no tendency for individual foragers to move toward better plants over the short term. Thus, area restricted search is a pattern of individual behavior that leads to a taxis term at the population level.[54]

Underlying the flux term in Equation 1 are several subtleties pertinent to an experimentalist interested in extrapolating individual behavior to the level of populations. First, populations may tend to flow toward regions of higher plant quality even though an individual shows no immediate bias in its direction of movement when placed between a plant of low quality and a plant of high quality. Second, the flux term in Equation 1 varies in form depending on exactly what microscale behavior underlies the tactic response. If individuals bias their direction of movement toward regions of higher Q, then the taxis and the diffusion components of flux are independent of one another; on the other hand, if individuals engage in area restricted search, the taxis and diffusion components of flux are mathematically related and are not free to vary independently.[76] Third, in pinning down the proper form of Equation 1, it is not sufficient to merely note that the rate of random motion varies with plant quality without having some idea of the spatial scale over which variation in quality affects the rate of movement. In particular, if that rate varies depending on some regional average of quality, then the flux term will include a diffusion component but no tactic component; whereas if the movement rate is determined only by the quality immediately "underfoot", then the flux term will include both a diffusion and the tactic component.[76] Finally, notice that the taxis term in Equation 1 is a function of $\partial Q/\partial x$, which means the rate at which searching insects will be attracted to a high quality plant is proportional to the local gradient in plant quality, and not just the suitability of the plant on which the insect currently resides. These theoretical distinctions are more than mathematical nuances; they tell us something about the consequences of subtle variations in how insects respond to heterogeneous plant quality, and identify distinctions that should be resolved experimentally. These distinctions are discussed in a later section concerning specific simulations and the design of field experiments.

The model embodied by Equations 1 and 2 summarizes the movement behavior of foraging herbivores when faced with an array of plants whose quality varies in space. Because it omits insect population growth, the model is only useful when movement occurs on a time scale faster than that at which insects give birth and die. Thus, for sedentary insects possessing high intrinsic rates of population growth (e.g., aphids), one should not adopt Equations 1 and 2 without first adding a population growth function. A less serious simplification is the one-dimensional framework of the model (i.e., spatial position is given in terms of only one coordinate, x). A one-dimensional model such as Equations 1 and 2 can easily be modified to deal with movement in two dimensions; alternatively, a one-dimensional model can be directly applied to rows of crop plants or to data collected from experimentally contrived rows of plants.

In order to use the model to predict the outcome of herbivore search, one must supply the initial distributions of mobile and sessile insects, the spatial pattern of plant suitability $Q(x)$, and specific forms for the functions Φ (M,S,Q), D(Q), and Ψ (Q). In the following section the model is invested with particular functions and parameters, and then used as a heuristic tool to investigate the consequences of random and nonrandom movement.

IV. USING THE SEARCHING MODEL TO HIGHLIGHT THE CONSEQUENCES OF RANDOM AND NONRANDOM MOVEMENT

By examining the special cases of Equations 1 and 2, it is possible to determine how different forms and mixtures of nonrandom and random behavior influence the success with which insects place themselves on the best plants (i.e., plants with the highest Q). For many of the specific models explored, it is possible to obtain an analytic expression for the equilibrium distribution of insects in space, and hence the equilibrium average Q experienced by the foragers; however, interest is not based solely in this equilibrium, but also in the rate at which this equilibrium is attained. For example, a rapid approach to the equilibrium spatial distribution means the equilibrium will be achieved before mortality factors substantially reduce the population of foragers, and that the insects will be able to track changes in Q when the more realistic assumption is made that Q is a perpetually changing variable. Consequently, searching efficiency is quantified by keeping track of the average value of Q experienced by searching herbivores; if this value increases rapidly and attains a high level, then the corresponding searching behavior is efficient. The authors emphasize that the conclusions drawn depend critically on the nature of variation in plant quality, and represent specific numerical results rather than global generalizations.

A. NONRANDOM SETTLEMENT CAN BE AN EXTREMELY EFFECTIVE SEARCHING STRATEGY

Phytophagous insects can be remarkably adept at aggregating on suitable food plants, even though their motions are completely random. To illustrate this, the rate of random movement, D, is fixed as a constant and the coefficient of taxis Y is set equal to zero. This suite of assumptions depicts foraging insects whose movement is unaffected by spatial variation in host plants. The only effect of plant quality is to change the rate at which foragers cease their wanderings and settle down to a sedentary existence. Specifically, the general function F (M,S,Q) is replaced by the expression:

$$-\Phi(MQ-S(1-Q)) \tag{3}$$

This function has two desirable properties: (1) when the plant suitability Q = 1, mobile herbivores settle at the maximum rate determined by the parameter j; (2) mobile herbivores will not settle on plants that are completely unsuitable (i.e., Q = 0). The authors reiterate that this mechanism of host plant choice requires minimal sophistication on the part of the insect herbivore. The outcome of this mechanism of host plant search is illustrated in Figure 1. Initially, both mobile and sessile herbivores are distributed at a uniform density of ten per plant. As mobile herbivores begin to settle on suitable plants, small peaks appear in the distribution of sessile herbivores. After a short interval, the initially uniform distribution of sessiles has become highly aggregated on the most suitable plants. This result can be summarized by plotting the average quality experienced by herbivores as their spatial distribution changes through time (Figure 2). Note that when the maximum rate of transition from mobile to sessile behavioral classes is low (e.g., j = 0.1, equivalent to a rate of 9.5%/d), the average quality experienced by herbivores increases slowly through time and is relatively insensitive to the diffusion coefficient of the herbivore, even if the diffusion coefficient varies by two orders of magnitude (Figure 2c vs. 2d). The diffusion coefficent does influence the rate at which herbivores attain high quality plants when the settling rate for those herbivores is high (Figure 2a vs. 2b), but the magnitude of the change is still comparatively small. It appears that the source of nonrandomness, the settling rate, dominates the effectiveness of this searching process, with some modulation by rates of random motion.

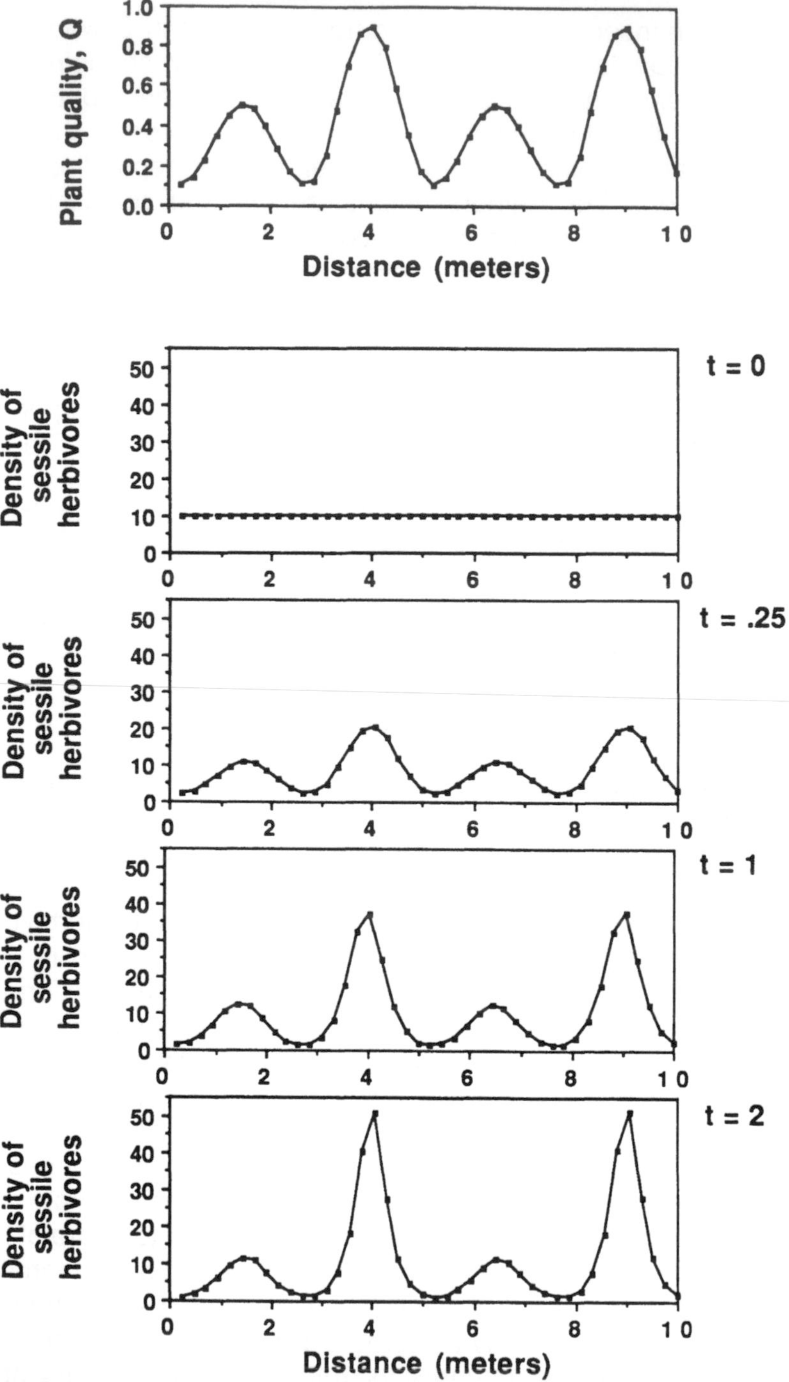

FIGURE 1. Nonrandom settlement model: the spatial distribution of plant quality (upper panel) and the changing distribution of sessile herbivores through time, measured in days. Parameter values for Equations 1 and 2: $\varphi = 10$, $D = 1$, $\Psi = 0$. Compare with Figure 2, curve b.

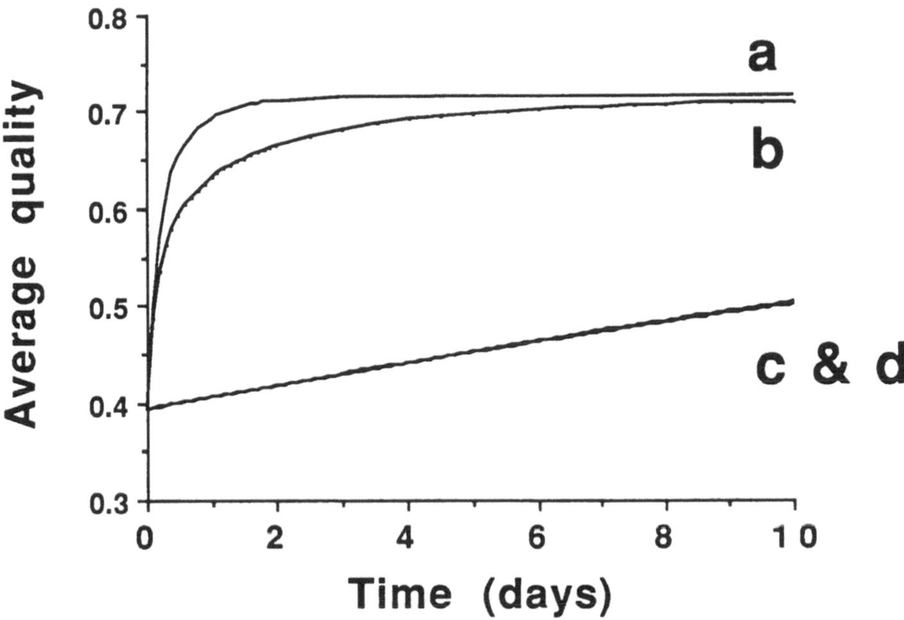

FIGURE 2. Nonrandom settlement model: average plant quality experienced by sessile herbivores through time. (a) $\varphi = 10$, $D = 100$; (b) $\varphi = 10$, $D = 1$; (c) $\varphi = 0.1$, $D = 100$; (d) $\varphi = 0.1$, $D = 1$.

B. THE EFFECTIVENESS OF AREA RESTRICTED SEARCH DEPENDS ON AN INTERPLAY BETWEEN MEAN PLANT QUALITY AND THE SENSITIVITY OF MOVEMENT TO CHANGES IN QUALITY

Imagine insects that are always mobile, always move in a random direction (i.e., that move left or right with equal probability, even if plants in one direction have a much higher Q), but are more likely to turn around or to slow down when they find themselves on a plant of high Q. Such individual behavior can be described at the population level by the partial differential equation[54]

$$\frac{\partial M}{\partial t} = \frac{\partial}{\partial x}[D(Q)\frac{\partial M}{\partial x} + \frac{\partial D}{\partial Q}\frac{\partial Q}{\partial x}M]$$

Notice that this equation bears a strong resemblance to the general model (Equation 1), where $F = 0$ and $\partial D/\partial Q$ takes the place of the coefficient of taxis Y (Q). An important consequence of the equivalence between Y and $\partial D/\partial Q$ in Equation 4 is that if nonrandomness enters through the rate of random movement depending on Q, then it is impossible to change diffusion without also changing taxis in a coordinated fashion. As pointed out earlier, this is not the case when insects actively orient toward plants of higher suitability.

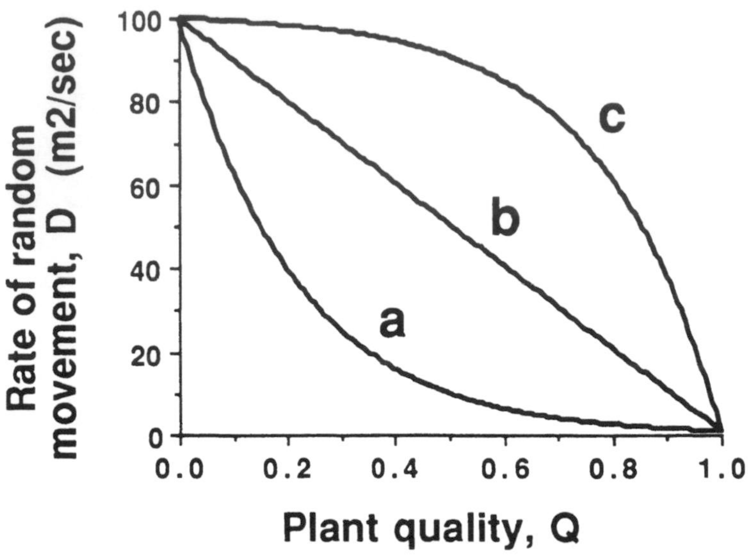

FIGURE 3. Area-restricted search model: three strategies by which herbivores can alter their rate of random movement in response to plant quality.(a) "nonchoosy" herbivores; (b) "intermediate" herbivores; (c) "choosy" herbivores (see text).

Assuming insects engage in area restricted search, there are three fundamental ways in which their rates of movement might vary with plant quality (Figure 3). "Choosy herbivores" would be expected to exhibit D(Q)'s that increased rapidly as quality fell below 1.0 (curve c in Figure 3); "nonchoosy herbivores" would tolerate large declines in plant quality before their diffusion rate rose (curve a in Figure 3); and "intermediate herbivores" can be represented by a linear change in diffusion rate with changing plant quality (curve b in Figure 3). All three types of herbivores will aggregate on high quality plants, but the strategy that produces the most efficient searching depends on the mean quality of plants available. Consider herbivores searching in two spatially heterogeneous environments posessing the same amplitude and wavelength of variation in plant quality, but differing in the mean quality (Figure 4). When quality is generally high choosy herbivores are most effective at congregating on the best plants (Figure 5a), whereas when quality is low, so-called nonchoosy herbivores are the most effective searchers (Figure 5b). This result suggests two important points: (1) one must know the range over which plant quality varies in the field in order to predict the effectiveness of different search strategies, and (2) the way in which insects alter their movement in response to Q may represent an adjustment to the distribution of plant quality they regularly encounter. In particular, species inhabiting environments with generally high-quality plants might be expected to exhibit diffusion coefficients that varied, as shown in curve c of Figure 3, whereas species inhabiting environments with generally low-quality plants could be expected to exhibit convex curves (such as curve a in Figure 3).

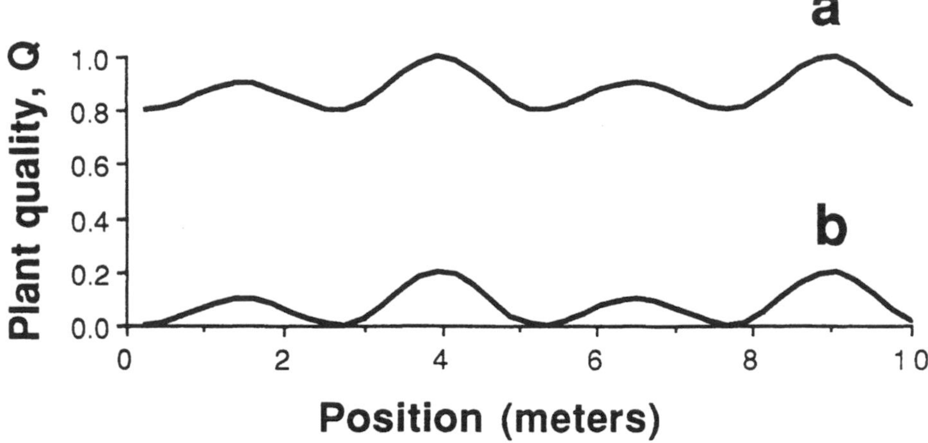

FIGURE 4. Two spatial distributions of plant quality with the same variance but different ranges in Q.

C. THE OUTCOME OF ORIENTED SEARCH REPRESENTS A BALANCE BETWEEN RANDOM AND NONRANDOM MOVEMENT

Consider the situation in which searching insects jump about at random, but are more likely to move to the left if the plant on the left is of higher quality than the plant on the right, and vice versa. Such movement represents a bias in the direction of higher quality plants and gives rise to a partial differential equation of the form[88]

$$\frac{\partial M}{\partial t} = \frac{\partial}{\partial x}[D\frac{\partial M}{\partial x} + \Psi M\frac{\partial Q}{\partial x}] \qquad (5)$$

where D and Ψ are constants (i.e., they do not depend on Q). Note that although Equation 5 is superficially similar to Equation 4, in the above model the coefficient of taxis Ψ is independent of the diffusion coefficient D.

The ability of searchers to aggregate on high quality plants using oriented movement depends on the ratio of diffusive to tactic movement (as given by D/Y). For instance, if the rate of diffusive movement is large relative to taxis, herbivores will wander away from patches of high quality plants and may never accumulate on the better plants (see Figure 6). Thus, as seen in the previous models of movement, the effectiveness of searchers represents a balance between random, diffusive movement and movement that is responsive to variation in plant quality.

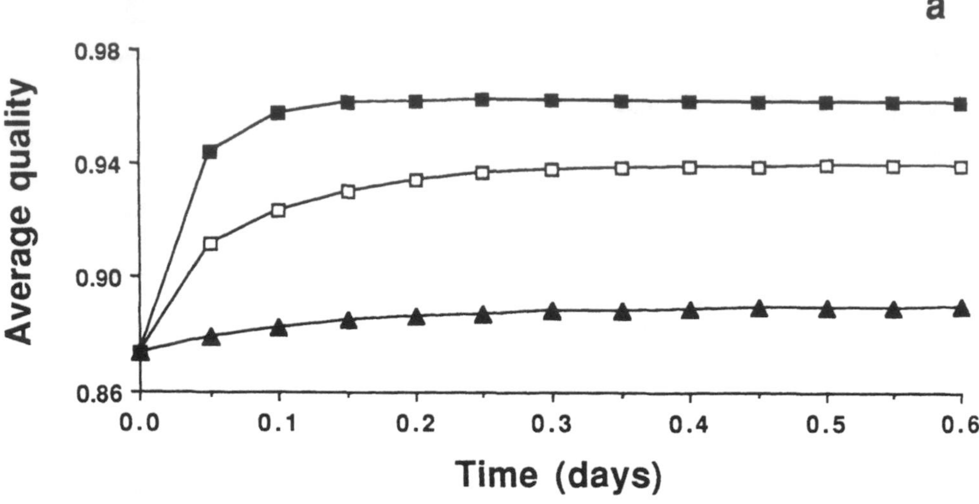

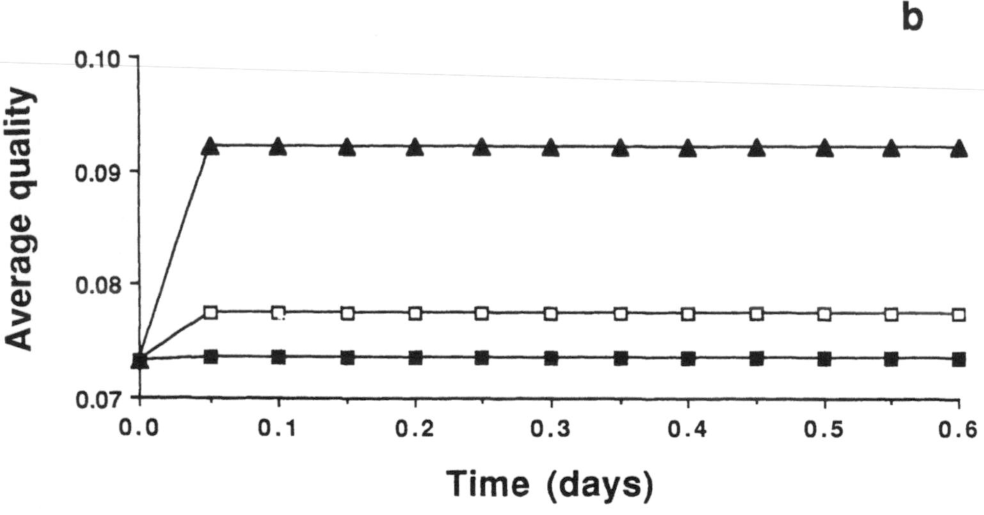

FIGURE 5. Area-restricted search model: the effectiveness of the three strategies illustrated in Figure 3 when herbivores forage on the two plant quality distributions shown in Figure 4 (indicated by the letter at upper right). Closed squares = "choosy" herbivores, open squares = "intermediate" herbivores, closed triangles = "nonchoosy" herbivores.

D. IN SPITE OF STRONGLY DIRECTED MOVEMENT, RANDOM MOTION MAY BE NECESSARY FOR EFFECTIVE SEARCH

At first glance, directed movement such as that embodied in Equation 5, would seem to be superior to any searching process that included a large degree of random wandering. One way of assessing this is to use the authors' model to contrast the outcome of search involving purely random motion and nonrandom settlement to the outcome of purely tactic movement (always

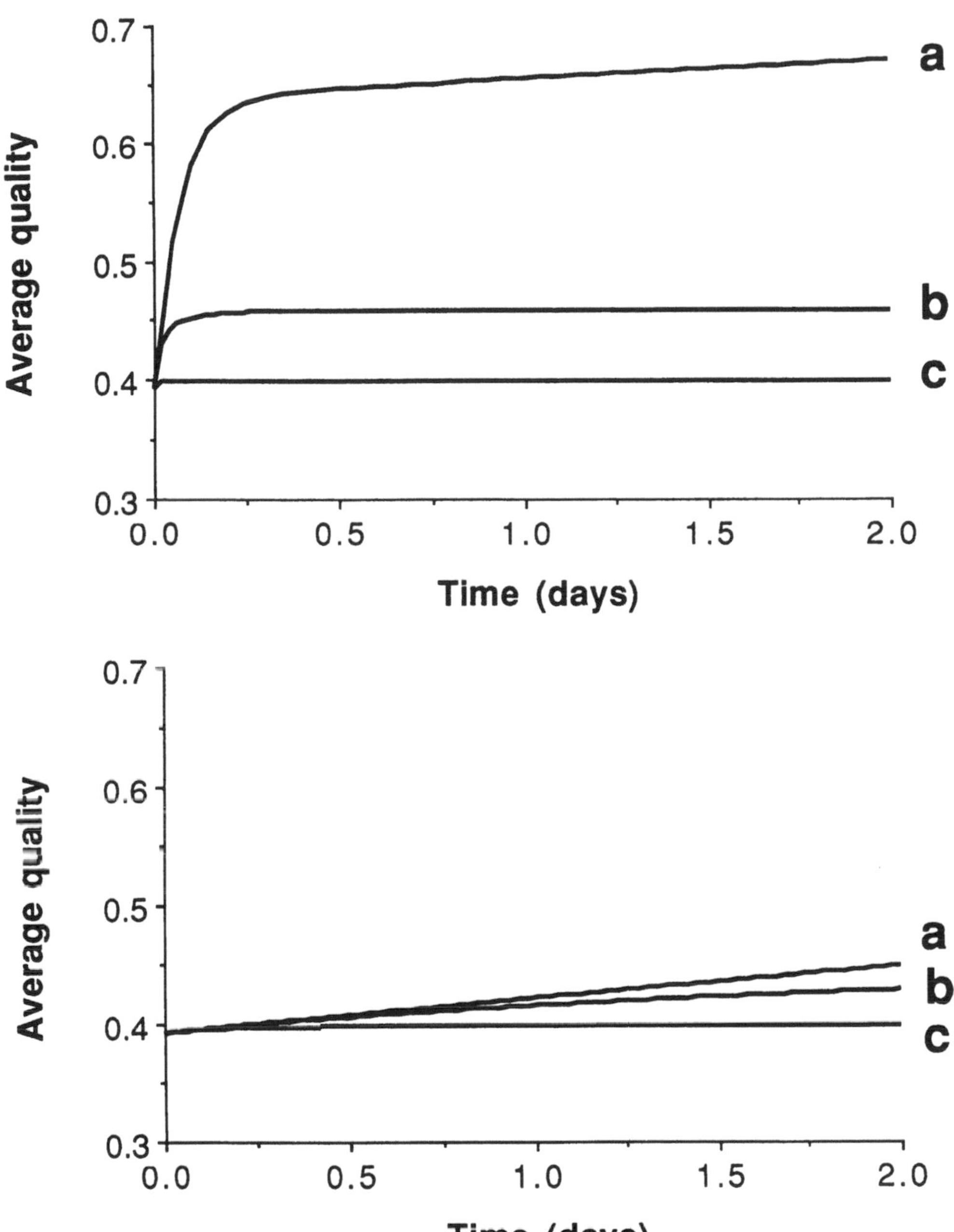

FIGURE 6. Oriented search model: the outcome of oriented search for various combinations of diffusion (D) and taxis (Ψ) in Equation 5. In the upper panel, $\Psi = 10$; in the lower panel, $\Psi = 0.1$. Curves a: $D = 0.1\ \Psi$; curves b: $D = \Psi$; curves c: $D = 10\ \Psi$. The distribution of plant quality is the same as in Figure 1.

toward regions of higher Q), also with nonrandom settlement. Specifically, one can numerically solve Equations 1 and 2 with constant coefficients while varying the relative strengths of diffusion, D, and taxis, Ψ. Somewhat surprisingly, insects behaving in accordance with the taxis model do not fare dramatically better than insects lacking any tactic response (Figure 7). In fact, under some arrangements of plant quality, the addition of random motion can actually increase foraging efficiency as measured by the maximum average Q attained (Figure 8). Thus, even though random motion means that individuals move without regard to plant quality, and directed motion means that individuals move toward regions of higher plant quality, random motion can enhance a searcher's success. The advantage of random motion is that it fosters exploratory search, such that individuals adandoning a region of relatively high plant quality might stumble across a region of even higher plant quality. In contrast, an insect that relied only on directed motion would never leave any "peaks in Q", since to do so would imply moving toward a region of lower Q over the short run. In other words, too strong a tactic response, or directed movement in the absence of randomness, can lead searching herbivores to be caught on minor peaks in plant quality, thereby preventing them from discovering the major "peaks in Q, just over the next ridge" (contrast Figure 9b to 9c).

It is important to realize that the above advantage to random movement arises only because the model assumes searching individuals cannot respond to distant "peaks" in plant quality when there are intervening "hills" in plant quality. This assumption probably applies to most insects — it certainly applies to those many herbivores that can only perceive host plant cues when they approach within a few centimeters of the plant.[23] Moreover, although numerous phytophagous insects are endowed with relatively long-range sensory abilities,[1] doubt remains that even these species would move toward a distant favorable patch of food plants in preference to a nearby, slightly less favorable patch of food plants. If this conjecture is correct, then a substantial component of random motion, even to the point of abandoning what seem to be perfectly suitable patches of food plants, could be interpreted as a behavioral mechanism for searching widely. The adaptive value of such wide-ranging random search will, of course, depend on the details of how plant quality varies in space.

V. A REVIEW OF PREVIOUS STUDIES OF INSECT HERBIVORE MOVEMENT: HOW OFTEN DO ECOLOGISTS QUANTIFY THE CRITICAL COMPONENTS OF MOVEMENT?

By examining a simple model of insect search, it can be seen that the success with which insects avail themselves of the best food plants depends upon how the rates of settlement (F in Equations 1 and 2), of random motion (D in Equations 4 and 5), and of taxis (Y in Equations 4 and 5) vary in response to host plant suitability. Moreover, the effectiveness of each component of nonrandom search generally depends on the nature and magnitude of all other components in the searching process. To what extent are field investigations of foraging behavior generating data that would allow us to predict the likely searching effectiveness of different insects? Table 1 summarizes published reports of movement in insect herbivores, indicating whether each study provides information on the rates of arriving and leaving host plants or host

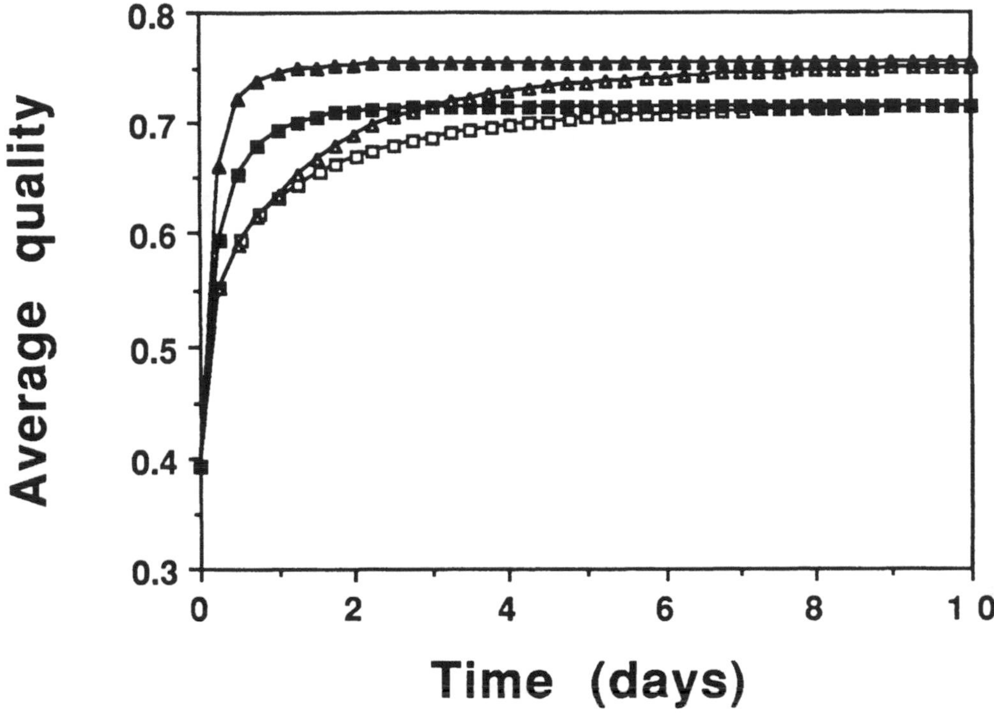

FIGURE 7. Effectiveness of nonrandom settlement when herbivores move randomly (squares — open: D = 1, Ψ = 0; closed: D = 100, Ψ = 0) vs. when herbivores use oriented movement (triangles — open: D = 0, Ψ = 1; closed: D = 0, Ψ = 10). In all cases, the rate of nonrandom settlement, φ, is 10. The distribution of plant quality is the same as in Figure 1.

plant patches, diffusion rates, or rates of oriented movement. Table 1 includes only those studies that provide measurements that can be used to estimate the parameters in movement models such as those presented above (for example, the positions of marked insects relative to some initial release point can be used to estimate a diffusion coefficient[52]). Most of the studies listed in the table provide quantitative data on only a single component of movement. For example, 23 of the 72 studies (32%) measured only the rate of emigration or immigration (the most frequently measured component of movement); however, as noted above, it is difficult to predict the outcome of host plant search if the rates of immigration and emigration are known but no information exists about the rate of diffusive movement (see Figure 2). For only 2 of the 45 species in the table (i.e., *Tetranychus urticae* and *Pieris rapae*) could quantitative measurements pertaining to all three components of movement be found. In addition, very few studies have quantified how insects change their movement behavior in response to inter- or intraspecific differences in plant suitabilility. Thus, in spite of much data concerning foraging movements among herbivorous insects, there is an almost total lack of descriptions that would allow evaluation, much less the prediction, of which components of behavior contrib-

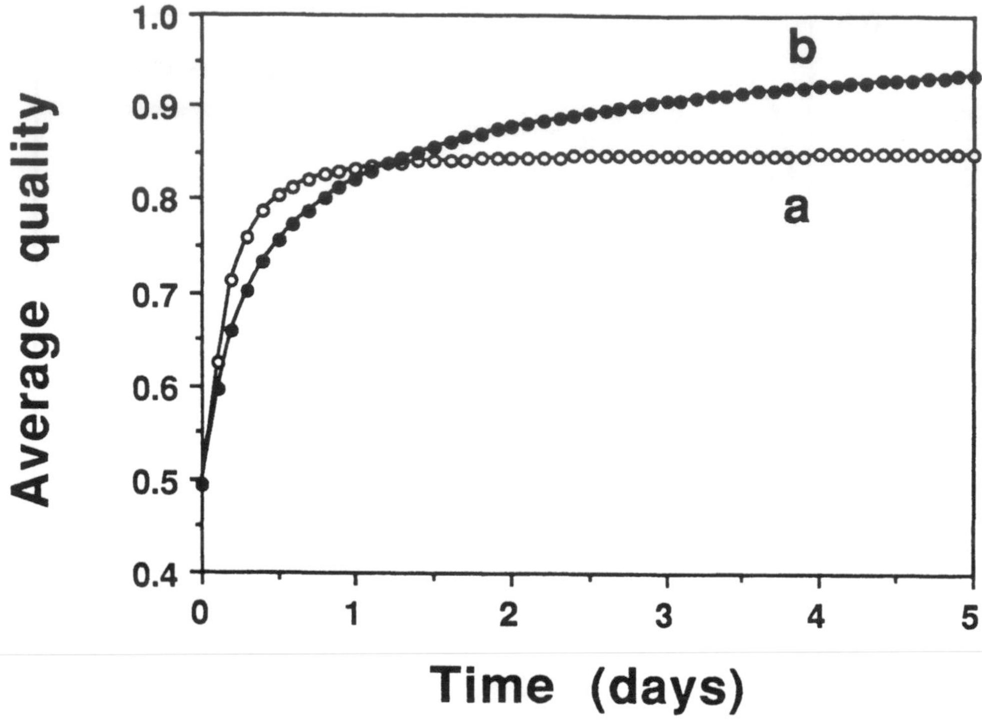

FIGURE 8. Distribution of sessile herbivores attained after 5 d from an initially uniform distribution of herbivores foraging on the plant quality distribution shown in panel a. Note that herbivores which have a random component to their movement (shown in panel c; $\varphi = 10$, $D = 100$, $\Psi = 10$) become more concentrated on the highest peaks in Q than do purely tactic herbivores (shown in panel b; $\varphi = 10$, $D = 0$, $\Psi = 10$).

ute to the searching success of various insects. In particular, there is insufficient data to determine whether phytophagous insects sucessfully find good food plants because of highly tuned sensory abilities or because of wide-ranging motion combined with minimal sensitivity to local food quality. The intent of the authors is not to criticize past studies of insect movement (undertaken to address a wide range of questions), but rather to highlight a major gap in knowledge about herbivorous insects. If the process by which insects locate superior food plants is not understood, one is severely handicapped when addressing questions about host race formation, evolution of diets, plant-herbivore coevolution, the implications of different cropping patterns, the consequences of changing food quality for herbivore populations, etc.

VI. GUIDELINES FOR MAKING QUANTITATIVE PREDICTIONS ABOUT THE OUTCOME OF HOST PLANT SEARCH

Most studies of insect searching have focused either on the outcome of the searching process (quantifying where eggs and individuals end up at the

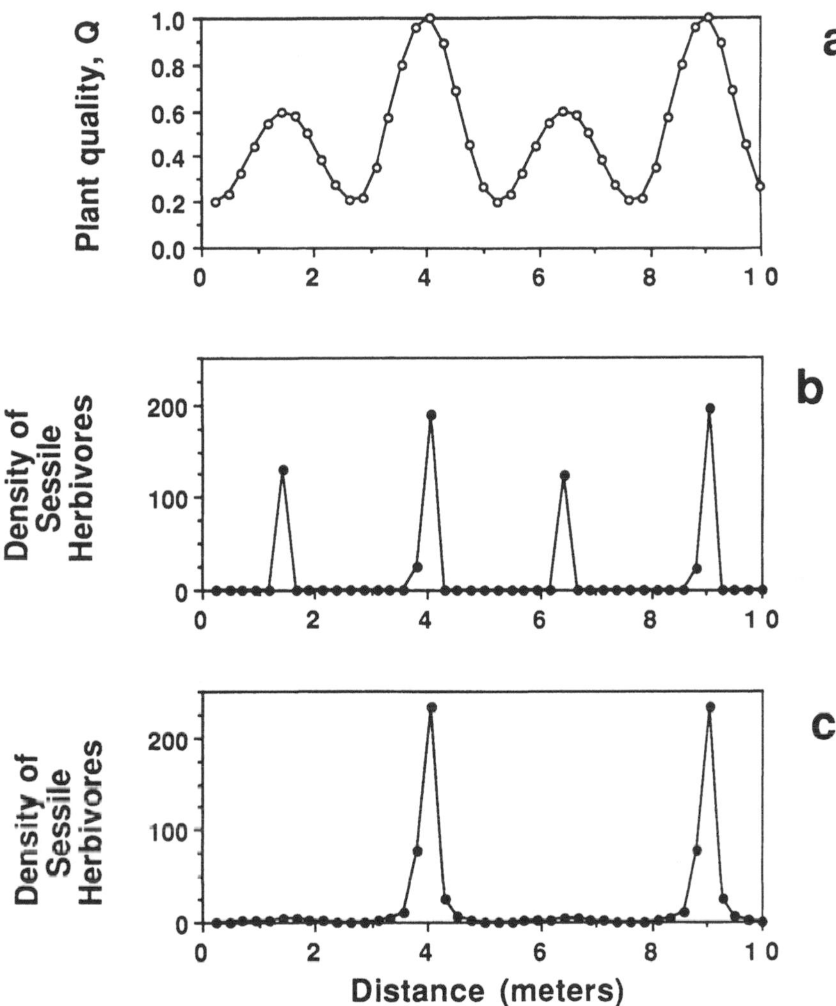

FIGURE 9. Effectiveness of oriented movement alone (curve a, same parameter values as Figure 8a) vs. oriented movement plus diffusion (curve b, same parameter values as Figure 8b).

termination of search), or on the ability of insects to sense differences among plants. Neither of these approaches provide enough information on the mechanics of search to develop predictions about the relative efficiencies of different species when faced with heterogeneous arrays of food plants. To predict when "species A" is more likely to be effective at finding the best food plants than is "species B", a model of the searching process is needed. Even if the research is concerned with only one herbivore, a model is needed to generate explicit hypotheses about how changes in the degree of host plant heterogeneity will alter that herbivore's success at finding high quality food. Without searching or foraging models, the best that can be done is to build up an encyclopedic collection of anecdotes about how herbivores distribute themselves in a variety of circumstances. Although the need for models is clear, no single superior method exists to model the foraging or searching

Table 1
A SUMMARY OF EMPIRICAL STUDIES THAT PROVIDE QUANTITATIVE INFORMATION ON 3 COMPONENTS OF MOVEMENT (RATES OF EMIGRATION/IMMIGRATION (E/I), RATES OF RANDOM SPREAD (D), AND ORIENTED MOVEMENT (O)) IN ARTHROPOD HERBIVORES

Species	Movement component	Adult/ imm.	Quantity measured	Field or Lab?	Sensitivity to Q measured?	Ref.
ACARINA						
Tetranychus urticae	E/I	A	Emigration rate	L	Y	12
	D	A	Distance moved/time	L	N	47
		A	Turning frequency	L	Y	24
	O	A/I	% Moving toward damaged vs. undamaged plants	L	Y	43
ORTHOPTERA						
Hesperotettix viridis	E/I	A/I	Emigration rate	F	N	78
Myrmeleotettix maculatus	D	I	Distance moved/time, diffusion coefficient	F	N	2
Schistocerca gregaria	O	I	% Moving toward odor source	L	N	44,58,71
HOMOPTERA						
Acyrthosiphon pisum	D	A/I	Distance moved/time	L,F	N	85

Aphis fabae	E/I	A	Emigration rate	L	N	42
Dalbulus maidis	D	A	Diffusion coefficients	F	Y	79
Drepanosiphum platanoides	E/I	A	Emigration rate	L	Y	25
Eucallipterus tiliae	E/I	A	Emigration rate	L	Y	59
Pubilia concava	D	A	Distance moved/time, diffusion coefficient	F	N	52,69
Rhopalosiphum padi	E/I	A	Emigration rate	L	Y	26
HEMIPTERA						
Lygus lineolaris	D	A	Distance moved/time, diffusion coefficient	F	N	38
Murgantia histrionica	E/I	A	Emigration rate	F	Y	33
Nezara viridula	D	A/I	Distance moved/time, diffusion coefficient	F	N	52,60
	D	A	Average move length, frequency of movement	L	N	3
Oncopeltus fasciatus	E/I	A	Emigration rate	F	Y	81
Tingis ampliata	D	A	Distance moved/time, diffusion coefficient	F	N	31,52

Table 1 (continued)
A SUMMARY OF EMPIRICAL STUDIES THAT PROVIDE QUANTITATIVE INFORMATION ON 3 COMPONENTS OF MOVEMENT (RATES OF EMIGRATION/IMMIGRATION (E/I), RATES OF RANDOM SPREAD (D), AND ORIENTED MOVEMENT (O)) IN ARTHROPOD HERBIVORES

Species	Movement component	Adult/ imm.	Quantity measured	Field or Lab?	Sensitivity to Q measured?	Ref.
COLEOPTERA						
Acalymma innubum	E/I	A	Emigration rate	F	Y	5
	D	A	Move length	F	N	5
Acalymma vittatum	E/I	A	Emigration rate	F	Y	4,6,67
Chauliognathus pennsyl-vanicus	D	A	Distance moved/time	F	N	14
Diabrotica balteata	E/I	A	Emigration rate	F	Y	84
Diabrotica undecimpunctata howardi	E/I	A	Emigration rate	F	Y	6,67
Epilachna sparsa orientalis	D	A	Distance moved/time	F	N	48
Epilachna varivestis	E/I	A	Emigration rate, immigration rate	F	Y	93,94
Gastrophysa viridula	D	A	Distance moved/time, diffusion coefficient	F	N	52,89

Leptinotarsa decemlineata	D	A	Distance moved/time	F	N	101
	O	A	% Orienting toward host odor	L	N	98
Oryctes rhinoceros	D	A	Distance moved/time	L	N	21
Pantorhytes szentivanyi	D	A	Diffusion coefficient	F	Y	61
Phyllotreta cruciferae	E/I	A	Emigration rate, immigration rate	F	Y	32,51,53
and *P. striolata*	D	A	Diffusion coefficient	F	N	51
Tetraopes tetra-opthalmus	E/I	A	Emigration rate	F	N	66
	D	A	Distance moved/time	F	N	65,68
Tetraopes femoratus	D	A	Distance moved/time	F	N	65
DIPTERA						
Delia (Erioischia) brassicae	D	A	Distance moved/time	F	N	45
	O	A	% Orienting toward odor	L,F	N	45
Hylemya antiqua	D	A	Distance moved/time	F	N	97
LEPIDOPTERA						
Aporia crataegi	D	A	Distance moved/time	F	Y	99
Battus philenor	D	I	Distance moved/time, diffusion coefficient	F	N	52,83

Table 1 (continued)
A SUMMARY OF EMPIRICAL STUDIES THAT PROVIDE QUANTITATIVE INFORMATION ON 3 COMPONENTS OF MOVEMENT (RATES OF EMIGRATION/IMMIGRATION (E/I), RATES OF RANDOM SPREAD (D), AND ORIENTED MOVEMENT (O)) IN ARTHROPOD HERBIVORES

Species	Movement component	Adult/ imm.	Quantity measured	Field or Lab?	Sensitivity to Q measured?	Ref.
Cidaria albulata	D	A	Move lengths, turning angles	F	Y	28
Coleophora seratella	D	I	Distance moved/time	F	N	12
Colias philodice eriphyle	D	A	Move lengths, diffusion coefficient	F	Y	90,91
Danaus plexippus	D	A	Move lengths, turning angles	F	Y	102
Euphydryas anicia	D	A	Distance moved/time	F	N	74
	D	A	Move lengths, turning angles	F	Y	75
Lymantria dispar	E/I	I	Emigration rate	L,F	Y	11,17, 61–63
	D	I	Distance moved/time	L	N	100
Melitaea harrisii	D	I	Distance moved/time	F	N	22
Phthorimaea operculella	E/I	I	Emigration rate	L	Y	96

Pieris rapae	E/I	A	Immigration rate	F	Y	35,36
	D	I	Crawling velocity, turning frequency	L	N	49
	D	A	Move lengths, turning angles, distance moved/time	F	Y	50,86
	O	I	% of moves toward plant;	L	N	49
		A	% of moves toward host patch	F	N	36
Pieris virginiensis	D	I	Distance moved/time	F	N	18
		A	Distribution of turning angles	F	N	18

process. The suite of partial differential equations presented above is one of many mathematical frameworks that can generate precise predictions about searching effectiveness, and can focus research on pertinent behavioral distinctions. One advantage of the partial differential equation structure is that it compresses into a compact notation in which concrete parameters or terms can be directly related to behavioral data. What follows is a description of the sorts of data and observations that go into building these models. Since many empiricists will be uninterested in modeling for its own sake, the authors also comment on problems of experimental inference that are highlighted by a consideration of searching models such as Equations 1 and 2.

A. A HIERARCHY OF KEY QUESTIONS

For any foraging insect, the first question to ask is whether it should be divided into mobile and sedentary subpopulations, or can be treated as one pool of mobile individuals. If a sedentary stage is evident, then there is a good possibility that the key to foraging is nothing more than the rate at which the transition is made from mobile to sessile stages. In a sense, oviposition of eggs by a mobile female is one extreme example of the sedentary/mobile compartmentalization, and it is widely recognized that by engaging in postalightment selectivity, randomly moving herbivores can exhibit highly selective patterns of host usage.

A second key question is whether a herbivore alters its rate of movement in accord with spatial variation in the quality of its potential host plants. It is suspected that all herbivores reduce their rates of movement in patches of high quality food plants, and the authors know of no examples to the contrary. A subtler facet of this question is determining the spatial scale over which an herbivore is sensitive to changes in resource quality. Some insects, such as flea beetles, may adjust their rate of movement in response to each individual plant they encounter[51], whereas other insects, such as butterflies[86], may adjust rates of movement in accord with some average quality integrated over scores of meters. If experimentalists misidentify the scale at which herbivores respond to variation in plant quality, then they may erroneously conclude searching is random when in fact the insect does alter its movement in reference to plant variation.

Finally, to characterize the foraging behavior of an herbivore, one needs to determine whether individuals orient and move in a directed manner toward plants whose quality stands above the quality of surrounding vegetation. The presence of such a tactic response could generate a strong tendency for individuals to rapidly collect on the nearest patch of higher-than-average quality plants. Although both variation in the rate or direction of movement can generate nonrandom spatial distributions in insects, it is important to distinguish between these mechanisms when attempting to model host plant search.

1. Do Purely Experimental Studies of Herbivore Foraging Have Anything to Learn from Models of Movement?

The experimental investigation of movement behavior is not as facile as casual reasoning might suggest. For example, the basic question of whether a species engages in directed movement cannot be answered by simply recording the ultimate distribution of individuals — individuals could appear to have "flowed toward" the best plants, even though over the short-term they are no more likely to move toward a high quality plant than a low quality plant. Second, there is a tendency in the literature to associate randomness in movement patterns with inefficient search; however, simple models such as the one delineated here indicate that with only a very slight tendency to settle down on "better plants", herbivores that wander about in a random fashion can be very effective searchers. Indeed, the magnitude of "nonrandomness", as measured by a change in probability of movement as a function of plant quality could be so small that it is difficult to establish experimentally, yet when combined with high mobility such behavior produces efficient search.

2. How Does One Use Experiments to Build A Model of Movement?

The most straightforward approach to modeling herbivore search is to quantify an insect's instantaneous pattern of movement when placed in experimentally contrived arrays of plants that differ in quality. "Instantaneous pattern of movement" means rates of departure, settlement, turning frequencies, and directional biases over the time scale of minutes. While this might appear to be an overwhelming task, it may be possible to forego many of these measurements for some species. For example, if an insect forages by making short hops from plant to plant, one probably need only place a population of individuals on a central plant amidst a short row of plants whose qualities have been experimentally manipulated. In order to distinguish among the many possible ways in which movement might be nonrandom, the authors advocate a three-way factorial experiment in which the quality of plants on which individuals are released is varied as one experimental treatment and the quality of plants to the left and to the right of the release plant are manipulated as the two remaining treatments. By combining these three factors, and using three different levels of quality for each factor, or position, it should be possible to determine the extent to which herbivores alter their movement in repsonse to local conditions, to gradients in plant quality, or to "average" quality (see Figure 10). A straightforward analysis of variance with a test for interactions will allow one to identify which aspects of nonrandomness produce "significant" effects. To actually build a model from such data, however, one must relate instantaneous rates of movement or biases in rates of movement to parameters in partial differential equations. This is a standard exercise and there are several examples of this approach as applied to field populations of insects.[54,95]

Sometimes it may not be possible to follow individual movements. In such cases, the experimentalist will still want to manipulate the spatial patterning of plant quality in a fashion analogous to Figure 10. Rather than recording individual trajectories, however, large numbers of insects can be released on the central plant and their distribution at successive time periods recorded. The appropriate model is selected by then finding the parameters or functions in Equations 1 and 2 that produce the best match to observed population distributions.[8-10] This approach is analogous to a multiple regression approach, except that instead of using linear or polynomial regression equations, the

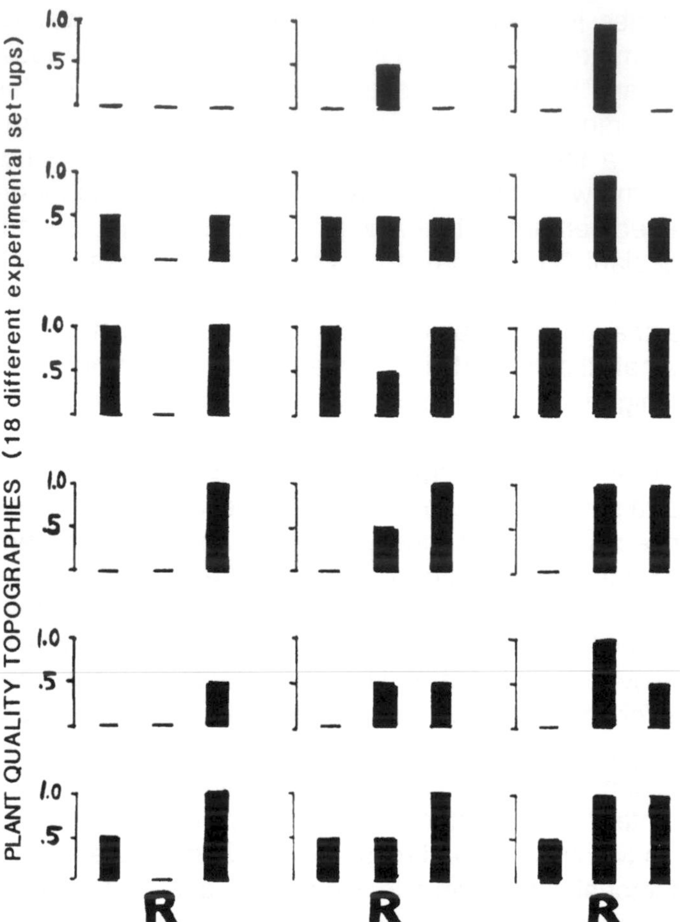

FIGURE 10. The complete suite of plant quality manipulations required to determine the dependence of movement on local conditions, average conditions, and/or spatial gradients in plant quality. "R" indicates the position at which insects are released. By varying the quality of plants at "R" and at neighboring positions in a systematic fashion, it is possible to identify the exact form of the "flux terms" in Equations 1 and 2. Above are three qualities arbitrarily selected as 0, 0.5, and 1.0; other selections are possible. Indeed, one can imagine behavior that depends on the absolute values of qualities and not just their relative variation.

researcher is fitting the solution of a partial differential equation to field data. Recently, maximum likelihood statistics have been developed so that this analysis can be used to identify what aspects of nonrandom behavior explain a significant proportion of the spatiotemporal changes in a population distribution.[7]

The major problem with following either individuals or masses of insects is that one must first determine whether the insects alter their movement behavior as a result of interaction with one another. If there is some sort of density-dependent variation in movement, the diffusion and taxis terms in Equations 1 and 2 become functions of herbivore density as well as plant quality. Although such equations pose no special mathematical problems,[95] the number of experimental permutations required to develop the appropriate model is large. It should be emphasized, however, that many patterns attributed to density effects in herbivore populations may actually be mediated through damage to food plants (i.e., high density herbivore populations damage plants, yielding high rates of movement away from lower quality patches of food); thus, although one's field experience suggests density is important to herbivore movement over the long run, the appropriate model of movement may include only plant quality effects on the instantaneous time scale. In addition, although density effects may be detectable under extreme conditions, a large proportion of the variation in an insect's distribution may still be explained without incorporating density as an immediate factor.

Of course, several other logistical and biological complications are likely to complicate movement investigations. First, handling insects prior to such observations may perturb their natural behaviors. In addition, the feeding history of an insect could alter its behavioral response to enviromental heterogeneity.[78] Both hazards can be overcome by building appropriate controls into the experimental design. A much more difficult complication is the fact that the degree of nonrandomness in the movement of any species will almost inevitably depend on a combination of the absolute quality of plants and the magnitude of any spatial gradient in plant quality. Thus, one insect may respond to differences in quality only in the range of Q between 0.1 and 0.3, whereas another species is sensitive to variation in quality in the range 0.5 to 0.7, etc. (see Figure 3). Nonetheless, framing the investigation in terms of movement of insects placed on differing topographies of plant quality at least draws explicit attention to the importance of the details of spatial variation in Q. In contrast, wind tunnel experiments or laboratory choice assays skirt the importance of quality topography in determining movement patterns. More importantly, as discussed in regards to the derivation of taxis models, it is possible for a species on the short term to always move in a random direction with respect to plant quality, yet as a population (or probability density) they tend to flow toward regions of high quality plants.

The key point is that to understand search, observations must be made in arenas where the topography of plant quality has been manipulated. Although several researchers have experimentally altered plant quality and quantified effects on herbivore movement,[33,51,78] these studies have failed to explicitly manipulate the topography of quality. Thus, while such studies provide evidence as to whether searching herbivores respond to plant quality, they do not facilitate the construction of a model of the searching process.

VII. CONCLUDING REMARKS

The models examined make it clear that a wide variety of behaviors, ranging from nonrandom settlement to changes in rates of movement to actual orientation, all represent mechanisms by which herbivores may concentrate on patches of plants of higher-than-average quality. While this has been widely recognized by behaviorists, these different mechanisms have not been explored in the context of a unified model. The virtue of such a model is that it allows the evaluation of the relative importance of different behaviors to an herbivore's success (as measured by increases in the average quality of plants it experiences in a heterogeneous environment).

A second critical point that emerges from the examination of searching models is that the rate of random motion can be of overwhelming importance to the effectiveness of nonrandom search. Thus, although an investigator is interested in "nonrandom behavior" or "nonrandom search", the effectiveness of such behavior is modulated by the rate of random movement. For this reason, it is essential that rates of random movement are quantified in searching herbivores (standard methods are readily available[52,55]), and the nonrandom features of the searching process are recorded.

A final conclusion that emerges from this analysis is that a complete picture of any herbivore's searching behavior is lacking. This lack is starkly evident when the question is posed: to what extent have empiricists collected the data necessary to build a model of search? (Table 1). The authors hope that by putting forth their search model, they can help direct experimental work toward providing the sort of data needed for a predictive model of herbivore movement in variable environments.

They believe that until studies of herbivore search meld experimental manipulations of plant quality and dispersion with mechanistic models, one is left with only anecdotes about the abilities of different species at discovering food plants. Although such data are valuable, by themselves they do not yield predictions or an understanding of how evolution might shape searching efficiency. With a model, it is possible to note, for example, that a mere 10% increase in rate of motion produces a dramatic increase in the average quality of plants on which an herbivore feeds.

Finally, the authors reiterate that the models presented here are far from the last word in models of herbivore search. Their models omit two important biological features of insect search: the intrinsic dynamics of the herbivore population (i.e., birth and death of searchers) and temporal changes in plant quality (e.g., declining food suitability caused by accumulated herbivore damage). Nevertheless, such models serve as a jumping-off point for more complex models of interacting plant and herbivore populations. Also, models such as the those described here identify the key experiments needed to distinguish among different forms of nonrandom search (Figure 10). It is the authors' belie that researchers must start with explicit models of the searching process if they are to understand both the short-term redistribution of searching insects and the long-term dynamics of herbivore populations.

ACKNOWLEDGMENTS

The authors wish to thank G. Odell for assistance in using his Biograph numerical methods software for solving partial differential equations, and Sigma Xi for a Grant-in-Aid of Research to W. Morris.

REFERENCES

1. **Ahmad, S., Ed.,** *Herbivorous Insects: Host-Seeking Behavior and Mechanisms,* Academic Press, New York, 1983.
2. **Aikman, D. and Hewitt, G.,** An experimental investigation of the rate and form of dispersal in grasshoppers, *J. Appl. Ecol.,* 9, 807, 1972.
3. **Andow, D. A. and Kiritani, K.,** Fine structure of trivial movement in the green rice leafhopper *Nephotettix cinctipes* (Homoptera: Cicadellidae), *Appl. Entomol. Zool.,* 19, 306, 1984.
4. **Bach, C. E.,** Effects of plant density and diversity on the population dynamics of a specialist herbivore, the striped cucumber beetle, *Acalymma vittata, Ecology,* 61, 1515, 1980.
5. **Bach, C. E.,** Plant spatial pattern and herbivore population dynamics: plant factors affecting the movement patterns of a tropical cucurbit specialist (*Acalymma innubum*), *Ecology,* 65, 175, 1984.
6. **Bach, C. E.,** Effects of host plant patch size on herbivore density: underlying mechanisms, *Ecology,* 69, 1103, 1988.
7. **Banks, H. and Fitzpatrick, B.,** Inverse problems for distributed systems: statistical tests and ANOVA, in *Mathematical Approaches to Problems in Resource Management and Epidemiology,* Castillo-Chavez, C. and Levin, S., Eds., Springer-Verlag, New York, 1990.
8. **Banks, H. and Kareiva, P.,** Parameter estimation techniques for transport equations with application to population dispersal and tissue bulk flow models, *J. Math. Biol.,* 17, 253, 1983.
9. **Banks, H., Kareiva, P., and Murphy, K.,** Parameter estimation techniques and redistribution models of species interactions: a predator-prey example, *Oecologia,* 74, 356, 1987.
10. **Banks, H., Kareiva, P., and Zia, L.,** Analyzing field sutdies of insect dispersal using two-dimensional transport equations, *Environ. Entomol.,* 17, 815, 1988.
11. **Barbosa, P., Cranshaw, W., and Greenblatt, J. A.,** Influence of food quantity and quality on polymorphic dispersal behaviors in the gypsy moth, *Lymantria dispar, Can. J. Zool.,* 59, 293, 1981.
12. **Bergelson, J., Fowler, S., and Hartley, S.,** The effects of foliage damage on casebearing moth larvae, *Coleophora serratella,* feeding on birch, *Ecol. Entomol.,* 11, 241, 1986.
13. **Bernstein, C.,** Prey and predator emigration responses in the acarine system *Tetranychus urticae-Phytoseiulus persimilis, Oecologia,* 61, 134, 1984.
14. **Brown, L. and Brown, J.,** Dispersal and dispersion in a population of soldier beetles, *Chauliognathus pennsylvanicus* (Coleoptera: Cantharidae), *Environ. Entomol.,* 13, 175, 1984.
15. **Cain, M. L.,** Random search by herbivorous insects: a simulation model, *Ecology,* 66, 876, 1985.
16. **Cain, M., Eccleston, J., and Kareiva, P.,** The influence of food plant dispersion on caterpillar success, *Ecol. Entomol.,* 10, 1, 1985.
17. **Capinera, J. L. and Barbosa, P.,** Dispersal of first-instar gypsy moth larvae in relation to population quality, *Oecologia,* 26, 53, 1976.
18. **Cappucino, N. and Kareiva, P.,** Coping with a capricious environ-

ment: a population study of a rare pierid butterfly, *Ecology,* 66, 152, 1985.

19. **Carter, M. and Dixon, A. F. G.**, Foraging behavior of coccinellid larvae: duration of intensive search, *Entomol. Exp. Appl.,* 36, 133, 1984.
20. **Chandler, A.**, Locomotory behaviour of first instar aphidophagous Syrphidae (Diptera) after contact with aphids, *Anim. Behav.,* 17, 673, 1969.
21. **Cumber, R. A.**, Ecological Studies of the Rhinoceros Beetle *Oryctes rhinoceros* L. in Western Samoa, Technical Paper No. 107, South Pacific Commission, Noumea, New Caledonia, 1966.
22. **Dethier, V. G.**, Foodplant distribution and density and larval dispersal as factors affecting insect populations, *Can. Entomol.,* 91, 581, 1959.
23. **Dethier, V. G.**, Patterns of locomotion of polyphagous arctiid caterpillars in relation to foraging, *Ecol. Entomol.,* 14, 375, 1989.
24. **Dicke, M.**, Volatile spider-mite pheromone and host-plant kairomone, involved in spaced-out gregariousness in the spider mite *Tetranychus urticae, Physiol. Entomol.,* 11, 251, 1986.
25. **Dixon, A. F. G.**, Population dynamics of the sycmore aphid *Drapanosiphum platanoides* (Schr.) (Hemiptera: Aphididae): Migratory and trivial flight activity, *J. Anim. Ecol.,* 38, 585, 1969.
26. **Dixon, A. F. G. and Glen, D. M.**, Morph determination in the bird cherry-oat aphid *Rhopalosiphum padi* L., *Ann. Appl. Biol.*, 68, 11, 1971.
27. **Doucet, P. and Wilschut, A.**, Theoretical studies of animal orientation III. A model for kinesis, *J. Theor. Biol.*, 127, 111, 1987.
28. **Douwes, P.**, Host-selection and host-finding in egg-laying female *Cidaria albulata* L. (Lepidoptera: Geometridae), *Opusc. Entomol.,* 33, 233, 1971.
29. **Edelstein-Keshet, L.**, Mathematical theory for plant-herbivore systems, *J. Math. Biol.,* 24, 25, 1986.
30. **Edelstein-Keshet, L. and Rausher, M. D.**, The effects of inducible plant defenses on herbivore populations. I. Mobile herbivores in continuous time, *Am. Nat.,* 133, 787, 1989.
31. **Eguagie, W. E.**, An analysis of movement of adult *Tingis ampliata* (Heteroptera: Tingidae) in a natural habitat, *J. An. Ecol.,* 43, 521, 1974.
32. **Elmstrom, K., Andow, D., and Barclay, W.**, Flea beetle movement in a broccoli monoculture and diculture, *Environ. Entomol.,* 17, 299, 1988.
33. **English-Loeb, G .M. and Collier, B. D.**, Nonmigratory movement of adult harlequin bugs *Murgantia histrionica* (Hemiptera: Pentatomidae) as affected by sex, age and host plant quality, *Am. Midl. Nat.,* 118, 189, 1987.
34. **Evans, H. F.**, The searching behaviour of *Anthocoris confusus* (Reuter) in relation to prey density and plant surface topography, *Ecol. Entomol.,* 1, 163, 1976.
35. **Fahrig, L. and Paloheimo, J.**, Interpatch dispersal of the cabbage butterfly, *Can. J. Zool.,* 65, 616, 1987.
36. **Fahrig, L. and Paloheimo, J.**, Effect of spatial arrangement of habitat patches on local population size, *Ecology,* 69, 468, 1988.
37. **Feeny, P. P., Rosenberry, L., and Carter, M.**, Chemical aspects of oviposition behavior in butterflies, in *Herbivorous Insects: Host Seeking Behavior and Mechanisms,* Ahmad, S., Ed., Academic Press, New York, 1983, 27.
38. **Fleisher, S. J., Gaylor, M. J., and Hue, N. V.**, Dispersal of *Lygus*

lineolaris (Hemiptera: Miridae) adults through cotton following nursery host destruction, *Environ. Entomol.,* 17, 533, 1988.

39. **Fleschner, C.**, Studies on the searching capacity of the larvae of three predators of the citrus red mite, *Hilgardia,* 20, 233, 1950.
40. **Fromm, J. E. and Bell, W. J.**, Search orientation of *Musca domestica* in patches of sucrose drops, *Physiol. Entomol.,* 12, 297, 1987.
41. **Harborne, J. B.**, *Introduction to Ecological Biochemistry,* 2nd ed., Academic Press, London, 1982, chap. 5.
42. **Hardie, J.**, Behavioral differences between alate and apterous larvae of the black bean aphid, *Aphis fabae:* dispersal from the host plant, *Entomol. Exp. Appl.,* 28, 338, 1980.
43. **Harrison, S. and Karban, R.**, Behavioural response of spider mites (*Tetranychus urticae*) to induced resistance of cotton plants, *Ecol. Entomol.,* 11, 181, 1986.
44. **Haskell, P. T., Paskin, M. W. J., and Moorhouse, J. E.**, Laboratory observations on factors affecting the movements of hoppers of the desert locust, J. *Insect Physiol.,* 8, 53, 1962.
45. **Hawkes, C.**, Dispersal of adult cabbage rootfly (*Erioischia brassicae*) in relation to a brassica crop, *J. Appl. Ecol.,* 11, 83, 1974.
46. **Heinrich, B.**, Resource heterogeneity and patterns of movement in foraging bumblebees, *Oecologia,* 40, 235, 1979.
47. **Hussey, N. W. and Parr, W. J.**, Dispersal of the glasshouse red spider mite *Tetranychus urticae* Koch (Acarina, Tetranychidae), *Entomol. Exp. Appl.,* 6, 207, 1963.
48. **Iwao, S. and Machida, A.**, A marking- and recapture analysis of the adult population of a phytophagous lady-beetle, *Epilachna sparsa orientalis, Res. Pop. Ecol.,* 5, 107, 1963.
49. **Jones, R. E.**, Search behavior: a study of three caterpillar species, *Behavior,* 60, 237, 1977.
50. **Jones, R. E.**, Movement patterns and egg distribution in cabbage butterflies, *J. An. Ecol.,* 46, 195, 1977.
51. **Kareiva, P.**, Experimental and mathematical analyses of herbivore movement: quantifying the influence of plant spacing and quality on foraging discrimination, *Ecol. Monogr.,* 52, 261, 1982.
52. **Kareiva, P. M.**, Local movement in herbivorous insects: applying a passive diffusion model to mark-recapture field experiments, *Oecologia,* 57, 322, 1983.
53. **Kareiva, P. M.**, Finding and losing host plants by *Phyllotreta*: patch size and surrounding habitat, *Ecology,* 66, 1809, 1985.
54. **Kareiva, P. and Odell, G.**, Swarms of predators exhibit "preytaxis" if individual predators use area-restricted search, *Am. Nat.,* 130, 233, 1987.
55. **Kareiva, P. and Shigesada, N.**, Analyzing insect movement as a correlated random walk, *Oecologia,* 56, 234, 1983.
56. **Keller, E. and Segel, L.,** A model for chemotaxis, *J. Theor. Biol.,* 30, 225, 1971.
57. **Kennedy, J. S., Booth, C. O., and Kershaw, W. J. S.**, Host finding by aphids in the field. III. Visual attraction, *Ann. Appl. Biol.,* 49, 1, 1959.
58. **Kennedy, J. S. and Moorehouse, J. E.**, Laboratory observations of locust responses to wind-borne grass odour, *Entomol. Exp. Appl.,* 12, 487, 1969.
59. **Kidd, N. A. C.**, The influence of populaton density on the light behaviour of the lime aphid, *Eucallipterus tiliae, Entomol. Exp. Appl.,* 22, 251, 1977.

60. **Kiritani, K., Hokyo, N., and Iwao, S.**, Population behavior of the southern green stinkbug, *Nezara viridula*, with special reference to the developmental stages of early-planted paddy, *Res. Pop. Ecol.*, 8, 133, 1966.
61. **Lamb, K. P., Hassan, E., and Scotter, D. R.**, Dispersal of scandium-46-labelled Pantorhytes weevils in papuan cacao plantations, *Ecology,* 52, 178, 1971.
62. **Lance, D. R.**, Host-seeking behavior of the gypsy moth: the influence of polyphagy and highly apparent host plants, in *Herbivorous Insects: Host Seeking Behavior and Mechanisms,* Ahmad, S., Ed., Academic Press, New York, 1983, 201.
63. **Lance, D. R. and Barbosa, P.**, Host tree influences on the dispersal of first instar gypsy moths, *Lymantria dispar* (L.), *Ecol. Entomol.*, 6, 411, 1981.
64. **Lance, D. R. and Barbosa, P.**, Host tree influences on the dispersal of late instar gypsy moths, *Lymantria dispar* (L.), *Oikos*, 38, 1, 1982.
65. **Lawrence, W. S.**, Sexual dimorphism in between and within patch movements of a monophagous insect, *Tetraopes* (Coleoptera: Cerambycidae), *Oecologia,* 53, 245, 1982.
66. **Lawrence, W. S.**, Effects of sex ratio on milkweed beetle emigration from host plant patches, *Ecology,* 68, 539, 1987.
67. **Lawrence, W. S. and Bach, C. E.**, Chrysomelid beetle movement in relation to host-plant size and surrounding non-host vegetation, *Ecology,* 70, 1679, 1989.
68. **McCauley, D. E., Ott, J. R., Stime, A., and McGrath, S.**, Limited dispersal and its effect on population structure in the milkweed beetle *Tetraopes tetraophthalmus, Oecologia,* 51, 145, 1981.
69. **McEvoy, P. B.**, Adaptive Significance of Clumped Dispersion in the Treehopper, Publilia concava (Homoptera: Membracidae), Dissertation, Cornell University, Ithaca, NY, 1977.
70. **Miller, J. R. and Strickler, K. L.**, Finding and accepting host plants, in *Chemical Ecology of Insects,* Bell, W. and Carde, R., Eds., Sinauer Associates, Sunderland, MA, 1984.
71. **Moorehouse, J. E.**, Experimental analysis of the locomotor behaviour of *Schistocerca gregaria* induced by odour, *J. Insect Physiol.*, 17, 913, 1971.
72. **Murdie, G. and Hassell, M. P.**, Food distribution, searching success and predator-prey models, in *The Mathematical Theory of the Dynamics of Biological Populations,* Bartlett, M. S. and Hiorns, R. W., Eds., Academic Press, New York, 1973.
73. **Myers, J. H.**, Effect of physiological condition of the host plant on the ovipositional choice of the cabbage white butterfly, *Pieris rapae, J. An. Ecol.,* 54, 193, 1985.
74. **Odendaal, F. J., Turchin, P., and Stermitz, F. R.**, An incidental-effect hypothesis explaining aggregation of males in a population of *Euphydryas anicia, Am. Nat.,* 132, 735, 1988.
75. **Odendaal, F. J., Turchin, P., and Stermitz, F. R.**, Influence of host-plant density and male harassment on the distribution of female *Euphydryas anicia* (Nymphalidae), *Oecologia,* 78, 283, 1989.
76. **Okubo, A.**, *Diffusion and Ecological Problems: Mathematical Models,* Springer-Verlag, Berlin, 1980.
77. **Owens, E. D. and Prokopy, R. J.**, Relationship between reflectance spectra of host plant surfaces and

visual detection of host fruit by *Rhagoletis pomonella* flies, *Physiol. Entomol.*, 11, 297, 1986.

78. **Parker, M. A.**, Local food depletion and the foraging behavior of a specialist grasshopper, *Hesperotettix viridis, Ecology,* 65, 824, 1984.
79. **Power, A. G.**, Plant community diversity, herbivore movement, and an insect-transmitted disease of maize, *Ecology,* 68, 1658, 1987.
80. **Prokopy, R. J. and Owens, E. D.**, Visual detection of plants by herbivorous insects, *Annu. Rev. Entomol.,* 28, 337, 1983.
81. **Ralph, C. P.**, Search behavior of the large milkweed bug *Oncopeltus fasciatus* (Hemiptera: Lygaeidae), *Annu. Entomol. Soc. Am.,* 70, 337, 1976.
82. **Rausher, M. D.**, Coevolution in a Simple Plant-Herbivore System, Dissertation, Cornell University, Ithaca, NY, 1977.
83. **Rausher, M. D.**, Search image for leaf shape in a butterfly, *Science,* 200, 1071, 1978.
84. **Risch, S.**, Insect herbivore abundance in tropical monocultures and polycultures: an experimental test of two hypotheses, *Ecology,* 62, 1325, 1981.
85. **Roitberg, B. D., Myers, J. H., and Frazer, B. D.**, The influence of predators on the movement of apterous pea aphids between plants, *J. An. Ecol.,* 48, 111, 1979.
86. **Root, R. B. and Kareiva, P. M.**, The search for resources by cabbage butterflies (*Pieris rapae*): ecological consequences and adaptive significance of Markovian movements in a patchy environment, *Ecology,* 65, 147, 1984.
87. **Shaw, M. J. P.**, Effects of population density on alienicolae of *Aphis fabae* Scop. III. The effect of isolation on the development of form and behaviour of alatae in a laboratory clone, *Ann. Appl. Biol.,* 65, 205, 1970.
88. **Shigesada, N., Kawasaki, K., and Teramoto, E.**, Spatial segregation of interacting species, *J. Theor. Biol.,* 79, 83, 1979.
89. **Smith, R. W. and Whittaker, J. B.**, Factors affecting *Gastrophysa viridula* populations (Coleoptera: Chrysomelidae) in different habitats, *J. Anim. Ecol.,* 49, 537, 1980.
90. **Stanton, M. L.**, Searching in a patchy environment: foodplant selection by *Colias periphyle* butterflies, *Ecology,* 63, 839, 1982.
91. **Stanton, M. L.**, Spatial patterns in the plant community and their effects upon insect search, in *Herbivorous Insects: Host Seeking Behavior and Mechanisms,* Ahmad, S., Ed., Academic Press, New York, 1983, 125.
92. **Thorsteinson, A. J.**, Host selection in phytophagous insects, *Annu. Rev. Entomol.,* 5, 193, 1960.
93. **Turchin, P.**, Modelling the effect of host patch size on Mexican bean beetle emigration, *Ecology,* 67, 124, 1986.
94. **Turchin, P.**, The role of aggregation in the response of Mexican bean beetles to host-plant density, *Oecologia,* 71, 577, 1987.
95. **Turchin, P.**, Population consequences of aggregative movement, *J. An. Ecol.,* 58, 75, 1989.
96. **Varela, L. G. and Bernays, E. A.**, Behavior of newly hatched potato tuber moth larvae, *Phthorimaea operculella* Zell. (Lepidoptera: Gelechiidae), in relation to their host plants, *J. Insect Behav.,* 1, 261, 1988.
97. **Vernon, R. S. and Borden, J. H.**, Dispersion of marked-released *Hylemya antiqua* (Meigen) (Diptera: Anthomyiidae) in an onion field, *Environ. Entomol.,* 12, 646, 1983.

98. **Visser, J. H. and Nielson, J. K.**, Specificity in the olfactory orientation of the Colorado beetle, *Leptinotarsa decemlineata, Entomol. Exp. Appl.,* 21, 14, 1977.
99. **Watanabe, M.**, Adult movements and resident ratios of the black-veined white, *Aporia crataeyi,* in a hilly region, *Jpn. J. Ecol.,* 28, 101, 1978.
100. **Weseloh, R. M.**, Dispersal and survival of gypsy moth larvae, *Can. J. Zool.,* 65, 1720, 1987.
101. **Williams, C. E.**, Movement, dispersion, and orientation of a population of the Colorado potato beetle, *Leptinotarsa decemlineata* (Coleoptera: Chrysomelidae) in eggplant, *Great Lakes Entomol.,* 21, 31, 1988.
102. **Zalucki, M. P. and Kitching, R. L.**, The analysis and description of movement of adult *Danaus plexippus* L. (Lepidoptera: Danaidae), *Behaviour,* 80, 174, 1982.

Host Range Patterns of Hymenopteran Parasitoids Of Exophytic Lepidopteran Folivores

William Sheehan
Insect Biology and Population Management Research Laboratory
USDA/ARS
Tifton, Georgia

TABLE OF CONTENTS

I. INTRODUCTION

Host range (or diet) specialization presents a central challenge for evolutionary ecologists. Much attention has been focused on patterns of diet specialization in leaf-feeding insects in relation to plant attributes.[21,32] By contrast, little is known about carnivore host range in relation to folivore qualities.

One group of carnivores, parasitoids, is particularly well suited for such an investigation because folivore species are often attacked by multiple parasitoid species with diverse host ranges. Lack of extensive, reliable rearing records of parasitoids in natural systems has hindered evaluation of the possible role of parasitoid host range in structuring host communities.[4] Recently, categorizing parasitoid host range by attack strategy has spurred attempts to elucidate its role in community organization.[4,37] Idiobiont parasitoids kill or permanently paralyze hosts during attack, whereas koinobionts permit continued development of the host. Therefore, koinobionts are presumably under increased selection pressure, relative to idiobionts, to adapt to specific physiological host defenses.[4] Although more quantitative data are needed, available evidence suggests that idiobionts have wider host ranges, on average, than koinobionts.[4,34,79]

Most evidence for patterns of association between parasitoid host range and ecological characteristics of phytophagous hosts comes from studies of parasitoids of endophytic hosts, particularly leaf-miners and gall-makers.[2,4,86,93] Furthermore, such studies have typically focused on patterns of parasitoid species richness; quantitative information on relative importance of parasitoids varying in host range is extremely rare.

In this chapter I examine associations between host range of parasitoids in one superfamily Ichneumonoidea (Hymenoptera) and five characteristics of exophytic lepidopteran hosts: food plant range (number of plant taxa utilized by the host), food plant type (tree, shrub or herb), exposure (exposed or semiconcealed in webs, leaf ties, or leaf rolls), integument defenses (presence or absence of hairs or spines), and gregariousness (solitary, egg-clumping, or larval-gregarious). I also examine relationships between these variables and host abundance and phylogeny. *A priori* hypotheses are presented or developed from the literature regarding expected correlations between parasitoid host range and each host variable. There is little comparative experimental work on host selection behavior of a range of specialist and generalist parasitoids; however, limited data do exist for some insect and avian predators of Lepidoptera.[11,40] These results are used for comparison and to generate hypotheses, assuming—until evidence indicates otherwise—that generalist parasitoids are similar to generalist predators in their response to prey. I also assumes that the insect and avian predators in question are in fact relative generalists, and that learning produces in them, on average, only temporary individual specialization.

Hypotheses are tested using data from an extensive rearing program for parasitoids of exophytic caterpillars native to the northeastern U.S., undertaken between 1915 and 1933.[71] In addition to the size of the database, these data have two particular advantages for testing hypotheses about host range associations. First, a wide range of native hosts was collected and native parasitoids reared. In contrast, most ecological research on parasitoids (particularly parasitoids of exophytic hosts) has been host or host plant-oriented and has typically focused on introduced host or parasitoid species, often only under outbreak conditions.[4] Second, the data set includes collection frequencies of hosts and parasitoids, which allows estimation of relative abundance. Abundance data is critical in assessing host ranges of all but the best known insects.[46]

II. METHODS

The author analyzed data collected by the Melrose Highlands Laboratory (Massachusetts, U.S.) of the former federal Bureau of Entomology using a published summary[71] as well as original data summary records. During the project 11,416 collections of Macrolepidoptera larvae and pupae were reared, representing 340 species and 442,112 identified specimens. Approximately 2000 collections yielded hymenopteran parasitoids. Sampling was restricted to nine northeastern states (Maine, Massachusetts, New Hampshire, Vermont, Rhode Island, Connecticut, New York, New Jersey, and Pennsylvania) with the majority of collections coming from Massachusetts, Maine, and New Jersey. A "collection" includes all the individuals of a species taken at one locality and time. No attempt was made to sample randomly, but collections were made by numerous persons during the 19 years of the study. There is no *a priori* reason to suspect any sampling bias with respect to parasitoid host range. All parasitoids and hosts analyzed are native to the northeastern U.S.,except *Pieris rapae* and its parasitoid, *Cotesia glomerata*, both of which established at least half a century before the rearing program began.[76] I excluded the few intentionally introduced parasitoid and host species from analysis.

"Generalist" parasitoids are defined here as those reared from hosts in two or more families, while "specialists" were reared from hosts in one lepidopteran genus ("genus specialists") or family ("family specialists"; see also Methods). This usage is taxonomic and differs from "specialists" of Force[28] (K-selected species), Askew[2] (parasitoids of primary gall makers and leaf miners), and Askew and Shaw[4] (koinobionts); other aspects of parasitoid specialization are discussed by Zwölfer[95] and Sheehan.[77] Because idiobionts are considered generalists and koinobionts specialists by recent authors,[4,37] their findings are compared with those in this chapter. Lepidopteran hosts are classified as "oligophagous", "intermediate", and "polyphagous" by whether they feed on plants in one, two or three, or four or more families, respectively.

I examined two types of patterns of association between specialist and generalist parasitoids and their hosts:

1. Species richness of parasitoids in relation to host characteristics and parasitoid host range (species richness may indicate evolutionary advantages for parasitoids exploiting specific types of hosts[63])
2. Parasitoid prevalence in host collections was calculated in relation to the same variables

In the absence of detailed comparative population dynamics studies it is impossible to assess the importance of parasitoids as selective agents; however, prevalence in host collections (percent of host collections yielding a given parasitoid) provides a useful measure of parasitoid impact *relative* to other species of parasitoids attacking a particular host. Schaffner and Griswold,[71] original data summary records, and other published sources (see below) were used to characterize host ranges of parasitoids and their hosts. A separate paper will address associations between parasitoids and the food plants of their hosts.

A. LEPIDOPTERA

Parasitoids were reared from 98 species of Lepidoptera (Appendix 1). All host species feed exophytically on trees, shrubs, and herbs, including ten that feed or rest in webs and six that feed in rolled or tied leaves, at least in early instars. Almost all are common or abundant species;[24-27,96] these species represent only 29% of the 340 species collected during the study. Median collection frequency for the 98 species was 54 collections/species (range, 1 to 623). An additional 38 host species were collected 15 or more times and yielded either no ichneumonoid parasitoids (26 species) or ichneumonoid parasitoids represented in three or fewer total collections (12 species). These host species (marked with asterisks in Appendix 1) were included for comparison with the more frequently attacked species (see Section IV).

Host attribute determinations (Table 1) were based on data reported by Schaffner and Griswold[71] and on information in published manuals (see References 5, 17, 18, 22-27, 30, 60, and 74). In cases where sources did not agree on particular host attributes, the most often cited description was accepted or further confirmation was sought in the primary literature. Questionable or exceptional food plant records (e.g., plants used only during outbreaks) were excluded, as were plants not found in the study area. Plant type was determined from the principal food plants utilized. Host species that regularly feed on a mixture of trees, shrubs, and/or herbs ("mixed" in Appendix 1) were not used in plant type analyses. Exposure of caterpillars that make webs or leaf ties or rolls was listed as "web maker" or "leaf tier", respectively, even when such behavior is restricted to early instars (e.g., *Euphydras phaeton*); together these categories are called semiconcealed. In addition to "solitary", three categories of gregarious host species were distinguished: "egg-gregarious" (species that disperse from clumped eggs, e.g., *Alsophila pometaria*), "early-gregarious" (species gregarious only during early instars, e.g., *Euphydras phaeton*), and "late-gregarious" (species that continue gregarious behavior until later instars or throughout larval development, e.g., *Malacosoma americanum*). Host species were classified as "hairy/spiny" if they possess long, dense hairs or spines during most instars. "Smooth" caterpillars have short, minute, or sparse hairs and do not have spines.

Table 1
FREQUENCY OF HOST SPECIES AND COLLECTIONS WITHIN VARIABLES AND CATEGORIES

	No. of species	No. of collections
Food plant range		
1 family	40	2728
2—3 families	14	759
4+ families	44	5597
Plant type		
Trees	54	6058
Shrubs	15	408
Herbs	21	1810
Mixed	8	808
Exposure		
Exposed feeders	82	7432
Tiers and folders	6	353
Web makers	10	1299
Gregariousness		
Solitary	58	3898
Egg	10	848
Early	19	1711
Late	11	2627
Integument		
Smooth	51	2790
Hairy/spiny	47	6294
Total	98	9084

B. PARASITOIDS

Seventy-three identified ichneumonoid parasitoid species were represented in four or more collections per host species (36 Braconidae and 37 Ichneumonidae; Appendix 2). All but ten of these species are koinobionts. These species attack one larval instar and emerge from a later larval instar or from the host pupa, or, in the case of *Cotesia hyphantriae*, attack eggs and emerge from the larva.[56] Virtually all of the koinobionts were reared from hosts collected as larvae. By contrast, the ten idiobiont species were reared primarily from hosts collected and presumably attacked as pupae (some cratichneumonines were recorded as having been reared from hosts collected as larvae). All of the idiobionts are ichneumonids (four ichneumonines, four pimplines, and two phygadeuontines).

Host range determinations are sensitive to sampling effort,[3] and it is desirable to distinguish usual from extraordinary hosts.[41] I therefore used abundance information in Schaffner and Griswold,[71] supplemented by original data records and by published host records.[15,52] Parasitoids reared from ten or more host collections were categorized as genus or family specialists if at least 90% of collections came from hosts in one genus or one host family, respectively. For parasitoids collected fewer than ten times, hosts reported in Marsh[52] and Carlson[15] were also considered. Fifty-six species (77%) are apparently restricted to one host family (including 24 restricted to one host genus), and 17 species (23%) are reported from two or more host families. Most generalists in this data set are koinobiont ichneumonids. Two braconid species were classified as generalists (*Cotesia phobetri* and *Meteoris hyphantriae*), as were four of the ten idiobiont species. Of the five generalist species reared from five or more host families, only *Itoplectis conquistor*, is an idiobiont.

C. ANALYSIS

The collection was used as the basic measure of sampling effort for both hosts and parasitoids. Host collections include all individuals of a species taken at one locality and time. The individuals of one parasitoid species reared from one host collection constitute a "parasitoid collection". The mean number of host larvae per collection per host species ranged from 1 to 217 (median = 10.4) individuals; the mean number of host collections yielding parasitoids per host species ranged from 1 to 27.6 (median = 1.5). Gregarious host species are typically represented by more individuals than solitary species, and more individuals of a species are collected during outbreaks than during endemic phases. To facilitate comparisons and to provide conservative estimates of relative frequencies, collections were therefore used as the basic unit of analysis.[46]

Parasitoid species richness was analyzed by examining frequencies of parasitoid-host associations with respect to parasitoid host range and host variables. Patterns across all levels of parasitoid host range (three levels) and host variables (two, three, or four levels) were analyzed with χ^2 tests (Table 2, Appendix 3); subdivision of contingency tables was performed as described in Reference 84. All χ^2 analyses with one degree of freedom employ the conservative Yates continuity correction. Both parasitoid species richness and prevalence within host collections were further analyzed for specific comparisons of interest using weighted tests of proportions (Figures 1 to 5). Parasitoid species richness per host category is the sum of parasitoid-host associations for that host category, and parasitoid prevalence with respect to a host category is the sum of all collections of hosts of that category yielding parasitoids. Both frequencies were weighted by using the total number of hosts collected in each category. In these tests, the frequencies of parasitoids associated with alternative host characteristics (e.g., oligophagy vs. polyphagy) were compared within specialist and generalist parasitoids separately, since there are different numbers of specialist and generalist species. Associations between host variables and host abundance (as measured by collection frequency) were analyzed by ANOVA, using log-transformed values.

Statistical tests assume that parasitoid rearings were independent (i.e., the number of host collections yielding multiple species of parasitoids was insignificant). This assumption is approximately true because half (49/98) of all host

Table 2
PARASITOID-HOST ASSOCIATIONS BY HOST VARIABLE AND PARASITOID HOST RANGE, ALL LEVELS

	Host variable		
	df	χ^2	**P**[a]
Food plant range			
Food plant range • parasitoid host range	4	4.3	NS
1 vs. 2–3 families • parasitoid host range	2	1.3	NS
1–3 vs. 4+ families • genus vs. family specialists	1	1.7	NS
1–3 vs. 4+ families • specialists vs. generalists	1	0.4	NS
Plant type			
Plant type • parasitoid host range	4	14.1	**
Herbs vs. shrubs • parasitoid host range	2	0.6	NS
Herbs and shrubs vs. trees • genus vs. family specialists	1	4.4	*
Herbs and shrubs vs. trees • specialists vs. generalists	1	6.8	**
Exposure			
Exposure • parasitoid host range	4	4.9	NS
Tiers vs. web makers • parasitoid host range	2	1.6	NS
Exposed vs. semiconcealed • genus vs. family specialists	1	0.1	NS
Exposed vs. semiconcealed • specialists vs. generalists	1	2.8	NS
Gregariousness			
Gregariousness • parasitoid host range	4	35.0	***
Egg and early vs. late gregarious • parasitoid host range	2	5.1	NS
Solitary vs. gregarious • genus vs. family specialists	1	9.2	**
Solitary vs. gregarious • specialists vs. generalists	1	18.4	***
Integument			
Integument • parasitoid host range	2	21.7	**
Smooth vs. hairy • genus vs. family specialists	1	7.70	*
Smooth vs. hairy • specialists vs. generalists	1	11.4	**

Note: * = $p < 0.05$, ** = $p < 0.005$, *** = $p < 0.0005$.

[a] NS = not significant.

species yielded only one ichneumonoid parasitoid species, and because, for host species yielding more than one parasitoid species, parasitoids were reared from a minority of all host collections (the median percent of collections per host species yielding a given parasitoid species was only 3.2%).

III. HOST VARIABLES AND PARASITOID HOST RANGE

A. HOST FOOD PLANT RANGE

Several authors, extending theory developed for interactions between plants and folivores, have speculated that oligophagous folivores that feed on

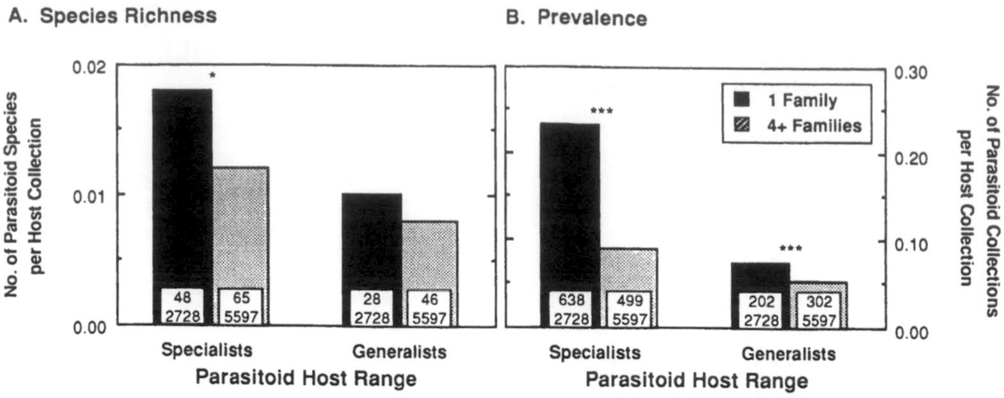

FIGURE 1. Host food plant range and parasitoid host range; effects on parasitoid species richness (A) and prevalence of parasitoids in host collections (B). Boxes at bases of bars show frequencies used in analyses. Top numbers give frequencies of unique host-parasitoid associations (A) or host collections yielding parasitoids (B); bottom numbers give total number of hosts collected having given attribute. Asterisks (*) show significance levels of comparisons: * = $p<0.05$, ** = $p<0.005$, *** = $p<0.0005$.

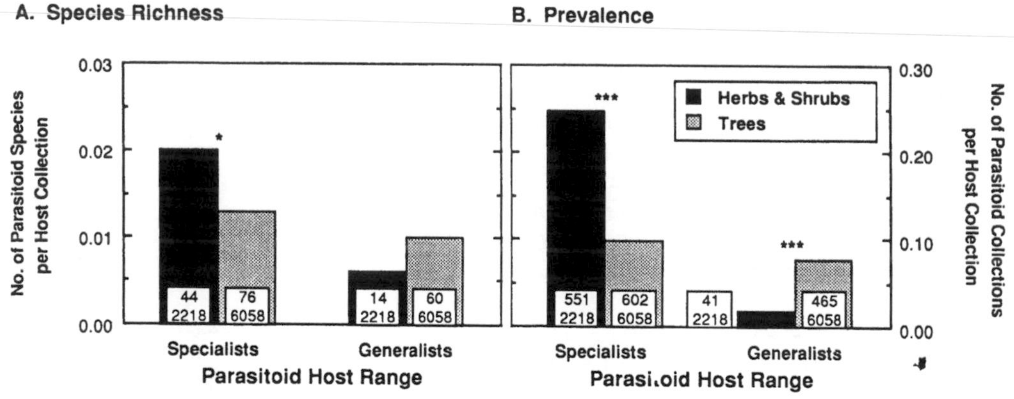

FIGURE 2. Host plant type and parasitoid host range; effects on parasitoid species richness (A) and prevalence of parasitoids in host collections (B). See legend to Figure 1 for details.

toxic plants should be relatively more vulnerable to predation by adapted specialist than by generalist enemies.[21,65] Edmunds[20] proposed that low predation pressure by generalist birds on aposematic (typically oligophagous) caterpillars selects for increased utilization and species richness of specialist parasitoids that can adapt to host toxins (and thereby avoid incidental consumption by generalist predators). Limited evidence, mostly anecdotal, also suggests that some insect predators are specialized on chemically defended, oligophagous prey.[62] However, Barbosa[6] summarized data from several experiments on effects of plant allelochemicals on the physiology and survival of six species of parasitoids. While he found that specialist parasitoids were indeed usually less susceptible than generalists to antibiotic effects of plant chemicals, susceptibility was poorly correlated with host food plant range.

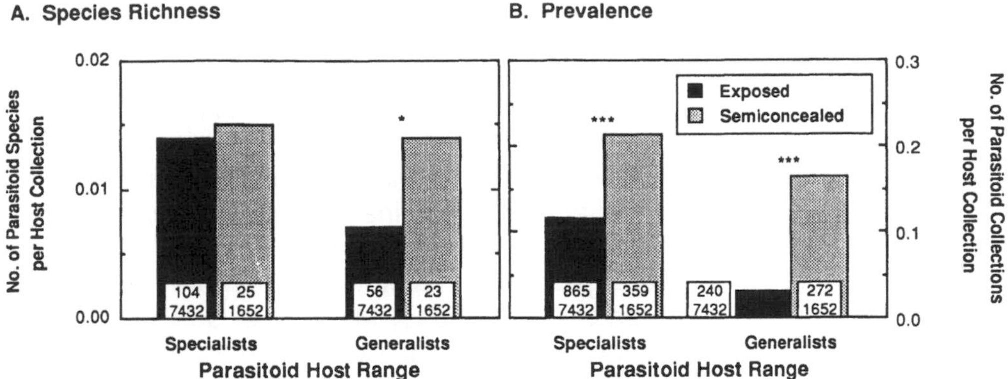

FIGURE 3. Host exposure and parasitoid host range; effects on parasitoid species richness (A) and prevalence of parasitoids in host collections (B). See legend to Figure 1 for details.

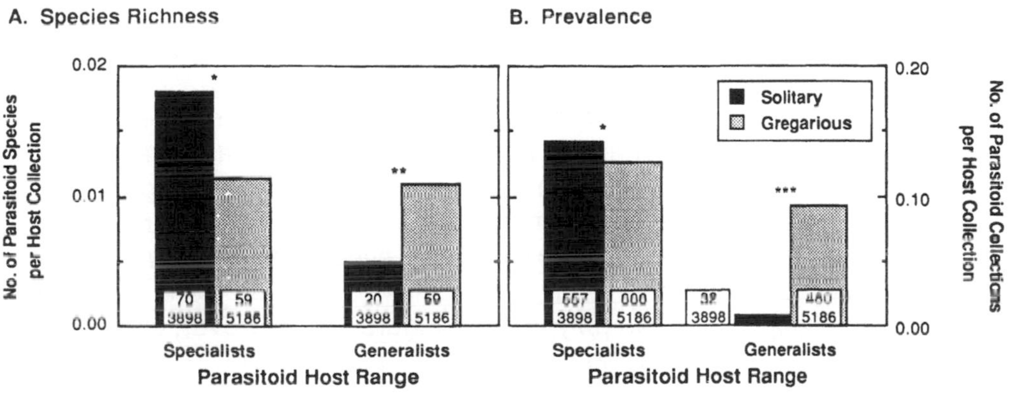

FIGURE 4. Host gregariousness and parasitoid host range; effects on parasitoid species richness (A) and prevalence of parasitoids in host collections (B). See legend to Figure 1 for details.

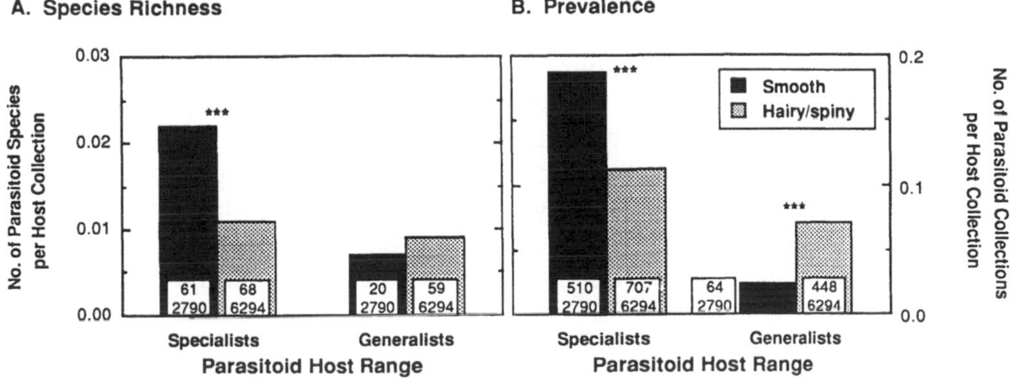

FIGURE 5. Host integument and parasitoid host range; effects on parasitoid species richness (A) and prevalence of parasitoids in host collections (B). See legend to Figure 1 for details.

Others have argued that generalist enemies are important selective agents affecting folivore food plant range.[13a] Lawton[44,49] argued that folivores are restricted from using certain food plants by generalist predators and parasitoids. Recently, Bernays and Graham[11] proposed that associations between generalist enemies and polyphagous folivores, in particular, may explain the observed predominance of narrow folivore food plant ranges. Using the vespid wasp, *Mischocyttarus flavitarsus*, and the argentine ant, *Iridomyrmex humilis*, Bernays[9,10] showed experimentally that these generalist predators preferred polyphagous to oligophagous Lepidopteran caterpillars when prey was presented in choice situations. On the other hand, Maclean et al.[51] examined preferences of generalist bird predators for adult Lepidopteran prey. In no-choice experiments they found that the proportion of prey taken for prey whose larvae are polyphagous (0.83; N=1,186) was similar to that for prey whose larvae are monophagous or oligophagous (0.77; N = 917 [p. 299 of Reference 51]). Comparable experimental data for parasitoids are unavailable.

Results

For species richness, there was no overall pattern of association between host ranges of parasitoids and plant ranges of hosts (Table 2, "Food plant range"). Oligophagous caterpillars did not have proportionally more specialist parasitoid species than polyphagous hosts (although there was a trend), and polyphages did not have more generalist parasitoid species (Figure 1A). Including caterpillar species of intermediate plant range with either oligophagous or polyphagous species had no significant effect. In terms of prevalence in host collections, collections yielding specialist parasitoids were more frequent among oligophages than polyphages (Figure 1B); however, contrary to predictions, generalist parasitoids were also more frequent in oligophage collections (Figure 1B).

B. HOST PLANT TYPE

Herbaceous plants are thought generally to have more toxic secondary plant compounds and to be less "apparent" to folivores than woody plants.[21] Although recent research has called into question the hypothesized digestibility-reducing function and nontoxic nature of tree-leaf secondary chemistry,[7,47] tree foliage may indeed be generally lower in nutrients than herbaceous foliage,[53] and folivores adapted to feeding on trees may typically grow more slowly than herb feeders.[88] In temperate regions herbs have a higher proportion of oligophagous Lepidoptera than do trees, and trees have proportionately more polyphagous species.[31,58,81] By extension from plant-folivore theory, folivores of herbaceous plants should be more prone to attack by specialist enemies that can find isolated, ephemeral host patches and adapt to sequestered plant toxins. Conversely, folivores of long-lived woody plants should, on average, be more vulnerable to generalist enemies. In support of the latter hypothesis, Maclean et al.[51] found that adult Lepidoptera whose larvae fed on woody plants were more acceptable to generalist bird predators than were prey whose larvae fed on herbaceous plants.

Other hypotheses predict that increasing complexity of plant architecture should influence the composition of enemy communities. For endophytic hosts, tree feeders have been found to support richer parasitoid assemblages than shrub and herb feeders.[3,4,36,38] Askew[2] found that endophytic folivores that gall or mine leaves on trees (cynipids and gracillariids) not only supported larger parasitoid complexes than those of shrubs and herbs, but also had a higher proportion of "generalist" species. For parasitoids of exophytic hosts, Hawkins[36] argued that host population fragmentation should decrease with increasing plant size and complexity (architecture), and that this should affect specialists more than generalists. As a result, exophytic tree-feeding folivores should have relatively higher specialist parasitoid loads than herb-feeding folivores. Hawkins et al.[37] examined parasitoid species richness patterns for 185 species of British holometabolous insects in 26 families. They reported an increase in the number and proportion of koinobiont (specialist) parasitoid species per host species on exophytic tree-feeding (vs. herb-feeding) folivores. This pattern is the opposite of that predicted by extension of plant-folivore theory.

Results

Records for parasitoids attacking caterpillars feeding on shrubs and herbs were combined, since they showed a similar pattern. The proportion of generalist parasitoid species was higher on trees than on herbs and shrubs (Table 2, "Plant type"). Specialist parasitoids were strongly associated with hosts on herbs and shrubs, in both species richness per host species and in prevalence in host collections (Figure 2). The weighted proportion of generalist species per host species was not lower on hosts on herbs and shrubs than on trees (Figure 2A); however, generalists were infrequent in collections of herb- and shrub-feeding Lepidoptera. Only 41 collections yielding generalist parasitoids came from herb-and shrub-feeding species, compared with 465 from tree-feeding species; the proportional difference in prevalence is highly significant (Figure 2B).

C. HOST EXPOSURE

Exposed-feeding, exophytic folivores may have different enemies than insects that feed exophytically in semiconcealed locations (leaf ties, leaf folds, and webs) or endophytically in more concealed locations (leaf mines, galls, stems, or roots).[35a] In an extensive survey of British phytophagous insects, Hawkins and Lawton[38] found that of the variables they examined host exposure ("niche") was the most important factor determining parasitoid species richness, accounting for 20.5% of the variance in number of parasitoid species per host species. Askew and Shaw[4] found that relatively more koinobiont than idiobiont parasitoid species attack exposed than concealed tenthredinid sawflies,[68] and Hawkins et al.[37] showed that this pattern holds for British holometabolous phytophages in general. Unfortunately, similar data were not reported for parasitoids of exposed and semiconcealed exophytic feeders (Hawkins, Askew, and Shaw lumped these categories because of insufficient sample size). Predator diet breadth may also be correlated with prey exposure.

Bernays and Cornelius[10] found that generalist Argentine ants took leaf tying caterpillars, irrespective of food plant range, more often than they took exposed-feeding oligophagous caterpillars. Together these results suggest that semiconcealed feeders may be more vulnerable than exposed feeders to generalist predators and parasitoids, and exposed feeders more vulnerable to specialists.

Results

Host exposure was independent of parasitoid host range in terms of frequency of association (Table 2, "Exposure"). Specialist species were not associated with exposed feeders (Figure 3A), and collections of exposed feeders yielded proportionally fewer specialists than did semiconcealed feeders (Figure 3B). Generalist parasitoids appeared proportionately more associated with semiconcealed than with exposed hosts (Figure 3), but this pattern was due to only two species of web makers and did not hold for leaf tiers or other web makers (see Section IV, Discussion).

D. HOST GREGARIOUSNESS

Aggregation may dilute per capita predation pressure,[16] while at the same time increasing the risk of colony discovery by generalist predators. Stamp[85] found that specialist parasitoids spent more time attending larger colonies of *Euphydras phaeton*. Birds and social wasps preferentially attacked larger webs of fall webworm *Hyphantrea cunea*,[55] but percentage of colonies attacked by parasitoids was independent of colony size.[57] If discovery is a major limiting factor, and if generalists are less efficient than specialists in host location, we would expect gregarious species to be attacked relatively more by generalist parasitoids, and less conspicuous solitary species more by specialists.

Results

Generalist parasitoids are strongly and positively associated with gregarious caterpillars, and specialists with solitary hosts. Gregarious hosts had proportionately more genus than family specialist associations (Table 2, "Gregariousness"). Solitary hosts yielded proportionately more specialist parasitoid species and collections (Figure 4). The proportion of generalists attacking gregarious hosts was significantly higher than the proportion attacking solitary hosts, for both parasitoid species and collections (Figure 4). Generalists were infrequently reared from solitary hosts; all solitary species (58 species, 90 associations, 3,898 collections) yielded a total of only 32 collections of generalists. By comparison, gregarious species (40 species, 118 associations, 5186 collections) yielded 480 collections of generalists (Figure 4).

E. HOST INTEGUMENT

Host integument may indicate palatability or unpalatability to enemies. Bright, aposematic coloration often indicates distastefulness to visually hunting generalist predators.[19] Interestingly, Bernays found that aposematic caterpillars were relatively unpalatable to Argentine ants, which are not visual hunters,[10] as well as to vespid wasps, which use both visual and olfactory cues;[9] however, determining which prey are aposematic, at what range, and to which enemies is extremely difficult.

Many species of lepidopteran larvae that feed exophytically are covered for some or all instars with setae or spines, a character that is more easily assigned than aposematic appearance. Such hairy/spiny species are probably relatively conspicuous to visually hunting enemies, because hairs and spines are often dark colored or on dark larvae, and because such species often feed openly and diurnally.[39,78] If ability to find hosts or prey determines usage, and if generalists are less efficient than specialists at finding hosts, smooth (and in this sample, usually cryptic) hosts should be attacked more by specialists, and hairy/spiny hosts more by generalists.

Some evidence, however, for predators and parasitoids suggests the contrary. In aviary experiments, black-capped chickadees (generalists predators) accepted smooth-skinned cryptic species without hesitation, but always rejected bristly or spiny species; species with silky plumose hair were accepted only when there was no alternative and then the birds ate only internal organs.[40] Bernays[9] found that generalist vespid wasps preferred hairy caterpillars less than smooth species. In choice experiments, wasps preferred polyphagous species except when hairy, then smooth oligophages were taken. Edmunds[20] reviewed evidence for higher specialist parasitism by *Apanteles* sp. on conspicuous (often aposematic) than on cryptic caterpillars. He hypothesized that conspicuous caterpillars are unpalatable to birds but vulnerable to adapted (specialist) parasitoids. From the above studies, it can be predicted that specialists should preferentially atttack hairy/spiny host species, and generalists primarily smooth species.

Results

Contrary to the latter prediction, but in support of the role of host finding as a limiting factor in host usage, specialist parasitoids were strongly associated with smooth species, and generalists with hairy/spiny species (Table 2, "Integument"). Smooth hosts yielded proportionately more specialist parasitoid species and collections (Figure 5). The proportion of generalists attacking hairy/spiny hosts was significantly higher than the proportion attacking smooth hosts, for parasitoid collections (Figure 5B) but not for species (Figure 5A).

F. HOST ABUNDANCE

The analyses of parasitoid species richness and relative abundance presented to this point control for sampling intensity but do not reveal whether generalists are typically associated with abundant host species, or specialists with less abundant species. Host "apparency" and ease of location have been proposed as major determinants of parasitoid exploitation patterns.[4,34,90] High mean host abundance should increase apparency and therefore increase exploitation by generalists, assuming generalists are less efficient at locating hosts than specialists. Very low host abundance may also restrict exploitation by specialists,[43] but since virtually all hosts examined are relatively common

Table 3
HOST COLLECTION FREQUENCY AND PERCENT INCIDENCE OF GENERALIST PARASITOID ASSOCIATIONS AND COLLECTIONS

Host variable	Host collection frequency[a]	Generalist associations[b]		Generalist collections[c]	
		%	Frequency	%	Frequency
Food plant range					
1 family	68.2 (12.5) a	36.8	28/76	24.0	202/840
2–3 families	54.2 (22.0) a	23.8	5/21	9.1	8/88
4+ families	127.2 (19.2) b	41.4	46/111	36.3	302/801
Plant type					
Trees	112.2 (16.8) b	44.1	60/136	43.6	465/1067
Shrubs	27.2 (5.8) a	17.4	4/23	10.4	11/106
Herbs	86.2 (20.6) b	28.6	10/35	6.2	30/486
Exposure					
Exposed feeders	90.6 (11.8) a	35.0	56/160	21.7	240/1105
Tiers and folders	58.8 (23.9) a	33.3	4/12	4.0	4/99
Web makers	129.9 (42.2) a	52.8	19/36	51.0	268/525
Gregariousness					
Solitary	67.2 (9.7) a	22.2	20/90	5.4	32/589
Egg	84.8 (24.7) ab	31.2	5/16	21.3	27/127
Early	90.1 (19.8) a	44.4	20/45	38.3	110/287
Late	238.8 (56.7) b	59.6	34/57	47.2	343/726
Integument					
Smooth	54.7 (7.4) a	24.7	20/81	11.1	64/574
Hairy/spiny	133.9 (19.6) b	46.5	59/127	38.8	448/1155

[a] Mean number of collections per species, with standard error in parentheses. Means within host variables followed by the same letter are not significantly different (based on ANOVAs using log-transformed values).
[b] Unique associations between generalist parasitoid species and host species. (Example: of 76 associations between parasitoid species and host species feeding within one plant family, 28 [36.8%] were with generalist parasitoid species.)
[c] Host collections yielding generalist (vs. specialist) parasitoids (% of total).

(see Section II, Methods) resource fragmentation is unlikely to be an important factor.

Host abundance may be correlated with collection frequency; however, conspicuous behavior or morphology independently contributes to apparency and may also have caused certain species to be sampled more than others — both by human collectors and by some parasitoids (see Section IV, Discussion). Although host abundance and conspicuousness are therefore confounded in collection frequencies, analysis of collection data may yield insight into the relationship between host apparency and parasitoid host range.

Results

Polyphagous, tree- and herb-feeding, late-gregarious, and hairy/spiny caterpillars were collected significantly more often than their counterparts (Table 3). Some but not all of these host types yielded a higher proportion of generalist parasitoid collections and species per host species than their counterparts. Late-gregarious feeders were collected more often than either early-gregarious or solitary feeders and had proportionately more generalist associations ($Z = 4.12$, $p < 0.0001$), and more generalist collections ($Z = 14.1$, $p < 0.0001$; Table 3). Hairy/spiny caterpillars were collected more than smooth caterpillars and yielded proportionately more generalist species than smooth caterpillars ($Z = 3.53$, $p < 0.0001$), and more generalist collections ($Z = 12.76$, $p < 0.0001$).

Other associations provide less support for a link between generalist parasitoids and host abundance. While polyphages were collected more than oligophagous host species (contrary to Reference 33), polyphages did not yield more generalist species ($Z = 0.39$, $p < 0.70$; Table 3) — although they did yield fewer generalist collections ($Z = 5.13$, $p < 0.0001$). (Hosts of intermediate food plant range had fewer total parasitoid species per host species than either oligophagous or polyphagous species [1.5 vs. 1.9 and 2.5, respectively], but not fewer parasitoid species per host collection [0.028, 0.028, and 0.020, respectively]. A similar decrease in parasitoid species richness on British folivores feeding on plants in two to four families[38] may thus be due to rarity or crypticity of intermediate hosts.) Shrub feeders were rarer than either herb or tree feeders but yielded a similar proportion of generalist species as herb feeders ($Z = 0.74$, $p < 0.48$) and a slightly higher proportion of generalist collections compared with herb feeders ($Z = 2.15$, $p < 0.03$; Table 3). Web makers were not collected more than tiers or exposed feeders but had proportionately more generalist associations than tiers and exposed feeders ($Z = 2.15$, $p < 0.03$), and more generalist collections ($Z = 13.36$, $p < 0.0001$).

G. HOST PHYLOGENY

Host phylogeny appears to exert a strong influence on the relative susceptibility of exophytic Lepidoptera to parasitism by specialists or generalists, since frequency of generalist parasitism varies widely within host families (Table 4). Parasitism of butterflies (Papilionoidea) was almost entirely by specialists. Only 5 (1.6%) of 319 collections of 14 butterfly species yielded generalist parasitoids, despite the presence in the sample of spiny and gregarious nymphalids. Similarly, sphingids yielded no generalist parasitoids. At the other extreme, generalists dominated rearings of lasiocampids (94.6%) and lymantriids (69.6%), both families characterized by hairy/spiny caterpillars. These were the only families with most species yielding more generalist than specialist parasitoid species (Table 4).

IV. DISCUSSION

A. HOST VARIABLES

Food plant range had the least association with parasitoid host range of any of the five variables examined (Table 5). Only the increased prevalence of specialist parasitoids in oligophagous collections fit predictions of plant-folivore theory (Table 2, "Food plant range"; Figure 1). The fact that no

Table 4
DISTRIBUTION OF HOSTS BY FAMILY AND FREQUENCY OF PARASITISM BY GENERALISTS

Host taxon	No. of host species	More generalist parasitoid species	No. of host collections	Collections yielding generalists (%)[a]
Hesperioidea				
Hesperiidae	1	1	157	9.1 (2/22)
Papilionoidea				
Papilionidae	3	0	316	0 (0/72)
Pieridae	1	0	201	0 (0/110)
Nymphalidae	10	0	888	3.6 (5/137)
Geometroidea				
Geometridae	9	2	664	27.6 (43/156)
Bombycoidea				
Lasiocampidae	2	2	603	94.6 (139/147)
Saturniidae	8	1	810	6.3 97/112)
Sphingoidea				
Sphingidae	17	0	518	0 (0/154)
Noctuoidea				
Notodontidae	15	4	1587	43.4 (86/198)
Arctiidae	12	5	2001	33.3 (155/466)
Lymantriidae	5	3	608	69.6 (55/79)
Noctuidae	16	3	859	16.0 (12/75)
Total	98	19	9084	29.2 (504/1728)

[a] Collection frequencies in parentheses (no. yielding generalists/no. yielding parasitoids).

generalist parasitoid species was restricted to oligophagous hosts may be significant; however, only 4 of the 15 generalist species were reared exclusively from polyphagous hosts. Most examples of insects that manufacture or sequester plant toxins are of oligophagous insects.[13] Recently, Jones et al.[45] showed that experimentally restricting the diet of the polyphagous lubber grasshopper, *Romalea guttata*, increased the amount of plant toxins it sequestered and reduced its palatability to ant predators. It may be that local oligophagy by polyphagous insect populations is more common than suspected[29,89] and affects suitability to parasitoids.

The increased proportion of specialist parasitoids on herbs and shrubs relative to trees (Table 2, "Plant type"; Figure 2) is consistent with plant-folivore theory; however, these data are contrary to the pattern reported by Hawkins et al.[37] for koinobionts vs. idiobionts on trees and herbs. Parasitoid communities of exophytic Lepidoptera may differ from the overall pattern for exophytic, holometabolous insects, or the pattern may differ within a sample, as occurs here, that consists primarily of koinobionts.

Host exposure had little effect on parasitoid host range (Table 2, "Exposure"; Figure 3). The association of generalists with semiconcealed hosts is due

Table 5
SUMMARY OF PREDICTED AND OBSERVED RELATIONSHIPS BETWEEN HOST VARIABLES AND PARASITOID HOST RANGE

Host variable[a] (prediction)	Species richness: Overall assoc.[b]	Species richness: Parasitoid species /host collection[c]	Prevalence: Parasitoid collections /host collection[c]
Gregariousness	0.0001		
More specialists on solitary		0.009	0.03
More generalists on gregarious		0.002	0.0001
Integument	0.0007		
More specialists on smooth		0.0001	0.0001
More generalists on hairy/spiny		NS	0.0001
Plant type	0.01		
More specialists on herb/shrub feeders		0.01	0.0001
More generalists on tree feeders		NS	0.0001
Exposure	NS		
More specialists on exposed		NS	(0.0001, on semiconcealed)
More generalists on semiconcealed		0.01	0.0001
Food plant range	NS		
More specialists on oligophagous		0.03	0.0001
More generalists on polyphagous		NS	(0.0004, on 1 family)

[a] Host variables are ranked in descending order of overall significance.
[b] P-values from chi square tests of the null hypothesis of no association (see Table 2).
[c] P-values from nonparametric tests of the null hypothesis of equal proportions (see Figures 1 to 5).

primarily to web makers, as is evident in the large partial chi square values for generalists reared from web makers (Appendix 3), and primarily to only two web maker species. *Hyphantrea cunea* and *Malacosoma americanum*, two of the most frequently collected species, accounted for 13 of 19 associations of web makers with generalists and 78% of 268 web maker collections yielding generalists. Without these species there is no pattern of association between generalists and semiconcealed feeders. The association between idiobionts and concealed feeders reported by Askew and Shaw[4] and Hawkins et al.[37] may reflect selection against parasitoids that kill or paralyze exposed hosts because of subsequent discovery by other predators.[34] Thus, generalist koinobionts may respond differently than generalist idiobionts in this regard, and differently than generalist insect predators.[10]

Associations between generalist parasitoids and gregarious and hairy/spiny hosts were among the strongest of all associations (Table 5); they are discussed in the next section. Specialists were associated with solitary hosts, although somewhat less than generalists with gregarious hosts (Table 2, "Gregariousness"; Figure 4). Contrary to predictions based on data of Bernays[9] and Edmunds[20], smooth hosts were strongly associated with specialist parasitoids (Table 2, "Integument"; Figure 5).

Some data are consistent with the hypothesis that generalist parasitoids attack more abundant host species (Table 3); however, conspicuousness is confounded with host abundance as measured by collection frequency, and other data do not support the hypothesis (Table 3). The fact that the strongest associations of generalist parasitoids with abundant hosts are also with late-gregarious, hairy/spiny hosts indeed suggests that conspicuousness may be more important than abundance per se; however, Nothnagle and Schultz[59] also found that gregariousness and hairy/spiny morphology are typical of Lepidoptera with eruptive populations; and these species have more generalist parasitoids and parasitism than other species collected equally frequently.[78] Host phylogeny (family) also appears to have a major influence on the proportion of generalists or specialists in a host"s parasitoid load (Table 4).

B. WHY ARE GENERALISTS REARED MOSTLY FROM HAIRY, GREGARIOUS TREE FEEDERS?

Associations between generalist parasitoids and hairy/spiny, gregarious and tree-feeding hosts suggest the possibility of a common denominator among these traits. Gregariousness and integument type were most highly associated of all host variables ($c^2 = 30.0$, df = 1, $p < 0.0001$), and both were just significantly associated with plant type (trees vs. herbs and shrubs; $c^2 = 4.2$, df = 1, $p = 0.04$ and $c^2 = 4.2$, df = 1, $p = 0.04$, respectively). Other host variables were not significantly associated. Selection for behaviors that increase growth rate may make these hosts more susceptible to generalist predation. Tradeoffs between growth rate and predation risk are posited for folivores generally[21,54,65,73] and for conspicuous[39] and gregarious[48,83] species in particular. I have argued elsewhere that hairy/spiny morphology facilitates adoption of behaviors, including aggregation and conspicuous, diurnal feeding, that enhance growth rate.[78] Here, it is suggested that conspicuous feeding behavior on low quality foliage (such as mature tree foliage) may increase host vulnerability to generalist, relative to specialist, enemies.

Hairy/spiny, gregarious caterpillars are typically more conspicuous than smooth, solitary caterpillars.[78] Conspicuous caterpillars feed readily during daylight as well as at night, at least during early instars,[39,48] while cryptic species tend to feed primarily at night or in very brief feeding bouts during the day.[42,72] Contrasting feeding patterns have been elegantly studied by Heinrich and Collins,[40] who noticed that cryptic species are "clean" feeders, in that they trim feeding damage within leaves and often chew off partially eaten leaves. These species apparently devote considerable time and energy to protecting themselves from discovery. Spiny and/or aposematic species are "messy" feeders that leave midribs and large veins and do not prune partially eaten leaves. Heinrich and Collins suggested that clean feeding is an adaptation to avoiding bird predation, "since all caterpillars are subject to insect parasitism, but only species palatable to birds . . . minimize their feeding damage" (Reference 40, see p. 601); however, messy feeders are attacked preferentially by generalist parasitoids. Of the folivore species studied by Heinrich and Collins and analyzed here, generalist parasitoids accounted for 61% (23/38) of associations with, and 62% (194/315) of collections of, messy feeders, but only 16% (3/19) of associations with, and 6% (7/126) of collections of, clean feeders (p <0.001 and p <0.0001, respectively, Fisher's exact test, two-tailed). Furthermore, messy feeding behavior is strongly associated with hairy/spiny morphology and gregariousness.[78] Thus, conspicuous feeding behavior by hairy/spiny, gregarious caterpillars appears to result in lower predation by birds,[40] but in higher parasitism by generalist parasitoids.

One reason generalist parasitoids are more successful exploiting hairy/spiny and gregarious hosts may be that the latter are conspicuous and therefore easy to find. In addition to messy feeding behavior and typically noncryptic morphology (relative to feeding sites), high mean population abundance may increase conspicuousness[59] (Table 3). The very elaborateness of cryptic disguises of many caterpillars implies a selective cost of conspicuousness.[64] Some evidence suggests that large colonies are more vulnerable to predation and generalist parasitism than small colonies.[55,85,94] Such studies seldom address discovery rate directly. It is commonly assumed that host finding is limiting to natural enemies and that specialists are more efficient at locating hosts than generalists;[90] however, generalist idiobiont ichneumonid parasitoids typically attack host pupae that are concealed[04] and therefore presumably relatively inconspicuous. Furthermore, host range need not be limited by host finding ability; a variety of ecological, behavioral, physiological and phylogenetic factors may constrain parasitoid host range [29,34,80]

A related possibility is that conspicuous feeding is selected for most strongly in species that feed on low quality foliage,[67,70b,78] and that feeding on low quality leaves results in increased vulnerability to generalist parasitoids. Mature, mid-, or late-season tree leaves tend to be tough and low in water, nitrogen and minerals, compared with herbaceous and early-season tree leaves.[53,88] Feeding on mature, mid-, or late-season leaves generally decreases growth rate and fitness in insects.[53,69,75] Thus, although possibly increasing growth rate by increasing the time available for feeding (a hypothesis yet to be tested), conspicuous feeding apparently does not compensate for inherently slow growth rates on old relative to young leaves. Reduced levels of toxins in mid- or late-season tree leaves[12,53] may increase suitability of hosts to generalist parasitoids, which may be more affected than specialists by plant-sequestered chemicals.[6,14] Host nutrition has also been shown to directly affect suitability of caterpillars to parasitoids,[35,82,91] but differential effects on specialists and generalists have not been examined.

The rarity of generalist parasitoids in rearings of herb-feeding hosts (Table 2, "Plant type" and Table 4) is consistent with the hypothesis that folivores feeding on low quality foliage are most vulnerable to generalist parasitoids;[37] however, among abundant hosts, the species that had the highest proportion of generalist parasitoids were *M. americanum* and *M. disstria*, both feeders on early-season foliage. If plant quality effects are indeed important, they may be overridden by phylogenetic constraints. Comparative experiments and extensive field data are needed to determine whether conspicuous consumptive behavior increases caterpillar growth rates, whether either growth rate or plant quality affects susceptibility to parasitoids, and whether specialists and generalists respond differently.

C. ROLE OF GENERALISTS IN THE EVOLUTION OF FOLIVORE FOOD PLANT RANGE

The proposition that folivore species compete via shared enemies for "enemy-free space" presumes the importance of generalist enemies in restricting folivore niches.[44,49] Bernays[8,9,11] further predicted that folivore niche restriction is primarily due to generalist enemies preferentially attacking polyphagous prey. Polyphagous hosts were not attacked more by generalist than by specialist ichneumonoid parasitoids (Table 2, "Food plant range" and Table 3). However, ichneumonoid parasitoids probably are, as a group, the most host-specific biotic mortality agents affecting exophytic Lepidoptera. Several lines of evidence suggest that other natural enemies of these hosts are likely to be even less host-specific, and in at least some cases cause more mortality:

1. Rare ichneumonoid parasitoids were not more host specific than those analyzed here (Appendix 2). Ichneumonoids reared from three or fewer (total) collections included a similar proportion of specialists (62% [8/13]), based solely on records in Carlson[15] and Marsh[52] as ichneumonoids reared from four or more host collections (77% [56/73]); Fisher's exact test, two-tailed, $p > 0.30$).
2. Among the chalcidoid parasitoids reared from hosts in the full data set, 85% (11 of 13 species) are host generalists, again based solely on records in Carlson[15] and Marsh[52]; none is a genus specialist.
3. Parasitoids that emerge from host eggs, which were not sampled by Schaffner and Griswold,[71] and parasitoids of host pupae, which were little sampled, are usually more host generalist than koinobiont larval parasitoids.[34,97]
4. Tachinids (Diptera), the group of parasitoids other than Hymenoptera recorded by Schaffner and Griswold,[71] are widely thought to have considerably wider host ranges, on average, than ichneumonoids.[1] Tachinids were reared from approximately 4500 host collections, more than twice the approximately 2000 collections that yielded hymenopteran parasitoids. Thus, tachinids may be more host generalist and kill more exophytic Lepidoptera than hymenopteran parasitoids.
5. Birds and social wasps, which are known to be important generalist mortality agents,[39,87] are also probably less host specialized than most ichneumonoid parasitoids. (Too little is known about microorganisms to permit useful speculation about their relative importance in relation to range of hosts attacked.)

These data imply that for common, temperate, exophytic macrolepidopterans the primary biotic mortality agents may be taxonomic generalists rather than specialists. This inference supports the importance of generalists posited by Lawton[49] and Bernays.[8,9] It remains to be shown, however, whether folivores typically compete for enemy-free space, whether generalist enemies other than ichneumonoid parasitoids preferentially attack polyphagous folivores, or whether natural enemies are even frequently significant selective factors.[64,66] Extensive experimental studies and direct measurements of mortality in the field will be necessary to answer these questions.

D. OTHER CORRELATES OF PARASITOID HOST RANGE

Potential ecological correlates with parasitoid host range other than those considered here are undoubtedly also important and deserve further study.

1. Host population size and variability may influence host range of colonizing parasitoid species. Theory suggests that resource predictability promotes diet specialization.[43,50,70] My finding[78] that species identified as outbreak pest Lepidoptera by Nothnagle and Schultz[59] have more generalist parasitoids and parasitism than other species collected equally frequently is in agreement with this theory; however, a full appraisal of population variability in these data awaits dynamic analysis of original data records.
2. Food plant quality is another, related candidate for further scrutiny. Schultz,[73] Price et al.,[67] and Rhoades[70a] have suggested ways plant quality might interact with predation and parasitism and this chapter contains speculation on possible consequences for parasitoid host range. Data are needed before conclusions can be drawn.
3. Quantification of host appearance may be useful in discerning patterns of association with parasitoids. Aposematism deserves a close look. Brightly colored, aposematic insects that are highly toxic may select for host range restriction in parasitoids, whereas hairy/spiny, conspicuous caterpillars are exploited preferentially by generalists (Table 2, "Integument" and Table 4). It will be necessary to quantify reaction distances with respect to specific natural enemies under realistic foraging conditions.
4. Host body size may correlate with parasitoid host range. Several studies have reported positive correlations between host range and body size of both arthropod folivores and predators;[33,61,92] however, koinobionts typically attack hosts before they are fully grown and many can regulate host growth.[82] Thus, it would be surprising to find strong correlations between parasitoid host range and host body size.

ACKNOWLEDGMENTS

T. M. Odell and W. E. Wallner of the U.S. Department of Agriculture, Forest Service, Hamden, CT, kindly made Schaffner and Griswold's original data summaries available to me. P. M. Marsh and D. Wahl graciously clarified taxonomic questions about parasitoids, and P. M. Marsh supplied several host record extensions and corrections for braconid species listed in Reference 52. I am grateful to the following for support, fruitful discussions, and/or critical comments on earlier drafts of this chapter: D. A. Andow, E. A. Bernays, C. Fox, P. Gross, A. M. Shelton, K. S. Hagen, B. A. Hawkins, J. P. Nyrop, F. L. W. Ratnieks, J. G. Franclemont clarified several details of lepidopteran natural history, and J. O. Washburn.

Appendix 1
BIOLOGICAL TRAITS AND COLLECTION FREQUENCIES OF LEPIDOPTERA ANALYZED

	Lepidopteran taxon[a]						
		Host range[b]	Plant type	Feeding exposure[c]	Gregariousness[d]	Host integument[e]	No. of collections
Hesperioidea							
Hesperiidae							
1	*Epargyreus clarus* (Cramer)	1	Mixed	Tie	Sol	Smooth	157
*2	*Erynnis icelus* (Scudder & Burgess)	2–3	Tree	Tie	Sol	Smooth	15
Papilionoidea							
Papilionidae							
3	*Papilio glaucus* (Linnaeus)	4+	Tree	Exp	Sol	Smooth	118
4	*Papilio polyxenes* (Fabricius)	1	Herb	Exp	Sol	Smooth	149
5	*Papilio troilus* (Linnaeus)	1	Shrub	Tie	Sol	Smooth	49
Pieridae							
6	*Artogeia rapae* (Linnaeus)	1	Herb	Exp	Sol	Smooth	201
Nymphalidae							
7	*Aglais milberti* (Godart)	1	Herb	Web	Early	H/S	56
8	*Basilarchia archippus* (Cramer)	1	Tree	Exp	Sol	H/S	191
*9	*Basilarchia arthemis* (Drury)	1	Tree	Exp	Sol	H/S	19
10	*Charidryas harrisii* (Scudder)	1	Herb	Web	Late	H/S	25
11	*Euphydryas phaeton* (Drury)	1	Herb	Web	Early	H/S	31
12	*Nymphalis antiopa* (Linnaeus)	2–3	Tree	Web	Early	H/S	323
*13	*Nymphalis vau-album* (Den. & Schiff.)	2–3	Tree	Exp	Late	H/S	16
14	*Polygonia comma* (Harris)	2–3	Mixed	Tie	Early	H/S	8
15	*Polygonia interrogationis* (Fabricius)	2–3	Mixed	Exp	Early	H/S	94

16	*Vanessa atalanta* (Linnaeus)	1	Herb	Tie	Sol	H/S	99
17	*Vanessa cardui* (Linnaeus)	4+	Mixed	Web	Sol	H/S	32
18	*Vanessa virginiensis* (Drury)	1	Herb	Tie	Egg	H/S	29
Danaidae							
*19	*Danaus plexippus* (Linnaeus)	1	Herb	Exp	Sol	Smooth	184
Geometroidea							
Geometridae							
20	*Alsophila pometaria* (Harris)	4+	Tree	Exp	Egg	Smooth	84
21	*Biston betularia* (Linnaeus)	4+	Tree	Exp	Sol	Smooth	185
22	*Cincilia catenaria* (Drury)	4+	Shrub	Exp	Early	Smooth	59
23	*Ennomos subsignaria* (Hübner)	4+	Tree	Exp	Early	Smooth	21
24	*Erannis tiliaria* (Harris)	4+	Tree	Exp	Sol	Smooth	100
25	*Hydria prunivorata* (Ferguson)	1	Tree	Web	Late	Smooth	116
26	*Itame ribearia* (Fitch)	1	Shrub	Exp	Sol	Smooth	7
*27	*Lycia ursaria* (Walker)	4+	Tree	Exp	Sol	Smooth	20
*28	*Nematocampa limbata* (Haworth)	4+	Tree	Exp	Sol	Smooth	26
29	*Phigalia titea* (Cramer)	4+	Tree	Exp	Sol	Smooth	76
30	*Prochoerodes transversata* (Drury)	4+	Tree	Exp	Sol	Smooth	16
Bombycoidea							
Lasiocampidae							
31	*Malacosoma americanum* (Fabricius)	1	Tree	Web	Late	H/S	364
32	*Malacosoma disstria* Hübner	4+	Tree	Exp	Early	H/S	239
*33	*Phyllodesma americana* (Harris)	4+	Tree	Exp	Sol	H/S	40
*34	*Tolype velleda* (Stoll)	4+	Tree	Exp	Early	H/S	21
Saturniidae							
35	*Actias luina* (Linnaeus)	4+	Tree	Exp	Sol	H/S	22
36	*Anisota senatoria* (J. E. Smith)	1	Tree	Exp	Early	H/S	64
37	*Anisota stigma* (Fabricius)	1	Tree	Exp	Sol	H/S	8
38	*Antheraea polyphemus* (Cramer)	4+	Tree	Exp	Sol	H/S	148
39	*Automeris io* (Fabricius)	4+	Tree	Exp	Early	H/S	162
40	*Callosamia promethea* (Drury)	4+	Tree	Exp	Egg	H/S	132

Appendix 1 (continued)
BIOLOGICAL TRAITS AND COLLECTION FREQUENCIES OF LEPIDOPTERA ANALYZED

	Lepidopteran taxon[a]	Host range[b]	Plant type	Feeding exposure[c]	Gregariousness[d]	Host integument[e]	No. of collections
41	*Dryocampa rubicunda* (Fabricius)	1	Tree	Exp	Early	H/S	114
*42	*Eacles imperialis* (Drury)	4+	Tree	Exp	Sol	H/S	24
43	*Hyalophora cecropia* (Linnaeus)	4+	Tree	Exp	Egg	H/S	160
Sphingoidea							
Sphingidae							
44	*Amphion floridensis* (B. P. Clark)	1	Shrub	Exp	Sol	Smooth	9
*45	*Ceratomia amyntor* (Geyer)	2–3	Tree	Exp	Sol	Smooth	28
46	*Ceratomia catalpae* (Boisduval)	1	Tree	Exp	Early	Smooth	30
47	*Ceratomia undulosa* (Walker)	2–3	Tree	Exp	Sol	Smooth	15
48	*Darapsa myron* (Cramer)	1	Shrub	Exp	Sol	Smooth	13
49	*Deidamia inscriptum* (Harris)	1	Shrub	Exp	Sol	Smooth	6
50	*Eumorpha pandorus* (Hübner)	1	Shrub	Exp	Sol	Smooth	20
51	*Hemaris thysbe* (Fabricius)	2–3	Shrub	Exp	Sol	Smooth	20
*52	*Hyles lineata* (Fabricius)	4+	Herb	Exp	Sol	Smooth	25
*53	*Laothoe juglandis* (J. E. Smith)	2–3	Tree	Exp	Sol	Smooth	20
54	*Lapara bombycoides* Walker	1	Tree	Exp	Sol	Smooth	16
55	*Manduca quinquemaculata* (Haworth)	1	Herb	Exp	Sol	Smooth	134
56	*Manduca sexta* (Linnaeus)	1	Herb	Exp	Sol	Smooth	9
57	*Paonias astylus* (Drury)	1	Shrub	Exp	Sol	Smooth	1
58	*Paonias excaecatus* (J. E. Smith)	4+	Tree	Exp	Sol	Smooth	40
59	*Paonias myops* (J. E. Smith)	1	Tree	Exp	Sol	Smooth	30
60	*Smerinthus jamaicensis* (Drury)	2–3	Tree	Exp	Sol	Smooth	43

61	*Sphecodina abbottii* (Swainson)	1	Shrub	Exp	Sol	Smooth	69
*62	*Sphinx chersis* (Hübner)	1	Tree	Exp	Sol	Smooth	38
*63	*Sphinx drupiferarum* (J. E. Smith)	2–3	Tree	Exp	Sol	Smooth	20
64	*Sphinx gordius* (Cramer)	4+	Shrub	Exp	Sol	Smooth	48
65	*Sphinx kalmiae* (J. E. Smith)	2–3	Shrub	Exp	Sol	Smooth	15
Noctuoidea							
Notodontidae							
66	*Clostera albosigma* Fitch	1	Tree	Tie	Sol	H/S	11
67	*Clostera inclusa* (Hübner)	1	Tree	Web	Early	H/S	75
68	*Clostera strigosa* (Grote)	1	Tree	Web	Early	H/S	18
69	*Datana angusii* Grote & Robinson	4+	Tree	Exp	Late	H/S	25
70	*Datana contracta* Walker	2–3	Tree	Exp	Sol	H/S	1
71	*Datana integerrima* Grote & Robinson	1	Tree	Exp	Late	H/S	158
*72	*Datana major* Grote & Robinson	1	Shrub	Exp	Late	H/S	17
73	*Datana ministra* (Drury)	4+	Tree	Exp	Late	H/S	392
74	*Datana perspicua* Grote & Robinson	1	Shrub	Exp	Late	H/S	51
*75	*Furcula occidentalis* (Lintner)	1	Tree	Exp	Sol	Smooth	38
*76	*Gluphisia septentrionis* Walker	1	Tree	Exp	Sol	Smooth	22
77	*Heterocampa guttivitta* (Walker)	4+	Tree	Exp	Sol	Smooth	77
78	*Lochmaeus manteo* Doubleday	4+	Tree	Exp	Early	Smooth	15
79	*Nadata gibbosa* (J. E. Smith)	1	Tree	Exp	Sol	Smooth	76
*80	*Notodonta scitipennis* Walker	1	Tree	Exp	Sol	Smooth	20
81	*Oligocentria lignicolor* (Walker)	2–3	Tree	Exp	Sol	Smooth	9
*82	*Pheosia rimosa* Packard	1	Tree	Exp	Sol	Smooth	56
83	*Schizura badia* (Packard)	1	Shrub	Exp	Sol	Smooth	4
84	*Schizura concinna* (J. E. Smith)	4+	Tree	Exp	Late	H/S	623
*85	*Schizura ipomoeae* Doubleday	4+	Tree	Exp	Sol	Smooth	28
86	*Schizura unicornis* (J. E. Smith)	4+	Tree	Exp	Sol	Smooth	52
*87	*Symmerista albifrons* (J. E. Smith)	4+	Tree	Exp	Late	Smooth	64
Arctiidae							
88	*Ctenucha virginica* (Esper)	4+	Herb	Exp	Egg	H/S	64

Appendix 1 (continued)
BIOLOGICAL TRAITS AND COLLECTION FREQUENCIES OF LEPIDOPTERA ANALYZED

	Lepidopteran taxon[a]	Host range[b]	Plant type	Feeding exposure[c]	Gregariousness[d]	Host integument[e]	No. of collections
89	*Cycnia inopinatus* (Henry Edwards)	1	Herb	Exp	Sol	H/S	12
90	*Cycnia tenera* Hübner	2–3	Herb	Exp	Early	H/S	29
91	*Estigmene acrea* (Drury)	4+	Herb	Exp	Sol	H/S	366
92	*Euchaetes egle* (Drury)	1	Herb	Exp	Late	H/S	243
93	*Halysidota harrisii* Walsh	1	Tree	Exp	Egg	H/S	18
94	*Halysidota tessellaris* (J. E. Smith)	4+	Tree	Exp	Egg	H/S	259
95	*Hyphantria cunea* (Drury)	4+	Tree	Web	Late	H/S	259
96	*Lophocampa caryae* Harris	4+	Tree	Exp	Early	H/S	183
97	*Lophocampa maculata* Harris	4+	Tree	Exp	Early	H/S	153
*98	*Phragmatobia fuliginosa* (Linnaeus)	4+	Herb	Exp	Sol	H/S	34
99	*Pyrrharctia isabella* (J. E. Smith)	4+	Herb	Exp	Sol	H/S	165
100	*Spilosoma virginica* (Fabricius)	4+	Mixed	Exp	Sol	H/S	250
Lymantriidae							
101	*Dasychira basiflava* (Packard)	2–3	Tree	Exp	Early	H/S	37
102	*Dasychira dorsipennata* (Barnes & McD.)	2–3	Tree	Exp	Egg	H/S	60
103	*Orgyia definita* (Packard)	4+	Tree	Exp	Egg	H/S	12
104	*Orgyia leucostigma* (J. E. Smith)	4+	Tree	Exp	Late	H/S	371
Noctuidae							
105	*Acronicta americana* (Harris)	4+	Tree	Exp	Sol	H/S	76
106	*Acronicta dactylina* (Grote)	2–3	Tree	Exp	Sol	H/S	82
*107	*Acronicta furcifera* (Guenée)	1	Tree	Exp	Sol	Smooth	92

*108	*Acronicta grisea* (Walker)	2–3	Tree	Exp	Sol	Smooth	20
*109	*Acronicta impleta* (Walker)	4+	Tree	Exp	Sol	Smooth	18
110	*Acronicta impressa* (Walker)	4+	Mixed	Exp	Sol	Smooth	105
*111	*Acronicta lanceolaria* (Grote)	4+	Shrub	Exp	Sol	Smooth	21
*112	*Acronicta lepusculina* (Guenée)	1	Tree	Exp	Sol	H/S	15
113	*Acronicta oblinita* (J. E. Smith)	4+	Mixed	Exp	Sol	Smooth	61
114	*Acronicta superans* (Guenée)	2–3	Tree	Exp	Sol	Smooth	23
115	*Agrotis fennica* (Tauscher)	4+	Herb	Exp	Sol	Smooth	3
*116	*Alypia octomaculata* (Fabricius)	1	Shrub	Exp	Sol	Smooth	93
*117	*Amphipyra pyramidoides* (Guenée)	4+	Tree	Exp	Sol	Smooth	73
118	*Calyptera canadensis* (Bethune)	1	Herb	Exp	Sol	Smooth	24
119	*Catocala neogama* (J. E. Smith)	1	Tree	Exp	Sol	Smooth	4
120	*Crymodes devastator* (Brace)	4+	Herb	Exp	Sol	Smooth	12
*121	*Cucullia convexipennis* (Grote & Rob.)	4+	Herb	Exp	Sol	Smooth	32
*122	*Eupsilia devia* (Grote)	4+	Mixed	Exp	Sol	Smooth	18
*123	*Heliothis zea* (Boddie)	4+	Herb	Exp	Sol	Smooth	19
*124	*Lacanobia legitima* (Grote)	4+	Mixed	Exp	Sol	Smooth	84
125	*Lithophane antennata* (Walker)	4+	Tree	Exp	Sol	Smooth	163
*126	*Melanchra adjuncta* (Guenée)	4+	Mixed	Exp	Sol	Smooth	41
127	*Melanchra picta* (Harris)	4+	Mixed	Exp	Sol	Smooth	101
*128	*Nephelodes minians* Guenée	4+	Herb	Exp	Sol	Smooth	28
129	*Orthosia hibisci* (Guenée)	4+	Tree	Exp	Sol	Smooth	9
*130	*Parallelia bistriaris* (Hübner)	1	Tree	Exp	Sol	Smooth	16
131	*Pseudaletia unipuncta* (Haworth)	4+	Herb	Exp	Sol	Smooth	20
132	*Pyreterra hesperidago* (Guenée)	1	Shrub	Exp	Sol	Smooth	37
*133	*Raphia frater* (Grote)	1	Tree	Exp	Sol	Smooth	30
134	*Simyra henrici* (Grote)	4+	Herb	Exp	Egg	Smooth	30
135	*Trichoplusia ni* (Hübner)	4+	Herb	Exp	Sol	Smooth	109

Appendix 1 (continued)
BIOLOGICAL TRAITS AND COLLECTION FREQUENCIES OF LEPIDOPTERA ANALYZED

[a] Asterisk (*) indicates species collected 15 or more times by Schaffner and Griswold (7), but without ichneumonoid parasitoids listed in Appendix 2 (i.e., those reared from 4 or more collections). Nomenclature follows Hodges et al. (4); arrangement is alphabetical.

[b] 1 = one food plant family, 2—3 = two-three families, 4+ = four or more families.

[c] "Exp" = exposed, "Tie" = feeds in leaf ties, folds or rolls during some or all instars, "web" = web-maker during some or all instars.

[d] "Sol" = solitary, "Egg" = eggs laid in clusters (larvae disperse), "Early" = early-gregarious, "Late" = late-gregarious

[e] "H/S" = hairy and/or spiny, "Smooth" = not hairy or spiny.

Appendix 2
BIOLOGICAL TRAITS AND COLLECTION FREQUENCIES OF PARASITOID SPECIES

Parasitoid taxon[a]	Parasitoid host range[b]	Koinobiont/ idiobiont	Collections[c]	Hosts[d]
Braconidae				
Rogadinae				
Aleiodes hyphantriae (Gahan)	1	K	7	95 (7)
Aleiodes stigmator (Say)	2	K	42	106 (28), 110 (7), 105 (3), 134 (2), 113 (1), 114 (1)
Microgastrinae				
Apanteles edwardsii (Riley)	2	K	4	16 (3), 7 (1)
Apanteles sarrothripae (Weed)	1	K	18	67 (16), 66 (1), 68 (1)
Cotesia acauda (Provancher)	1	K	60	25 (60)
Cotesia ammalonis (Musebeck)	1	K	5	90 (5)

Cotesia anisotae (Musebeck)	1	K	8	36 (5), 41 (2), 37 (1)
Cotesia argynnidis (Riley)	1	K	20	1 (20)
Cotesia atalantae (Packard)	2	K	76	16 (44), 7 (32)
Cotesia cingiliae (Musebeck)	2	K	4	22 (4)
Cotesia clisiocampae (Ashmead)	1	K	4	31 (4)
Cotesia congregata (Say)	2	K	113	55 (54), 46 (15), 51 (11), 61 (9), 50 (7), 48 (5), 56 (2), 64 (2), 58 (2), 54 (1), 57 (1), 49 (1), 44 (1)
Cotesia delicata (Howard)	1	K	9	104 (9)
Cotesia euchaetis (Ashmead)	1	K	103	92 (103)
Cotesia fiskei (Viereck)	1	K	10	102 (7), 101 (3)
Cotesia glomerata (Linnaeus)	2	K	110	6 (110)
Cotesia gordii (Musebeck)	2	K	6	64 (6)
Cotesia halisidotae (Musebeck)	1	K	35	97 (35)
Cotesia hyphantriae (Riley)	2	K	43	95 (43)
Cotesia limenitidis (Riley)	1	K	8	8 (8)
Cotesia lyciae (Musebeck)	1	K	9	21 (9)
Cotesia murtfeltae (Ashmead)	2	K	6	30 (2), 21 (1), 29 (1)
Cotesia nemoriae (Ashmead)	2	K	4	24 (2), 20 (1), 22 (1)
Cotesia phobetri (Rohwer)	3	K	26	94 (17), 93 (7), 104 (2)
Cotesia schizurae (Ashmead)	2	K	5	86 (3), 83 (2)
Cotesia smerinthi (Riley)	2	K	20	60 (10), 58 (5), 47 (2), 49 (1)
Diolcogaster schizurae (Musebeck)	2	K	4	84 (2), 78 (1), 81 (1)
Glyptapanteles nigricornis (Musebeck)	1	K	11	88 (11)
Microplitis ceratomiae (Riley)	2	K	10	64 (8), 65 (1), 59 (1)
Microplitis mamestrae (Weed)	1	K	8	127 (8)
Protapanteles paleacritae (Riley)	2	K	5	20 (4), 25 (1)
Protapanteles phigaliae (Musebeck)	1	K	7	29 (7)
Euphorinae				
Meteorus bakeri (Cook & Davis)	2	K	4	95 (4)
Meteorus cingiliae (Musebeck)	1	K	17	22 (17)

Appendix 2 (continued)
BIOLOGICAL TRAITS AND COLLECTION FREQUENCIES OF PARASITOID SPECIES

Parasitoid taxon[a]	Parasitoid host range[b]	Koinobiont/ idiobiont	Collections[c]	Hosts[d]
Meteorus communis (Cresson)	2	K	7	131 (4), 31 (2), 129 (1)
Meteorus datanae (Musebeck)	1	K	5	71 (4), 32 (1), 69 (1)
Meteorus hyphantriae (Riley)	3	K	79	95 (30), 104 (20), 31 (12), 32 (3), 92 (2), 125 (2), 15 (1), 16 (1), 74 (1), 86 (1), 118 (1),119 (1) 26 (1), 29 (1)
Meteorus leviventris (Wesmael)	2	K	13	125 (2), 115 (1), 120 (1)
Ichneumonidae				
Pimplinae				
Iseropus coelebs (Walsh)	3	I	15	104 (13), 96 (1), 103 (1)
Itoplectis conquistor (Say)	3	I	54	104 (8), 31 (8), 22 (8), 23 (7), 1 (1) 12 (1), 17 (1), 32 (1), 37 (1), 95 (1)
Pimpla pedalis (Cresson)	3	I	5	31 (3), 77 (1), 104 (1)
Scambus nigrifrons (Viereck)	3	I	5	104 (2), 1 (1), 94 (1)
Tryphoninae				
Netelia sayi (Cushman)	2	K	5	132 (3), 125 (2)
Phygadeuontinae (Gelinae)				
Gambrus extrematis (Cresson)	2	I	11	38 (7), 43 (3)
Gambrus nuncius (Say)	2	I	21	40 (21)
Ichneumoninae				
Conocalama brullei (Cresson)	2	K	7	59 (7)
Cratichneumon sublatus (Cresson)	3	I	5	77 (5)

Ichneumon caliginosus (Cresson)	2	I	14	16 (12), 15 (2)
Stenichneumon culpator (Cresson)	3	I	6	135 (6)
Thyrateles lugubrator (Gravenhorst)	2	I	23	17 (16), 18 (6), 14 (1)
Trogus fulvipes (Cresson)	1	K	17	3 (17)
Trogus pennator (Fabricius)	1	K	55	4 (42), 5 (8), 3 (5)
Banchinae				
Apophua simplicipes (Cresson)	3	K	4	31 (2), 23 (2)
Anomaloninae				
Agrypon anale (Say)	3	K	90	31 (54), 32 (23), 67 (11), 90 (1), 104 (1)
Aphanistes hyalinus (Norton)	3	K	4	84 (2), 96 (1)
Barylypa elongata (Davis)	1	K	13	97 (13)
Barylypa relicta (Fabricius)	1	K	10	73 (6), 71 (4)
Heteropelma datanae (Riley)	1	K	34	71 (16), 73 (10), 74 (7), 70 (1)
Therion morio (Fabricius)	2	K	18	91 (7), 100 (6), 89 (1), 97 (1)
Therion sassacus (Viereck)	2	K	22	95 (19), 90 (2), 89 (1)
Campopleginae				
Benjaminia euphydryadis (Viereck)	2	K	7	11 (6), 10 (1)
Campoletis argentifrons (Cresson)	3	K	4	127 (2), 16 (1)
Casinaria genuina (Norton)	2	K	4	91 (2), 95 (2)
Casinaria orgyiae (Howard)	3	K	6	104 (4), 95 (1), 101 (1)
Dusonia vitticollis Norton	2	K	4	29 (3), 32 (1)
Hyposoter fugitivus (Say)	3	K	75	84 (31), 95 (14), 31 (13), 36 (10), 92 (4), 37 (1), 39 (1), 73 (1)
Hyposoter rivalis (Cresson)	2	K	49	95 (43), 100 (3), 92 (3)
Hyposoter rubiginosus (Cushman)	1	K	5	102 (3), 101 (2)
Phobocampe clisiocampae (Weed)	3	K	25	31 (11), 32 (8), 84 (2), 104 (2), 20 (1), 74 (1)
Phobocampe flavipes (Provancher)	3	K	4	29 (2), 23 (1), 24 (1)
Sinophorus validus (Cresson)	3	K	118	95 (61), 67 (23), 25 (19), 92 (9), 84 (6), 90 (4), 68 (2), 31 (1), 99 (1), 104 (1)

Appendix 2
BIOLOGICAL TRAITS AND COLLECTION FREQUENCIES OF PARASITOID SPECIES

Parasitoid taxon[a]	Parasitoid host range[b]	Koinobiont/ idiobiont	Collections[c]	Hosts[d]
Ophioninae				
Enicospilus americanus (Christ)	2	K	61	40 (35), 39 (9), 38 (8), 43 (7), 35 (1)
Enicospilus arcuatus (Felt)	2	K	35	84 (33), 79 (1), 77 (1)

[a] Rearing records from Schafner and Griswold (7). Nomenclature follows Carlson (1) and Marsh (5), as modified by (2, 3, 6).

[b] 1 = one genus, 2 = one family, 3 = two or more families.

[c] Number of host collections yielding parasitoid (includes records from hosts identified to genus or family only, thus may exceed sum of collections under "Hosts" column).

[d] Host species are identified by number in Appendix 1, with number of collections yielding parasitoid in parentheses.

Appendix 3
PARASITOID-HOST ASSOCIATIONS BY HOST VARIABLE AND PARASITOID HOST RANGE, ALL LEVELS

Host variable		Parasitoid-host associations			Parasitoid collections		
	N	Genus specialists	Family specialists	Generalists	Genus specialists	Family specialists	Generalists
Food plant range							
1 family	40	17	31	28	302	336	202
2 – 3 families	14	6	10	5	21	59	8
4+ families	44	15	50	46	156	343	302
Plant type							
Herbs	21	4	21	10	161	295	30
Shrubs	15	3	16	4	32	63	11
Trees	54	29	47	60	258	344	465
Exposure							
Exposed	82	30	74	56	362	503	218
Tiers	6	3	5	4	29	66	4
Web makers	10	5	12	19	88	169	268
Gregariousness							
Solitary	58	12	58	20	127	430	32
Egg-clustering	10	3	8	5	21	79	27
Early-gregarious	19	11	14	20	100	77	110
Late-gregarious	11	12	11	34	231	152	343
Integument							
Smooth	51	10	51	20	193	317	64
Hairy	47	28	40	59	286	421	448

REFERENCES FOR APPENDICES

1. **Carlson, R. W.,** Ichneumonidae, in *Catalog of Hymenoptera in America North of Mexico, Vol. 1. Symphyta and Apocrita (Parasitica),* Krombein, K. V., Hurd, R. D., Smith, D. R., and Burk, B. D., Eds., Smithsonian Institution, Washington, D.C., 1979, 315–740.
2. **Fitton, M. G. and Gauld, I. D.,** The family-group names of the Ichneumonidae (excluding the Ichneumoninae) (Hymenoptera), *System. Entomol.,* 1, 247–258, 1976.
3. **Fitton, M. G. and Gauld, I. D.** Further notes on family-group names of Ichneumonidae (Hymenoptera), *System. Entomol.,* 3, 245–247, 1978.
4. **Hodges, R. W., Dominick, T., Davis, D. R., Ferguson, D. C., Franclemont, J. G., Munroe, E. G. and Powell, J. A.,** *Check List of the Lepidoptera of America North of Mexico,* E. W. Classey, London, 1983.
5. **Marsh, P. M.,** Braconidae, in *Catalog of Hymenoptera in America North of Mexico,* Krombein, K. V., Hurd, R. D., Smith, D. R., and Burk, B. D., Eds., Smithsonian Institution, Washington, D.C., 1979, 144–295.
6. **Mason, W. R. M.,** The polyphyletic nature of *Apanteles* Foerster (Hymenoptera: Braconidae): a phylogeny and reclassification of Microgastrinae, *Mem. Entomol. Soc. Can.,* 115, 1981.
7. **Schaffner, J. V., Jr. and Griswold, C. L.,** Macrolepidoptera and Their Parasites Reared from Field Collections in the Northeastern Part of the United States, U.S. Department of Agriculture Misc. Publ. No. 188, 1934.

REFERENCES

1. **Arnaud, P. H.,** A Host-Parasite Catalog of North Am. Tachinidae (Diptera), U.S. Department of Agriculture Misc. Publ. 1319, 1978.
2. **Askew, R. R.,** The diversity of insect communities in leaf mines and plant galls, *J. Anim. Ecol.*, 49, 817–829, 1980.
3. **Askew, R. R. and Shaw, M. R.,** An account of the Chalcidoidea (Hymenoptera: Pteromalidae) parasitising leaf-mining insects of deciduous trees in Britain, *Biol. J. Linn. Soc.*, 6, 289–335, 1974.
4. **Askew, R. R. and Shaw, M. R.,** Parasitoid communities: their size, structure and development, in *Insect Parasitoids*, Waage, J. and Greathead, D., Eds., Academic Press, New York, 1986, 225–264.
5. **Baker, W.,** Eastern Fores tInsects, U.S. Department of Agriculture Forest Service Misc. Publ. 1175, 1972.
6. **Barbosa, P.,** Natural enemies and herbivore-plant interactions: influence of plant allelochemicals and host specificity, in *Novel Aspects of Insect-Plant Interactions*, Barbosa, P. and Letourneau, D. K., Eds., Wiley-Interscience, New York, 1988, 201–229.
7. **Bernays, E. A.,** Plant tannins and insect herbivores: an appraisal, *Ecol. Entomol.*, 6, 353–360, 1981.
8. **Bernays, E. A.,** Host specificity in phytophagous insects: selection pressure from generalist predators, *Entomol. Exp. Appl.*, 49, 131–140, 1988.
9. **Bernays, E. A.,** Host range in phytophagous insects: the potential role of generalist predators, *Evol. Ecol.*, 3, 299–311, 1989.
10. **Bernays, E. A. and Cornelius, M. L.,** Generalist caterpillar prey are more palatable than specialists for the generalist predator *Iridomyrmex humilis*, *Oecologia*, 79, 427–430, 1989.
11. **Bernays, E. A. and Graham, M.,** On the evolution of host specificity in phytophagous arthropods, *Ecology*, 69, 886–892, 1988.
12. **Bernays, E. A. and Janzen, D. H.,** Saturniid and sphingid caterpillars: two ways to eat leaves, *Ecology*, 69, 1153–1160, 1988.
13. **Blum, M. S.,** *Chemical Defense in Arthropods*, Academic Press, New York, 1985.

13a. **Brower, L. P.,** Bird predation and foodplant specificity in closely related procryptic insects, *Am. Nat.*, 92, 183–187, 1958.

14. **Campbell, B. C. and Duffey, S. S.,** Tomatine and parasitic wasps: potential incompatibility of plant-antibiosis and biological control, *Science*, 205, 700–702, 1979.
15. **Carlson, R. W.,** Ichneumonidae, in *Catalog of Hymenoptera in America North of Mexico, Vol. 1, Symphyta & Apocrlta (Parasitica)*, Krombein, K. V., Hurd, R. D., Smith, D. R., and Burk, B. D., Eds., Smithsonian Institution, Washington, D.C., 1979, 315–740.
16. **Cott, H. B.,** *Adaptive Coloration in Animals*, Methuen, London, 1940.
17. **Covell, C. V., Jr.,** *A Field Guide to the Moths of Eastern North America*, Houghton Mifflin, Boston, 1984.
18. **Crumb, S. E.,** The Larvae of the Phalaenidae, U.S. Department of Agriculture Tech. Bull. 1135, 1956.
19. **Edmunds, M.,** *Defence in Animals*, Longmans, Essex, U.K., 1974.
20. **Edmunds, M.,** Larval mortality and population regulation in the butterfly *Danaus chrysippus* in Ghana, *Zool. J. Linn. Soc.*, 58, 129–145, 1976.
21. **Feeny, P.,** Plant apparency and chemical defense, *Recent Adv. Phytochem.*, 10, 1–40, 1976.
22. **Ferguson, D. C.,** *The Moths of America North of Mexico,* Fascicle

20.2, *Bombycoidea saturniidae*, E. W. Classey, London, 1972.

23. **Ferguson, D. C.,** *The Moths of America North of Mexico,* Fascicle 22.2, *Noctuoidea lymantriidae*, E. W. Classey, London, 1978.
24. **Forbes, W. T. M.,** Lepidoptera of New York and neighboring states. I. Primitive forms, Microlepidoptera, Pyraloids, Bombyces, *Cornell Univ. Agric. Exp. Stn. Mem.* No. 68, 1923.
25. **Forbes, W. T. M.,** Lepidoptera of New York and neighboring states. II. Geometridae, Lymantriidae, Notodontidae, Sphingidae, *Cornell Univ. Agric. Exp. Stn. Mem.* No. 274, 1948.
26. **Forbes, W. T. M.,** Lepidoptera of New York and neighboring states. III. Noctuidae, *Cornell Univ. Agric. Exp. Stn. Mem.* No. 329, 1954.
27. **Forbes, W. T. M.,** Lepidoptera of New York and neighboring states. IV. Agaristidae through Nymphalidae, including butterflies, *Cornell Univ. Agric. Exp. Stn. Mem.* No. 371, 1960.
28. **Force, D.,** Succession of r and K strategists in parasitoids, in *Evolutionary Strategies of Parasitic Insects and Mites*, Price, P. W., Ed., Plenum, New York, 1975, 112–129.
29. **Fox, L. R. and Morrow, P. A.,** Specialization: species property or local phenomenon?, *Science*, 211, 887–893, 1981.
30. **Françlemont, J. G.,** *The Moths of America North of Mexico,* Fascicle 20.1, *Mimallonoidea mimallonidae and Bombycoidea apatelodidae, bombycidae, lasiocampidae*, E. W. Classey, London, 1973.
31. **Futuyma, D. J.,** Food plant specialization and environmental predictability in Lepidoptera, *Am. Nat.*, 110, 285–292, 1976.
32. **Futuyma, D. J. and Moreno, G.,** The evolution of specialization, *Annu. Rev. Ecol. System.,* 19, 207–233, 1988.
33. **Gaston, K. J. and Lawton, J. H.,** Patterns in body size, population dynamics, and regional distribution of bracken herbivores, *Am. Nat.*, 132, 662–680, 1988.
34. **Gauld, I. D.,** Evolutionary patterns of host utilization by ichneumonoid parasitoids (Hymenoptera: Ichneumonidae and Braconidae), *Biol. J. Linn. Soc.*, 35, 351–377, 1988.
35. **Greenblatt, J. A. and Barbosa, P.,** Effects of host's diet on two pupal parasitoids of the gypsy moth: *Brachymeria intermedia* (Ness) and *Coccygomimus turionellae* (L.), *J. Appl. Ecol.*, 18, 1–10, 1981.

35a. **Gross, P., and Price, P. W.,** Plant influences on parasitism of two laefminers: a test of enemy-free space, *Ecology,* 69, 1506–1516, 1988.

36. **Hawkins, B. A.,** Species diversity in the third and fourth trophic levels: patterns and mechanisms, *J. Anim. Ecol.*, 57, 137–162, 1988.
37. **Hawkins, B. A., Askew, R. R., and Shaw, M. R.,** Influences of host feeding biology and plant architecture on generalist and specialist parasitoids, *Ecol. Entomol.*, 1990.
38. **Hawkins, B. A. and Lawton, J. H.,** Species richness for parasitoids of British phytophagous insects, *Nature*, 326, 788–790, 1987.
39. **Heinrich, B.,** Foraging strategies of caterpillars: leaf damage and possible predator avoidance strategies, *Oecologia*, 42, 325–337, 1979.
40. **Heinrich, B. and Collins, S. L.,** Caterpillar leaf damage, and the game of hide-and-seek with birds, *Ecology*, 64, 592–602, 1983.
41. **Heinrich, G. H.,** Synopsis of nearctic Ichneumoninae Stenopneusticae with particular reference to the northeastern region (Hymenoptera), *Can. Entomol.*, Suppl. 15, 1960.

42. **Herrebout, W. M., Kuyten, P. J., and de Ruiter, L.,** Observations in colour patterns and behaviour of caterpillars feeding on Scots pine, *Arch. Neerld. Zool.*, 3, 315–357, 1963.
43. **Janzen, D. H.,** The peak in North Am. ichneumonid species richness lies between 38° and 42°N, *Ecology*, 62, 532–537, 1981.
44. **Jeffries, M. J. and Lawton, J. H.,** Enemy free space and the structure of ecological communities, *Biol. J. Linn. Soc.*, 23, 269–286, 1984.
45. **Jones, C. G., Whitman, D. W., Compton, S. J., Silk, P. J., and Blum, M. S.,** Reduction in diet breadth results in sequestration of plant chemicals and increases efficacy of chemical defense in a generalist grasshopper, *J. Chem. Ecol.*, 15, 1811–1822, 1989.
46. **Karban, R. and Ricklefs, R. E.,** Host characteristics, sampling intensity, and species richness of lepidoptera larvae on broad-leaved trees in southern Ontario, *Ecology*, 64, 636–641, 1983.
47. **Karowe, D. N.,** Differential effect of tannic acid on two tree-feeding Lepidoptera: implication for theories of plant anti-herbivore chemistry, *Oecologia*, 80, 507–512, 1989.
48. **Knapp, R. and Casey, T. M.,** Thermal ecology, behavior, and growth of gypsy moth and eastern tent caterpillars, *Ecology*, 67, 598–608, 1986.
49. **Lawton, J. H.,** The effect of parasitoids on phytophagous insect communities, in *Insect Parasitoids*, Waage, J. and Greathead, D., Eds., Academic Press, New York, 1986, 265–287.
50. **Levins, R. and MacArthur, R.,** An hypothesis to explain the incidence of monophagy, *Ecology*, 50, 910–911, 1969.
51. **Maclean, D. B., Sargent, T. D., and Maclean, B. K.,** Discriminant analysis of lepidopteran prey characteristics and their effects on the outcome of bird-feeding trials, *Biol. J. Linn. Soc.*, 36, 295–311, 1989.
52. **Marsh, P. M.,** Braconidae, in *Catalog of Hymenoptera in America North of Mexico*, Krombein, K. V., Hurd, R. D., Smith, D. R., and Burk, B. D., Eds., Smithsonian Institution, Washington, D.C., 1979, 144–295.
53. **Mattson, W. J. and Scriber, J. M.,** Nutritional ecology of insect folivores of woody plants: nitrogen, water, fiber, and mineral considerations, in *Nutritional Ecology of Insects, Mites, Spiders and Related Invertebrates*, Slansky, F., Jr. and Rodriguez, J. G., Eds., John Wiley & Sons, New York, 1987, 105–146.
54. **Moran, N. and Hamilton, W. D.,** Low nutritive quality as defense against herbivores, *J. Theor. Biol.*, 86, 247–254, 1980.
55. **Morris, R.,** Predation by wasps, birds, and mammals on *Hyphantria cunea*, *Can. Entomol.*, 104, 1581–1591, 1972.
56. **Morris, R.,** Influence of genetic changes and other variables on the encapsulation of parasites by *Hyphantria cunea*, *Can. Entomol.*, 108, 673–684, 1976.
57. **Morris, R.,** Relation of parasite attack to the colonial habit of *Hyphantria cunea*, *Can. Entomol.*, 108, 833–836, 1976.
58. **Niemelä, P., Tahvanainen, J., Sorjonen, J., Hokkamen, T., and Neuvonen, S.,** The influence of host plant growth form and phenology on the life strategies of Finnish macrolepidopterous larvae, *Oikos*, 39, 164–170, 1982.
59. **Nothnagle, P. J. and Schultz, J. C.,** What is a forest pest?, in *Insect Outbreaks*, Barbosa, P. and Schultz, J. C., Eds., Academic Press, New York, 1988, 59–79.

60. **Opler, P. A. and Krizek, G. O.,** *Butterflies East of the Great Plains*, Johns Hopkins University Press, Baltimore, 1984.
61. **Owen, J. and Gilbert, F. S.,** On the abundance of hoverflies (Syrphidae), *Oikos*, 55, 183–193, 1989.
62. **Pasteels, J. M., Grégoire, J.-C., and Rowell-Rahier, M.,** The chemical ecology of defense in arthropods, *Annu. Rev. Entomol.*, 28, 263–290, 1983.
63. **Price, P. W.,** *Evolutionary Biology of Parasites*, Princeton University Press, Princeton, NJ, 1980.
64. **Price, P. W.,** The role of natural enemies in insect populations, in *Insect Outbreaks*, Barbosa, P. and Schultz, J. C., Eds., Academic Press, New York, 1987, 287–312.
65. **Price, P. W., Bouton, C. E., Gross, P., McPheron, B. A., Thompson, J. N., and Weis, A. E.,** Interactions among three trophic levels : influence of plants on interactions between insect herbivores and natural enemies, *Annu. Rev. Ecol. System.*, 11, 41–65, 1980.
66. **Price, P. W. and Clancy, K.,** Interactions among three trophic levels: gall size and parasitoid attack, *Ecology*, 67, 1593–1600, 1986.
67. **Price, P. W., Cobb, N., Craig, T. P., Fernandes, G. W., Itami, J. K., Mopper, S., and Preszler, R. W.,** Insect herbivore population dynamics on trees and shrubs: new approaches relevant to latent and eruptive species and life table development, in *Insect-Plant Interactions*, Bernays, E. A., Ed., CRC Press, Boca Raton, FL, 1990.
68. **Price, P. W. and Pschorn-Walcher, H.,** Are galling insects better protected against parasitoids than exposed feeders? A test using tenthredinid swaflies, *Ecol. Entomol.*, 13, 195–205, 1988.
69. **Raupp, M. J. and Denno, R. F.,** Leaf age as a predictor of herbivore distribution and abundance, in *Variable Plants and Herbivores in Natural and Managed Systems*, Denno, R. F. and McClure, M. S., Eds., Academic Press, New York, 1983, 91–124.
70. **Redfearn, A. and Pimm, S. L.,** Population variability and polyphagy in herbivorous insect communities, *Ecol. Monogr.*, 58, 39–55, 1988.
70a. **Rhoades, D. F.,** Offensive-defensive interactions between herbivores and plants: their relevance in herbivore population dynamics and ecological theory, *Amer. Nat.*, 125, 205–238, 1988.
71. **Schaffner, J. V., Jr. and Griswold, C. L.,** Macrolepidoptera and Their Parasites Reared from Field Collections in the Northeastern Part of the United States, U.S. Department of Agriculture Misc. Publ. 188, Washington, D.C., 1934.
72. **Schultz, J. C.,** Habitat selection and foraging tactics of caterpillars in heterogeneous trees, in *Variable Plants and Herbivores in Natural and Managed Systems*, Denno, R. F. and McClure, M. S., Eds., Academic Press, New York, 1983, 61–90.
73. **Schultz, J. C.,** Impact of variable plant defensive chemistry on susceptibility of insects to natural enemies, in *Plant Resistance to Insects*, Hedin, P. A., Ed., American Chemical Society, Washington, D.C., 1983, 37–54.
74. **Scott, J. A.,** *The Butterflies of North America*, Stanford University Press, Stanford, CA, 1986.
75. **Scriber, J. M. and Feeny, P.,** Growth of herbivorous caterpillars in relation to feeding specialization and to the growth form of their food plants, *Ecology*, 60, 829–850, 1979.
76. **Shapiro, A. M.,** Susceptibility of *Pieris napi* (Pieridae) to *Apanteles glomeratus* (Hymenoptera, Braconidae), *J. Lepidopterists' Soc.*, 35, 256, 1981.

77. **Sheehan, W.,** Response by specialist and generalist natural enemies to agroecosystem diversification: a selective review, *Environ. Entomol.*, 15, 456–461, 1986.
78. **Sheehan, W.,** Caterpillar defenses, conspicuous consumption and the evolution of gregariousness in Lepidoptera, manuscript in review.
79. **Sheehan, W. and Hawkins, B. A.,** Attack strategy as an indicator of host range in metopiine and pimpline Ichneumonidae (Hymenoptera), *Ecol. Entomol.*, 16, 129–131, 19??.
80. **Sheehan, W. and Shelton, A.,** Parasitoid response to concentration of herbivore food plants: finding and leaving plants by *Diaeretiella rapae* (Hymenoptera: Aphidiidae), *Ecology*, 70, 993–998, 1989.
81. **Slansky, F., Jr.,** Phagism relationships among butterflies, *J. NY Entomol. Soc.*, 84, 91–105, 1976.
82. **Slansky, F., Jr.,** Nutritional ecology of endoparasitic insects and their hosts: an overview, *J. Insect Physiol.*, 32, 255–261, 1986.
83. **Smiley, J. T. and Rank, N. E.,** Predator protection versus rapid growth in a montane leaf beetle, *Oecologia*, 70, 106–112, 1986.
84. **Snedecor, G. W. and Cochran, W. G.,** *Statistical Methods*, Iowa State University Press, Ames, IA, 1980.
85. **Stamp, N. E.,** Effect of group size on parasitism in a natural population of the Baltimore Checkerspot *Euphydryas phaeton*, *Oecologia*, 49, 201–206, 1981.
86. **Stary, P.,** On the strategy, tactics and trends of host specificity evolution in aphid parasitoids (Hymenoptera: Aphidiidae), *Folia Entomol. Hung.*, 32, 199–206, 1981.
87. **Steward, V. B., Smith, K. G., and Stephen, F. M.,** Predation by wasps on lepidopteran larvae in an Ozark forest canopy, *Ecol. Entomol.*, 13, 81–86, 1988.
88. **Van't Hof, H. M. and Martin, M. M.,** Performance of the tree-feeder *Orgyia leucostigma* (Lepidoptera: Liparidae) on artificial diets of different water content: a comparison with the forb-feeder *Manduca sexta* (Lepidoptera: Sphingidae), *J. Insect Physiol.*, 35, 635–641, 1989.
89. **Via, S.,** Ecology genetics and host adaptation in herbivorous insects: the experimental study of evolution in natural and agricultural systems, *Annu. Rev. Ecol. System.*, 35, 421–446, 1990.
90. **Vinson, S. B.,** Habitat location, in *Semiochemicals: Their Role in Pest Control*, Nordlund, D. A., Jones, R. L., and Lewis, W. J., Eds., Wiley, New York, 1981, 51–77.
91. **Vinson, S. B. and Barbosa, P.,** Interrelationships of nutritional ecology of parasitoids, in *Nutritional Ecology of Insects, Mites, and Spiders*, Slansky, F., Jr. and Rodriguez., J. G., Eds., John Wiley & Sons, New York, 1987, 673–695.
92. **Wasserman, S. S. and Mitter, C.,** The relationship of body size to breadth of diet in some Lepidoptera, *Ecol. Entomol.*, 3, 155–160, 1978.
93. **Whitfield, J. B. and Wagner, D. L.,** Patterns in host ranges within the nearctic species of the parasitoid genus *Pholetesor* Mason (Hymenoptera: Braconidae), *Environ. Entomol.*, 17, 608–615, 1988.
94. **Williams, W. S.,** The effects of group size and host species on development and survivorship of a gregarious caterpillar *Halisidota caryae* (Lepidoptera: Arctiidae), *Ecol. Entomol.*, 15, 53–62, 1990.
95. **Zwölfer, H.,** The structure and effect of parasite complexes attacking phytophagous host insects, in *Dynamics of Populations. Proc. Advanced Study Institute on Dynamics of Numbers in Populations*

(Oosterbeck, 1970), Den Boer, P. J. and Gradwell, G. R., Eds., PUDOC, Wageningen, The Netherlands, 1971, 405–418.

96. **Françlemont, J.,** personal communication.
97. **Strand, M.,** personal communication.

INDEX

A

B

C

D

G

H

I

J

K

L

M

N

O

P

Q

R

S

T

Milton Keynes UK
Ingram Content Group UK Ltd.
UKHW051933141024
449569UK00027B/1475